# Impacts of Extreme Weather on Hydrological Process, Water Quality and Ecosystem in Agricultural and Forested Watersheds under the Changing Climate

# Impacts of Extreme Weather on Hydrological Process, Water Quality and Ecosystem in Agricultural and Forested Watersheds under the Changing Climate

Guest Editors

**Ying Ouyang**
**Johnny M. Grace**
**Sudhanshu Sekhar Panda**

MDPI

Basel • Beijing • Wuhan • Barcelona • Belgrade • Novi Sad • Cluj • Manchester

*Guest Editors*

Ying Ouyang
Southern Research Station
USDA, Forest Service
Mississippi State, MS
USA

Johnny M. Grace
Southern Research Station
USDA, Forest Service
Tallahassee, FL
USA

Sudhanshu Sekhar Panda
Institute for Environmental
Spatial Analysis
University of North Georgia
Oakwood, GA
USA

*Editorial Office*
MDPI AG
Grosspeteranlage 5
4052 Basel, Switzerland

This is a reprint of the Special Issue, published open access by the journal *Climate* (ISSN 2225-1154), freely accessible at: https://www.mdpi.com/journal/climate/special_issues/ExtremeWeather_HydrologicalProcess.

For citation purposes, cite each article independently as indicated on the article page online and as indicated below:

| Lastname, A.A.; Lastname, B.B. Article Title. *Journal Name* **Year**, *Volume Number*, Page Range. |
|---|

**ISBN 978-3-7258-6792-9 (Hbk)**
**ISBN 978-3-7258-6793-6 (PDF)**
**https://doi.org/10.3390/books978-3-7258-6793-6**

# Contents

# About the Editors

**Ying Ouyang**

Ying Ouyang, Ph.D. in soil physics, registered Professional Hydrologist, Fellow of the Soil Science Society of America (SSSA), and recipients of the Robert G. Wetzel Award for Water Quality from the American Institute of Hydrology and the Don and Betty Kirkham Soil Physics Award from the SSSA. He is a research hydrologist at the United States Department of Agriculture, Forest Service, Southern Research Station. His main field of interests and activities include: (1) basin-scale modeling of surface and ground water hydrological processes and water quality (e.g., nutrients and sediment) in forested watersheds using HSPF, SWAT, and MODFLOW models and machine learning techniques; (2) multivariate statistical analysis of spatiotemporal variability and trends of hydrological processes and water quality using copula and Mann Kendall techniques; (3) field observation of streamflow, sap flow, and water quality using flowmeters, sensors, and YSI meters; and (4) developing models for simulating coupled streamflow and sediment erosion, biomass production, and $CO_2$ emission in forested watersheds using STELLA and HAWQS models.

**Johnny M. Grace**

Johnny M. Grace III, Ph.D. Biological and Agricultural Engineering. He is a General Engineer with the USDA Forest Service-Southern Research Station (SRS), Center for Forest Watershed Research in Tallahassee, Florida. He serves as Lead Scientist / Scientist-In-Charge (SIC) for the Chipola Experimental Forest and the Program Manager/Coordination Lead for the SRS-University Partnering Programs. He earned a B.S. in 1994 and M.S. in 1996 in Agricultural Engineering from Auburn University and earned a Ph.D. in 2004 in Biological and Agricultural Engineering with a minor in Water Resources from N.C. State University. His research program focuses on describing the influence of timber management, road management, BMPs, and fire on forest soil and water resources (quantity and quality) within a range of watersheds in the region. His research program aims to understand the benefits of forests and maintain or improve the health of watersheds under the changing conditions presented by management, disturbances, and land use.

**Sudhanshu Sekhar Panda**

Sudhanshu Panda is a professional engineer specializing in soil and water engineering, precision agriculture, GeoAI, and geospatial technologies for sustainable environmental management. He is a Professor of GIS and Environmental Science at the University of North Georgia's Institute for Environmental Spatial Analysis. He holds a Ph.D. in Biosystems and Agricultural Engineering from North Dakota State University, an M.S. in Environmental Remote Sensing from the Asian Institute of Technology, and a B.S. in Agricultural Engineering from Odisha University of Agriculture & Technology. Dr. Panda's career spans field engineering, industry, and academia, including roles in soil conservation, water resources, and geospatial modeling. His research integrates machine learning, environmental decision support systems, watershed management, climate change, and precision agriculture. A prolific modeler and developer, he has secured funding from federal, regional, and private agencies and collaborates internationally. Dr. Panda has authored numerous journal articles, book chapters, conference papers, and delivered multiple invited keynote lectures.

# Preface

Extreme weather such as hurricanes, severe storms, droughts, and tornadoes is not frequent but has disastrous events for human lives, social activities, and terrestrial ecosystems. Recent evidence confirms that the unnatural and unseasonal effects of climate variation are making extreme weather more frequent and destructive. Currently, insufficient efforts have been devoted to characterizing the impacts of extreme weather on hydrology, water quality, and ecosystems in agricultural and forested watersheds.

Inspired by the Hydrology-H030 Session of the 2021 American Geophysical Union Fall Meeting, this Reprint, entitled "Impacts of Extreme Weather on Hydrological Processes, Water Quality, and Ecosystems in Agricultural and Forested Watersheds under a Changing Climate", tackles the aspects of extreme weather-induced issues such as hydrological processes (e.g., streamflow, floods, drought, and water yield), water quality constituents (e.g., sediment and nutrients), and ecological services (e.g., wetlands and habitats) along with land use changes, best management practices, field observations, remote sensing, GIS analysis, computer modeling, and multivariate statistics. This Reprint also covers a wide range of geographical locations, including Africa, the Americas, and Asia. As such, the knowledge presented in this Reprint is relevant to other regions around the world.

We express our sincere appreciation to all the authors, reviewers, and editors who made valuable contributions to this Reprint. Without their input, this Reprint would never have been released.

**Ying Ouyang, Johnny M. Grace, and Sudhanshu Sekhar Panda**
*Guest Editors*

*climate* MDPI

*Article*

# Impacts of Multiple Hurricanes and Tropical Storms on Watershed Hydrological Processes in the Florida Panhandle

**Ying Ouyang [1,*], Johnny M. Grace [2], Prem B. Parajuli [3] and Peter V. Caldwell [4]**

1 USDA Forest Service, Center for Bottomland Hardwoods Research, Southern Research Station, 775 Stone Blvd., Thompson Hall, Room 309, Mississippi State, MS 39762, USA
2 USDA Forest Service, Center for Forest Watershed Research, Southern Research Station, 1740 S. Martin Luther King Jr. Blvd., Perry-Paige Bldg., Suite 303 North, Tallahassee, FL 32307, USA; johnny.m.grace@usda.gov
3 Department of Agricultural and Biological Engineering, Mississippi State University, Mississippi State, MS 39762, USA; pparajuli@abe.msstate.edu
4 USDA Forest Service, Center for Integrated Forest Science, Southern Research Station, 3160 Coweeta Lab Road, Otto, NC 28763, USA; peter.v.caldwell@usda.gov
* Correspondence: ying.ouyang@usda.gov

**Abstract:** Hurricanes and tropical storms (TS) are infrequent but disastrous events to human lives, social activities, and terrestrial ecosystems in coastal regions. Using the Environmental Protection Agency (US-EPA)'s Hydrologic and Water Quality System (HAWQS) model, principal component analysis (PCA), and principal factor analysis (PFA), we estimated impacts of multiple hurricanes and TS on hydrological processes in agricultural and forested watersheds. Five hurricanes and four TS that passed near or through the Apalachicola–Chattahoochee–Flint River basin (ACFRB) of the Florida panhandle from 1966 to 2018 were selected to estimate their impacts on rainfall, potential evapotranspiration (PET), evapotranspiration (ET), soil water percolation, surface runoff, stream discharge, groundwater recharge, and water yield (WYLD). Simulations showed that the category of hurricanes was not highly related to the amounts of rainfall, runoff, discharge, and WYLD. Based on PCA and PFA, PET and ET were highly and negatively, rainfall and discharge were highly and positively, and percolation, runoff, groundwater recharge and WYLD were moderately and positively affected by the hurricanes and TS at the ACFRB in the recent 50 years. This study provides water resource managers with critical insights into how multiple hurricanes and TS affected hydrological processes in agricultural and forested watersheds of the coastal region.

**Keywords:** Florida panhandle; hydrological processes; hurricane; tropical storm; HAWQS model; principal component analysis

## 1. Introduction

Hurricanes and tropical storms are disastrous natural events to human lives, economical activities, and natural environments in coastal regions [1,2]. Natural hazardous disasters and human-made catastrophes have resulted in human deaths and more than 11,000 missing as well as economic losses greater than USD 337 billion in 2017 [3]. Hurricanes Harvey, Irma and Maria alone caused a loss of USD 217 billion in North America [4]. Impacts of hurricanes and tropical storms on hydrological processes in coastal agricultural and forested watersheds primarily include high peak discharge, reduced evapotranspiration, overbank flooding and groundwater table rise for weeks to month due to storm surges and heavy precipitation [5,6]. Jayakaran et al. [7] investigated the impact of Hurricane Hugo in 1989 on two paired coastal watersheds in South Carolina, United States of America (USA). They found a significant change in rainfall–runoff relationships, lower vegetative water use, and increased stream outflow due to hurricane damage to forests. Torres et al. [8] simulated the hydraulic interactions of hurricane storm surge and rainfall runoff at the Houston-Galveston region, USA and reported that the peak flows from storm surge dominate those from rainfall runoff although rainfall runoff can constitute more than half of the total flood volume. Brena-Naranjo et al. [9] assessed the contributions of tropical storms and hurricanes to rainfall in Mexico from 1998 to 2013 using the satellite-derived precipitation dataset. These authors observed that the southern regions of Mexico can receive more than 2400 mm of cyclonic rainfall during years with significant tropical cyclone activity. They also found that the number of tropical cyclones increased significantly from 1998, yet their rainfall contributions did not increase, and wind speed and rainfall intensity during cyclones were not highly correlated. Hou et al. [10] investigated hydrological responses to extreme weather-induced forest disturbances in a tropical experimental watershed of Hainan Province, China using the single watershed and paired-year approaches. They reported that typhoon and cold wave-induced forest disturbances have a positive effect on stream discharge. Vidon et al. [11] synthesized the impacts of Hurricane Irene in 2011 and Tropical Storm Lee in 2011 on riparian hydrology, biogeochemistry, and water quality from North Carolina to Maine, USA. They found that in almost all cases, these storms generated unprecedented changes in hydrology and water quality.

All the above and many other studies provide very useful insights into the impacts of hurricanes and tropical storms on some hydrological processes in agricultural and forest watersheds. However, a thorough literature review reveals that few studies have been devoted to scrutinizing the impacts of hurricanes and tropical storms on hydrological processes before and after the hurricanes and tropical storms, which are essential to water resource management. The goal of this study was to quantify the impacts of multiple hurricanes and tropical storms on hydrological processes in agricultural and forested watersheds, using the US-Environmental Protection Agency (EPA)'s Hydrologic and Water Quality System (HAWQS) model and multivariate statistics. The Apalachicola–Chattahoochee–Flint River basin (ACFRB), draining into the northern Gulf of Mexico (GOM) region, was selected because hurricanes and tropical storms occur in this forested and agricultural dominated region more frequently than most other regions in the world, making this basin an ideal location for the purpose of this study. Our specific objectives were to: (1) create, calibrate, and validate the HAWQS-based ACFRB model; (2) apply the model to assess hydrological processes in agricultural and forested watersheds before-during-after the hurricanes and tropical storms that had passed through the ACFRB from 1966 to 2018; and (3) identify the hydrological variables most affected by hurricanes and tropical storms using principal component and factor analyses. Hydrological processes of interest in this study include rainfall, surface runoff, evapotranspiration (ET), potential ET (PET), stream discharge, soil water percolation, groundwater recharge, and water yield (WYLD).

## 2. Materials and Methods

### 2.1. Study Site, Hurricanes, and Tropical Storms

The ACFRB consists of three river basins, namely the Chattahoochee River basin, Flint River basin and Apalachicola River basin (50,701 km$^2$), with headwaters initiating in tributaries feeding the upper reaches of the Chattahoochee River, Georgia, continuing southward through west Georgia and east Alabama, and ending at the coast of the Florida panhandle (Figure 1). The Apalachicola River in the Florida panhandle is the confluence of the Chattahoochee River (flows through Georgia and Alabama) and Flint River (flows through Georgia). The Chattahoochee River, with a drainage area of 22,714 km$^2$, arises as a cold-water mountain stream at altitudes above 914 m, flows 692 km to its confluence with the Flint River, and is a heavily used water resource in Georgia. The Flint River is approximately 563 km in length and drains an area of 21,911 km$^2$ prior to its confluence with the Chattahoochee River to form the Apalachicola River. The Apalachicola River flows 171 km from Jim Woodruff Lock and Dam to the GOM. The river drains approximately 6734 km$^2$ in the Florida panhandle with a shallow estuary of approximately 539 km$^2$ [12]. ACFRB is an important ecological and economic region of tri-states (i.e., Florida, Alabama, and Georgia) in the southeastern USA [13]. There are four major land-resource areas, namely the Southern Piedmont, Georgia Sand Hills, Southern Coastal Plain, and Eastern Gulf Coast

Flatwoods, in the ACFRB. The Southern Piedmont area is dominated by ultisols [14], while the Southern Coastal Plain and the Georgia Sand Hills areas are derived from marine and fluvial sediments eroded from the Appalachian and Piedmont Plateaus. Ultisols are found throughout these areas with an exception of some areas in the Georgia Sand Hills with entisols locally. The Eastern Gulf Coast Flatwoods area, which covers most of the Apalachicola River basin, is dominated by spodosols [12]. The ACFRB spans approximately 5 degrees of latitude, with average annual temperatures ranging from 16 °C in the north to 21 °C in the south. Average daily temperatures range from to 13 °C in January and from 24 to 27 °C in July. Precipitation is greatest either in the mountains because of orographic effect or near the GOM because of high air moisture. Average annual precipitation is approximately 1295 mm, with a range from 1143 mm in the east-central portion of the ACFRB to 1524 mm in the Florida panhandle portion of the basin [12]. There is a total of 12 8-digit Hydrologic Unit Code (HUC-8) watersheds in the ACFRB, with four in the Chattahoochee River basin, six in the Flint River basin, and two in the Apalachicola River basin. Land uses and area of each watershed are shown in Figure 1b.

Five hurricanes (i.e., Alma, Agnes, Kate, Earl, and Michael) and four tropical storms (i.e., Becky, Allison, Bonnie, and Fay) that occurred from 1966 to 2018 were used in this study. These hurricanes and tropical storms were selected simply because their central paths (or storm tracks) passed through or near the ACFRB (Figure 1a). Alma was a Category 1 hurricane, its central path passed through the ACFRB, which occurred from 4 to 14 June 1966, with watershed impact times on 9 to 10 June and landfall date on 9 June. Agnes was a Category 1 hurricane, its central path passed through the ACFRB, which occurred from 14 to 23 June 1972, with watershed impact times on 19 to 20 June and landfall date on 19 June. Kate was a Category 2 hurricane, its central path passed through the ACFRB, which occurred from 15 to 23 November 1985, with watershed impact times on 21 to 22 November and landfall date on 21 November. Earl was a Category 1 hurricane, its central path passed through the ACFRB, which occurred from 31 August to 8 September 1998, and its watershed impact time and landfall date were both on 3 September. Michael was a Category 5 hurricane, its central path passed through the ACFRB, which occurred from 6 to 15 October 2018, with its watershed impact times on 10 to 11 October and landfall date on 10 October. The table embedded in Figure 1a lists the dates, years, categories, and watershed impact times of the hurricanes and tropical storms used in this study.

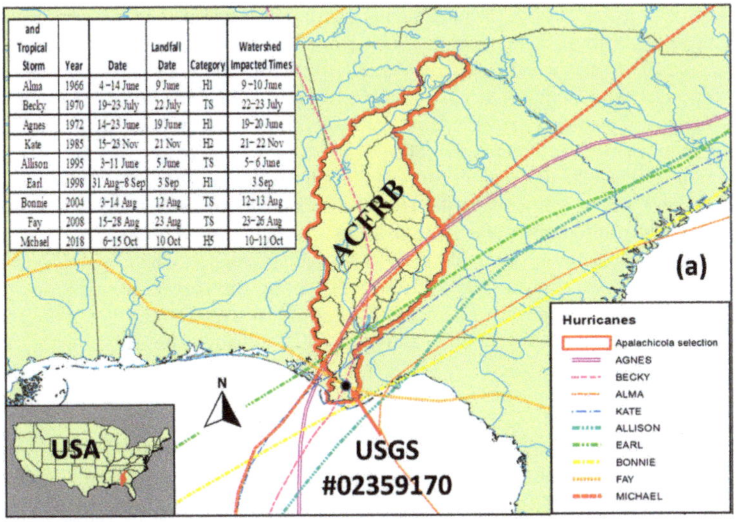


**Figure 1.** *Cont.*

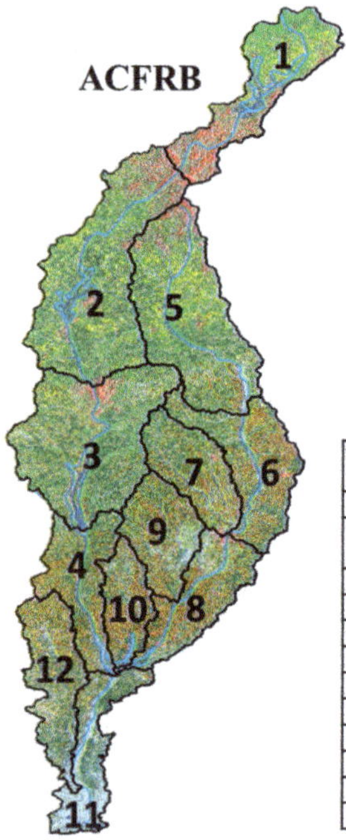

| Watershed ID | HUC Number | Watershed Name | Area ($km^2$) |
|---|---|---|---|
| 1 | 3130001 | Upper Chattahoochee | 4106.87 |
| 2 | 3130002 | Middle Chattahoochee-Lake Harding | 7875.21 |
| 3 | 3130003 | Middle Chattahoochee-Walter F | 7347.5 |
| 4 | 3130004 | Lower Chattahoochee | 3222.54 |
| 5 | 3130005 | Upper Flint | 6811.01 |
| 6 | 3130006 | Middle Flint | 4035.67 |
| 7 | 3130007 | Kinchafoonee-Muckalee | 2851.7 |
| 8 | 3130008 | Lower Flint | 3300.64 |
| 9 | 3130009 | Ichawaynochaway | 2859.48 |
| 10 | 3130010 | Spring | 2042.84 |
| 11 | 3130011 | Apalachicola | 2893.7 |
| 12 | 3130012 | Chipola | 3347.46 |

**Figure 1.** Location of the Apalachicola–Chattahoochee–Flint River basin (ACFRB) of the Florida panhandle, USA. (**a**) Hurricanes and tropical storms passed through or highly impacted the ACFRB from 1966 to 2018. (**b**) The name, area, and land use of the 12 watersheds at the ACFRB. USGS #02359170 is the US Geological Survey gage number.

### *2.2. The HAWQS Model and Statistical Analysis*

Using the Soil and Water Assessment Tool (SWAT) model as the core simulation engine, the HAWQS is a web-based modeling system applied to simulate water quantity and quality in watersheds of continental USA (https://hawqs.tamu.edu/, accessed on 6 March 2022). The SWAT model is a small watershed to river basin-scale model used to simulate the quality and quantity of surface and ground water and predict the environmental impact of land use, land management practices, and climate change. The SWAT model is widely used in assessing soil erosion prevention and control, non-point source pollution control and regional management in watersheds (https://swat.tamu.edu/, accessed on 6 March 2022). The HAWQS is a customized version of the SWAT model, which has interactive web interfaces and maps; online watershed-specific model development and execution; pre-loaded input data; and output data in tabulate and chart formats. Additionally, the past and future climate datasets are pre-loaded into the model for users to develop their own modeling projects [15]. Overall, the HAWQS is a time-saving and cost-effective modeling system for predicting water quantity and quality in complex watersheds.

The ACFRB model was developed with the HAWQS through the web interface at https://hawqs.tamu.edu/#/projects/ (accessed on 6 March 2022), using the following three major procedures: (1) Create a project. The ACFRB model was created at the data resolution of HUC-8 for the downstream HUC #03130011 (or Apalachicola River basin). After entering this HUC number into the Map Options of the HAWQS, the additional 11 upstream watersheds contributing to the Apalachicola River watershed were delineated. This generated the entire ACFRB model (Figure 1). (2) Set the Hydrologic Response Unit (HRU). A threshold level of 1% for HRUs was used to eliminate the effects of minor land uses, soils, and slopes in each watershed. (3) Create a scenario. Detailed procedures on how to develop a modeling project with the HAWQS can be found in the US-EPA's website (https://hawqs.tamu.edu/#/, accessed on 6 March 2022), whereas elaborated definitions of input parameters in the HAWQS are provided in the SWAT model user manual (https://swat.tamu.edu/docs/, accessed on 6 March 2022). After calibration and validation of the ACFRB model, a simulating scenario was created to assess hydrological processes under the stresses of nine hurricanes and tropical storms. The paths, durations and categories of the hurricanes and tropical storms are shown in Figure 1a.

Principal component analysis (PCA) and principal factor analysis (PFA) are multivariate statistics for identifying important components or factors that describe most of the variances in a system. They can be used to reduce explanatory variables to a small number of principal (or key) variables while preserving the relationship present in the original data [16,17]. PCA was first coined by Pearson (1901), who believed that this was the correct solution to some of the problems that were of interest to biometricians at that time, although he did not propose a practical method of calculation for more than two or three variables [16]. PFA seeks the fewest factors that can account for common variance in a set of variables. The major difference between the two methods is that PFA analyzes only the reliable common variance of data, while PCA analyzes all the variance of data. PCA is a linear combination of variables and PFA is a measurement model of a latent variable [18]. In recent decades, PCA and PFA techniques have widely been used in hydrology and water quality analysis [17,19–22]. In this study, PCA and PFA were applied to identify the hydrological variables affected most by the hurricanes and tropical storms in the R-statistics platform.

### 2.3. Model Calibration and Validation

The ACFRB model was manually calibrated by changing input parameter values within an acceptable range to match predicted discharges with observed discharges in the HAWQS. The discharge dataset was downloaded from the USGS Station #02359170 (Figure 1a) in Apalachicola River near Sumatra, FL. The input parameter values used for model calibration are given in Table 1. Figure 2a shows the predicted (Y) and observed (X) daily stream discharges for the near 10-year simulation period from 1 January 1983 to 31 December 1991. The values of $R^2$, the normalized root mean square error (nRMSE), the Nash–Sutcliffe Efficiency (NSE), and $p$-value were, respectively, 0.879, 0.159, 0.605, and 0, suggesting good agreement was obtained between the predicted and observed daily discharges during model calibration.

**Table 1.** Major input parameter values used for the ACFRB-HAWQS model.

| Parameter | Definition | Value | Unit/Method/Explanation | Reference |
|---|---|---|---|---|
| SFTMP | Snowfall temperature | 1 | °C | Local observation |
| SMTMP | Snow melt base temperature | 0.5 | °C | Local observation |
| SMFMX | Melt factor for snow on 21 June | 4.5 | mm $H_2O$/°C-day | Local observation |
| SMFMN | Melt factor for snow on 21 December | 4.5 | mm $H_2O$/°C-day | Local observation |
| TIMP | TIMP: Snow pack temperature lag factor | 1 | | Local observation |
| IPET | Potential evapotranspiration (PET) method | 0 | Priestley-Taylor method | |
| ESCO | Soil evaporation compensation factor | 1 | | Calibrated |
| EPCO | EPCO: Plant uptake compensation factor | 1 | | Calibrated |

**Table 1.** *Cont.*

| Parameter | Definition | Value | Unit/Method/Explanation | Reference |
|---|---|---|---|---|
| ICN | Daily curve number calculation method | 0 | Calculate daily CN value as a function of soil moisture | Calibrated |
| CNCOEF | Plant ET curve number coefficient | 2 | | |
| ICRK | Crack flow code | 0 | Do not model crack flow in soil | Local observation |
| SURLAG | Surface runoff lag time | 4 | days | Calibrated |
| CN2 | Subbasins curve number | −10% | CN2 decreased 10% for all subbasins | Calibrated |
| IRTE | Channel water routing method | 0 | Variable Storage Method | Calibrated |
| MSK_COL1 | Calibration coefficient used to control impact of the storage time constant for normal flow | 0 | | Calibrated |
| MSK_COL2 | Calibration coefficient used to control impact of the storage time constant for low flow | 3.5 | | Calibrated |
| MSK_X | Weighting factor controlling relative importance of inflow rate and outflow rate in determining water storage in reach segment | 0.2 | | Calibrated |
| TRNSRCH | Fraction of transmission losses from main channel that enter deep aquifer | 0 | | Calibrated |
| EVRCH | Reach evaporation adjustment factor | 1 | | |
| IDEG | Channel degradation code | 0 | Channel dimension is not updated as a result of degradation | Local observation |
| Dataset 1 | NCDC NWS/NOAA | Measured | Base scenario | Downloaded from HAWQS |

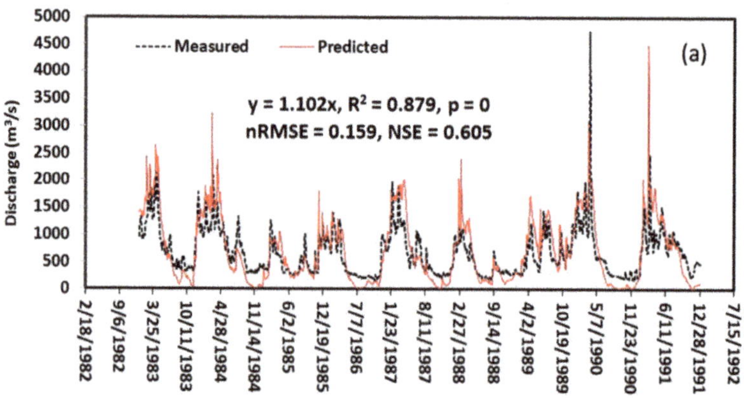


**Figure 2.** *Cont.*

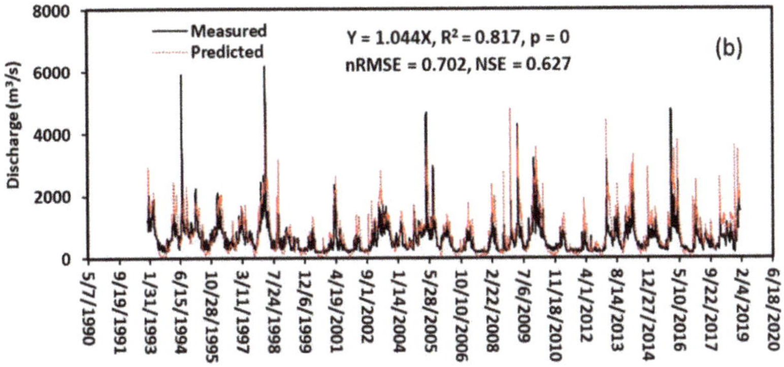


**Figure 2.** Comparison of the observed and predicted daily discharges during model calibration (**a**) and validation (**b**).

The ACFRB model was validated by comparing the daily stream discharges between model predictions and field observations for a 16-year simulation period from 1 January 1993 to 31 December 2016 (Figure 2b). During model validation, the input parameter values were the same as those used during model calibration. With values of $R^2$, nRMSE, NSE, and $p$ of 0.817, 0.702, 0.627, and 0, respectively (Figure 2b) that were comparable to those of model calibration, we demonstrated that very good agreement was achieved between the predicted and observed daily discharges during model validation.

## 3. Results and Discussion

### 3.1. Impact of Hurricane

Changes in daily observed rainfall and predicted stream discharge one month before, during, and one month after the hurricanes at the outlet of ACFRB are shown in Figure 3. In general, daily stream discharge increased with daily rainfall. For example, the average daily stream discharge was 787 $m^3/s$ (Figure 3g) during Agnes (watershed impact times or WIT: 19–20 June 1972) when the average daily rainfall was 48.9 mm (Figure 3b), while the average daily stream discharge was 2557 $m^3/s$ (Figure 3i) during Earl (WIT: 3 September 1993) when the average daily rainfall was 92.1 mm (Figure 3d). A 1.9-fold increase in daily rainfall increased the daily stream discharge by more than 3.2 folds. This occurred because rainfall is a major driving force for stream discharge. However, the amount of rainfall was not necessarily proportional to the amount of stream discharge during the hurricanes. For instance, the amounts of rainfall and stream discharge were, respectively, 60.7 mm (Figure 3b) and 1220 $m^3/s$ (Figure 3g) on 20 June 1972 for Agnes, whereas the amounts of rainfall and stream discharge were, respectively, 42.3 mm (Figure 3c) and 1943 $m^3/s$ (Figure 3h) on 22 November 1985 for Kate. Apparently, Kate had a higher stream discharge than that of Agnes although the opposite was true for the amount of rainfall. This occurred because there was a much higher amount of rainfall one month before Kate than before Agnes (Figure 3b,c), which resulted in a higher daily average soil water content (SWC) before Kate (SWC = 222 mm on 21 November 1985) than before Agnes (SWC = 110 mm on 18 June 1972). The higher antecedent SWC produced more stream discharge during the hurricanes. The results indicate that rainfall conditions preceding the event also played a role in stream discharge during the hurricanes. Figure 3 further reveals that the category of hurricane was not highly related to the amounts of rainfall and discharge during the hurricanes. The amounts of rainfall and discharge were, respectively, 42.3 mm and 1943 $m^3/s$ during Kate (Category 2) but were, respectively, 92.1 mm and 2461 $m^3/s$ during Earl (Category 1). Although Earl was of a weaker category than Kate, the amounts of rainfall and stream discharge during Earl were much higher than during Kate.

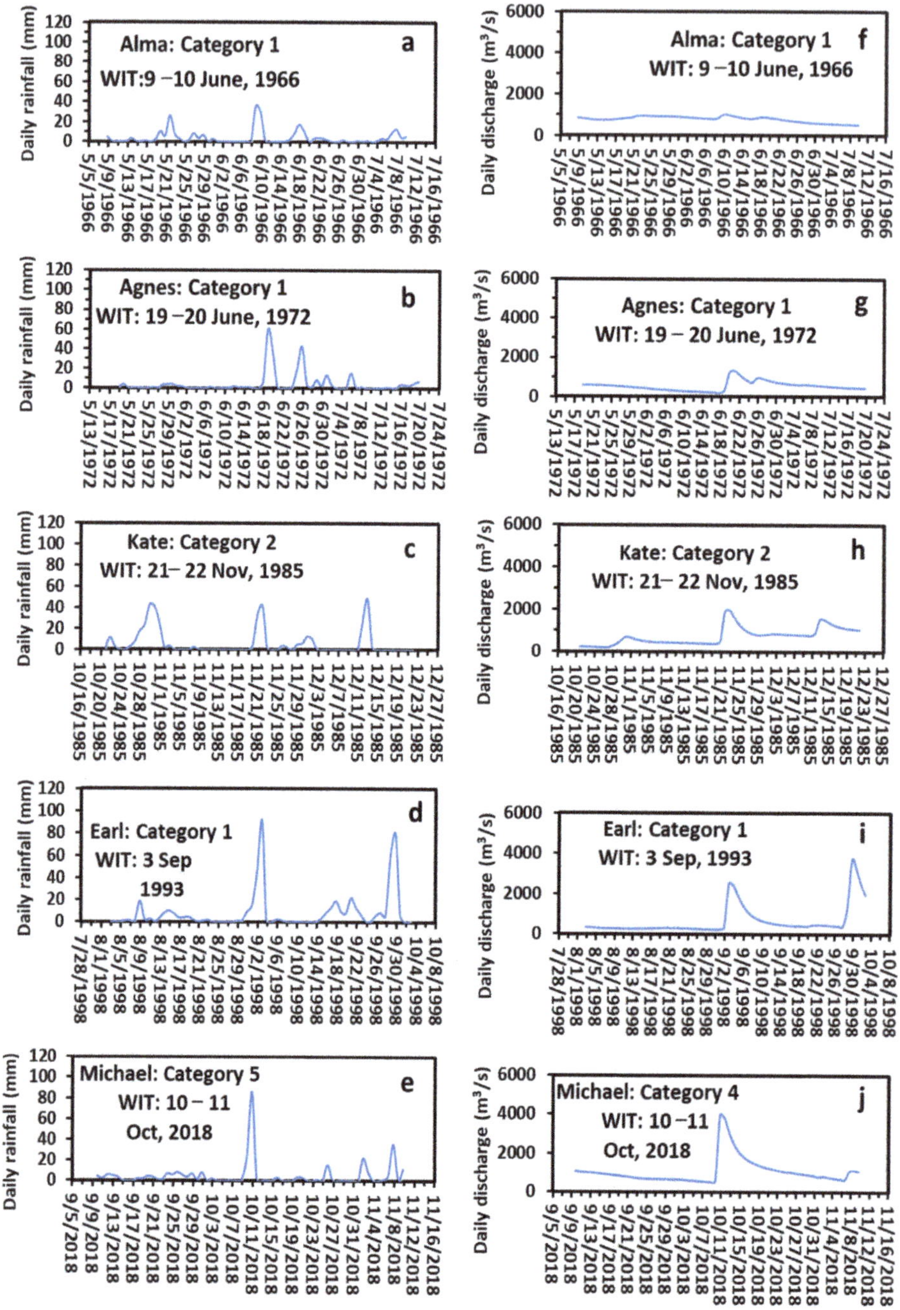


**Figure 3.** Daily rainfall and stream discharge under five hurricanes from 1966 to 2018. WIT is the watershed impact times.

In most cases, the rate of stream discharge was higher one month after the hurricanes than before the hurricanes. For Agnes, the rate of stream discharge was 206.2 $m^3/s$ on

18 June (just before the hurricane) and 450.1 $m^3/s$ on 20 July 1972 one month after the hurricane (Figure 3g)—approximately 2.18-fold (450.1 ÷ 206.2 = 2.18) higher. Similar results were also observed for Kate, Earl, and Michael (Figure 3h–j). That is, the rates of stream discharge were 2.14-, 5.48-, and 1.97-fold higher one month after Kate, Earl, and Michael than one month before. This occurred because more rainfall events were observed on the watersheds within one month after the hurricanes than before the hurricanes (Figure 3b–e). This also explained why much higher discharge was obtained almost one month after Earl (3751 $m^3/s$) than during Earl (2461 $m^3/s$), as shown in Figure 3i.

Similar variation patterns were observed between daily surface runoff and daily water yield (WYLD) one month before, during, and one month after the hurricanes (Figure 4). That is, an increase in surface runoff increased WYLD except that the amount of surface runoff was always lower than that of WYLD. For example, the amount of surface runoff was 8.75 mm on 22 November 1985 during Kate (Figure 4c) when the amount of WYLD was 13 mm (Figure 4h). The former was 48.57% lower than the latter. This occurred because WYLD includes not only surface runoff but also lateral flow and shallow groundwater discharge to reach.

Analogous to the case of stream discharge, the category of hurricane was not related to the amounts of runoff and WYLD during the hurricane. The amounts of runoff and WYLD were, respectively, 8.75 and 13 mm during Kate, which was a Category 2 hurricane (Figure 4c–h), but were, respectively, 29.7 and 32.5 mm during Earl, which was a Category 1 hurricane (Figure 4d–i). Although Earl was a lower category hurricane than Kate, the amounts of runoff and WYLD for Earl were higher than for Kate. It is apparent that the amount of rainfall rather than the category of hurricane was a controlling factor for changes in runoff and WYLD.

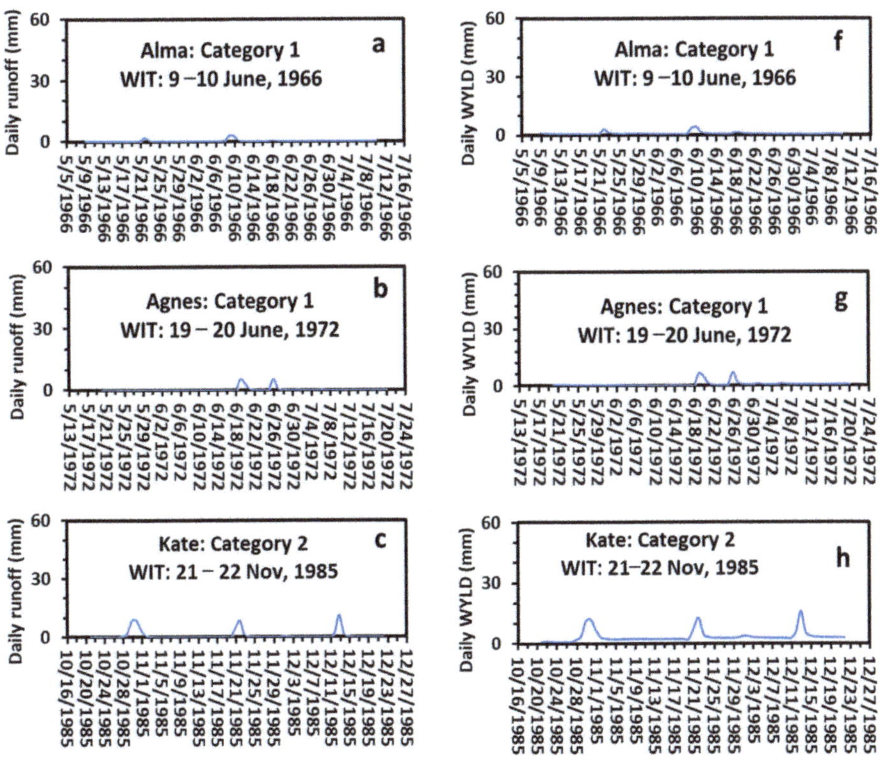


**Figure 4.** *Cont.*

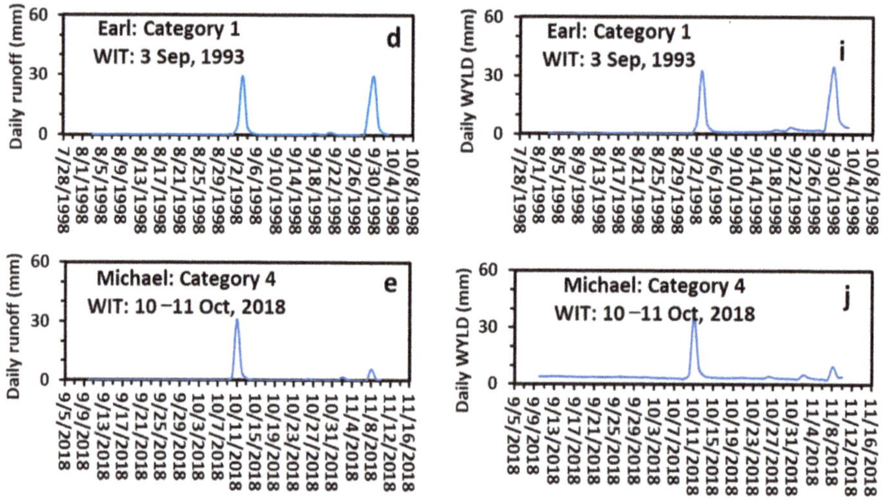


**Figure 4.** Simulated daily runoff and water yield (WYLD) under five hurricanes from 1966 to 2018. WIT is the watershed impact times.

### 3.2. *Impact of Tropical Storm*

Impacts of tropical storms on daily rainfall, stream discharge, runoff, and WYLD one month before, during, and one month after Barry and Fay are shown in Figure 5. Analogous to the impacts of the hurricanes, daily discharge, runoff, and WYLD increased with daily rainfall during tropical storms. Daily discharge was 899.2 $m^3/s$ on 6 August 2001 during Barry (Figure 5b) when the daily rainfall was 53.5 mm (Figure 5a), while daily discharge was 2956 $m^3/s$ on 23 August 2008 during Fay (Figure 5f) when the daily rainfall was 92.6 mm (Figure 5e). A 73% increase in rainfall resulted in a 2.3-fold increase in daily discharge. Comparable results were obtained for runoff and WYLD. That is, daily runoff and WYLD were, respectively, 13.8 mm and 18.3 mm on 6 August 2001 during Barry (Figure 5c,d) when the daily rainfall was 53.5 mm (Figure 5a), while daily runoff and WYLD were, respectively, 33.1 mm and 36.2 mm on 23 August 2008 during Fay (Figure 5g,h) when the daily rainfall was 92.6 mm (Figure 5e). A 73% increase in rainfall resulted in a 2.4- and 2.0-fold increase in daily runoff and WYLD, respectively. Similar results were also obtained for Becky and Bonnie (figures not shown). The results indicate that rainfall was a major driving force for stream discharge, runoff, and WYLD during tropical storms.

Average daily rainfall, runoff, WYLD, and stream discharge one month after the hurricanes compared with one month after the tropical storms are given in Figure 6. Overall, the average amounts of rainfall, runoff, WYLD, and stream discharge one month after the hurricanes were always higher than one month after the tropical storms. However, a Kolmogorov–Smirnov test revealed that the differences were not statistically significant at $\alpha = 0.05$. The results demonstrate that the hurricanes and tropical storms had similar impacts on daily rainfall, runoff, WYLD, and stream discharge.

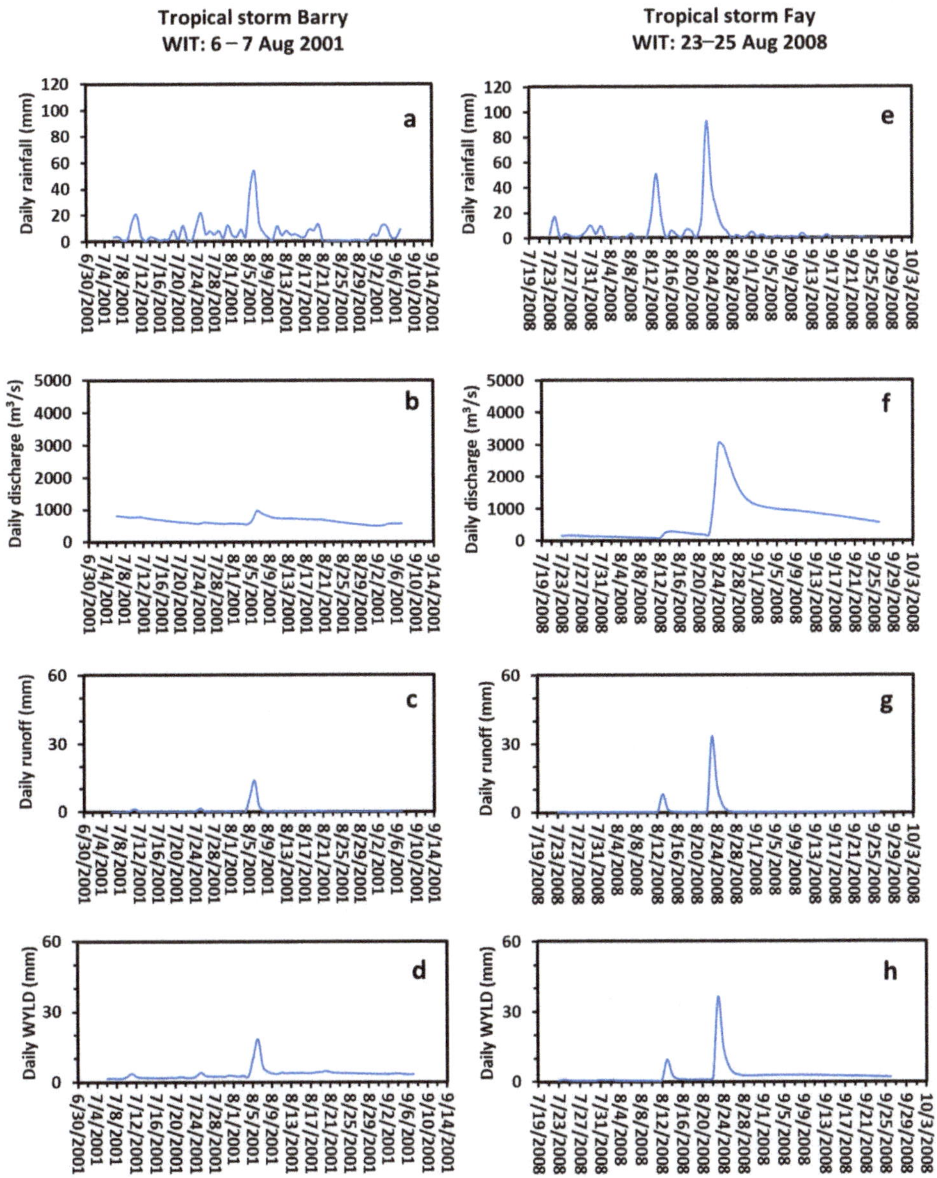


**Figure 5.** Daily rainfall, stream discharge, surface runoff and water yield (WYLD) under tropical storms Barry and Fay. WIT is the watershed impact times.

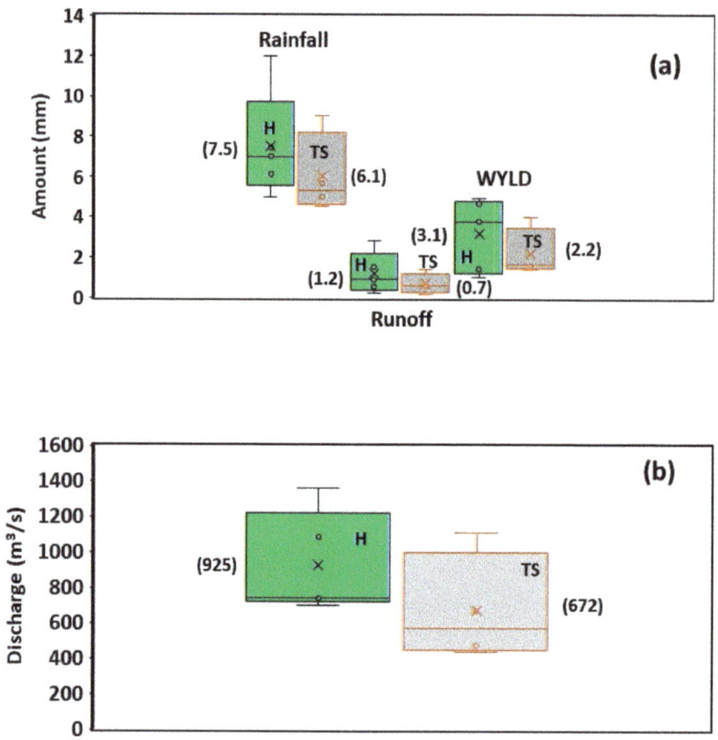


**Figure 6.** Comparison of (**a**) average daily rainfall, surface runoff and water yield (WYLD) and (**b**) stream discharge between the hurricanes (H) and the tropical storms (TS). The numbers in parentheses are the average values.

### *3.3. Identification of Most-Impacted Hydrological Variables*

Very few studies have been conducted to identify which hydrologic variables are affected most by hurricanes and tropical storms in the literature, which is critical for sustainable water resource and ecosystem management. In this study, eight hydrological variables, namely rainfall, PET, ET, percolation, runoff, groundwater discharge, WYLD and stream discharge, were analyzed with PCA and PFA to identify which hydrological variables were affected most during and one month after the hurricanes and tropical storms. In a PCA, the number of components is equal to the number of variables (or eight components in this case). However, a component is comprised of more than a single variable in a study [15]. Table 2 shows the component loadings of the hydrological variables during and one month after the hurricanes and tropical storms, which were obtained by PCA. Of the 8 components, the first component (PC1) accounted for 68% and the second component (PC2) accounted for 21.1% of the total variance in the dataset. These two components together accounted for 89% of the total variance and the rest of the components accounted for 11% of the total variance. Thus, we should focus on the first two components. For PC1, the component loadings of rainfall, percolation, runoff, groundwater recharge, WYLD, and stream discharge were positive (Table 2), indicating that the hurricanes and tropical storms resulted in increased amounts of these variables, whereas the component loadings of PET and ET were negative (Table 2), specifying that the hurricanes and tropical storms resulted in decreased amounts of these two variables. Apparently, PC1 represents changes in the hydrological variables. For PC2, the component loadings of rainfall, PET, ET, percolation, runoff, and WYLD were negative, while the component loadings of groundwater recharge and stream discharge were positive. This component seemed to measure the effects of the hurricanes and tropical storms on the hydrological variables. In other words, during and one month after the hurricanes and tropical storms, rainfall, PET, ET, percolation, runoff, and WYLD decreased but groundwater discharge and stream discharge increased.

**Table 2.** Component loading coefficients from principal component analysis. PC denotes principal component and the number refers to the number of components. For example, PC1 is principal component 1.

| Parameter | PC1 | PC2 | PC3 | PC4 | PC5 | PC6 | PC7 | PC8 |
|---|---|---|---|---|---|---|---|---|
| Precipitation | 0.2069 | −0.6412 | −0.2687 | 0.4357 | −0.1773 | 0.4809 | −0.1419 | 0.0334 |
| PET | −0.3386 | −0.3881 | 0.4167 | −0.0570 | −0.0406 | −0.3055 | −0.6789 | 0.0437 |
| ET | −0.3712 | −0.3020 | 0.2830 | 0.0231 | 0.6968 | 0.2615 | 0.3684 | −0.0268 |
| Percolation | 0.3835 | −0.2590 | −0.2982 | 0.1910 | 0.4423 | −0.6787 | 0.0400 | 0.0514 |
| Runoff | 0.3535 | −0.3865 | 0.1826 | −0.6682 | −0.1501 | 0.0260 | 0.2363 | 0.4084 |
| Groundwater discharge | 0.3724 | 0.3561 | 0.1336 | 0.0644 | 0.4328 | 0.3242 | −0.4548 | 0.4618 |
| Water yield (WYLD) | 0.4227 | −0.0482 | 0.1373 | −0.2963 | 0.1780 | 0.1581 | −0.2063 | −0.7830 |
| Stream discharge | 0.3396 | 0.0380 | 0.7177 | 0.4812 | −0.2101 | −0.1248 | 0.2765 | −0.0210 |
| Proportion of variance | 68.0% | 21.1% | 8.3% | 1.2% | 1.1% | 0.2% | 0.1% | 0.0% |
| Cumulative proportion | 68.0% | 89.0% | 97.4% | 98.6% | 99.6% | 99.9% | 100.0% | 100.0% |

While PCA identified that the first two components accounted for 89% of the total variance in the dataset, it did not determine which hydrological variables explained most of the variance. Therefore, PFA was employed to meet this need. Analogous to PCA, the number of factors is equal to the number of variables in PFA. Since there were eight hydrological variables in this study, there were eight factors. Principal factors with an eigenvalue above or close to one [17], a total of three factors, are discussed in our analysis. Table 3 shows the rotated factor correlation coefficients for all eight hydrological variables. The results show that factor 1 accounted for 38.8%, Factor 2 for 31.7%, and Factor 3 for 25.3% of the total variance in the dataset. These three factors together accounted for 95.8% of the total variance. In this study, the factor correlation coefficient is considered significant when it is greater than 0.85 (or 85%). This conservative criterion was used because the ACRFB is a complex river basin. Based on this criterion, rainfall, PET, ET, and stream discharge had coefficient values greater than 0.85 at least from one of the three factors. These hydrological variables were very important in explaining the total variance of the dataset, and were thereby highly affected by the hurricanes and tropical storms. In contrast, percolation, runoff, groundwater discharge and WYLD scored lower than 0.85 for all three factors and were considered less important in explaining the total variance of the dataset and were thereby relatively less affected by hurricanes and tropical storms.

**Table 3.** Rotated factor correlation coefficients from principal factor analysis.

| Parameter | Factor1 | Factor2 | Factor3 | Factor4 |
|---|---|---|---|---|
| Rainfall | −0.0561 | 0.9664 | 0.0332 | 0.0099 |
| PET | −0.9780 | −0.0910 | −0.1721 | −0.0288 |
| ET | −0.9094 | −0.1860 | −0.2789 | 0.0953 |
| Percolation | 0.5042 | 0.8049 | 0.2816 | 0.1173 |
| Runoff | 0.1895 | 0.7785 | 0.5517 | −0.2223 |
| Groundwater discharge | 0.7893 | 0.0360 | 0.6029 | 0.0913 |

Table 3. *Cont.*

| Parameter | Factor1 | Factor2 | Factor3 | Factor4 |
|---|---|---|---|---|
| Water yield | 0.5637 | 0.5173 | 0.6398 | −0.0590 |
| Stream discharge | 0.2959 | 0.1973 | 0.8702 | 0.0044 |
| Proportion of variance | 38.8% | 31.7% | 25.3% | 1.1% |
| Cumulative proportion | 38.8% | 70.5% | 95.8% | 96.9% |

## 4. Conclusions

The US-EPA's HAWQS model system was applied to develop a model for the Apalachicola–Chattahoochee–Flint River basin (ACFRB) of the Florida panhandle. There was good agreement between model predictions and field measurements during model calibration and validation. A simulation scenario was chosen to estimate the impact of hurricanes and tropical storms on hydrological processes in the last 52 years (from 1966 to 2018).

In general, daily stream discharge increased with daily rainfall and occurred because rainfall is a major driving force for stream discharge. However, the amount of rainfall was not necessarily proportional to the amount of stream discharge during the hurricanes and tropical storms. Rainfall conditions preceding the hurricanes also played a role in stream discharge during the events. The category of hurricane was not highly related to the amounts of daily rainfall, surface runoff, water yield (WYLD), and stream discharge during the hurricanes.

In most cases, the amounts of stream discharge, surface runoff, and WYLD were higher one month after the hurricanes and tropical storms than before the events at the ACFRB and occurred because more rainfall fell on the basin within one month after the events than before the events. It should be noted that there are no scientific reasons for more rainfall events after the hurricanes than before the hurricanes although this finding was observed in the Florida panhandle. Further study is therefore delineate the reasons for this.

The amount of surface runoff was always lower than that of WYLD during the hurricanes and tropical storms because WYLD includes not only surface runoff but also lateral flow and shallow groundwater discharge to reach.

The average amounts of rainfall, runoff, WYLD, and stream discharge one month after the hurricanes were higher than one month after the tropical storms. However, a Kolmogorov–Smirnov test revealed that the differences were not statistically significant at $\alpha = 0.05$. The results demonstrate that the hurricanes and tropical storms had similar impacts on daily rainfall, runoff, WYLD, and stream discharge.

Our PCA showed that rainfall, PET, ET, and stream discharge were highly influenced by the hurricanes and tropical storms, but percolation, runoff, groundwater discharge and WYLD were relatively less influenced by the hurricanes and tropical storms.

It should be noted that only one USGS gage station at the basin outlet and one hydrological variable (i.e., stream discharge) were used for model calibration and validation in this study. Further study is, therefore, warranted to include more gage stations and hydrological variables for such a purpose.

**Author Contributions:** Conceptualization, Y.O.; methodology, Y.O.; validation, Y.O.; formal analysis, Y.O.; investigation, Y.O.; resources, Y.O. and J.M.G.; data curation, Y.O.; writing—original draft preparation, Y.O.; writing—review and editing, Y.O., J.M.G., P.B.P. and P.V.C.; visualization. All authors have read and agreed to the published version of the manuscript.

**Funding:** This research received no external funding.

**Institutional Review Board Statement:** Not applicable.

**Informed Consent Statement:** Not applicable.

**Data Availability Statement:** Data will be made available by the authors upon request.

**Conflicts of Interest:** The authors declare no conflict of interest.

## List of Acronyms

| | |
|---|---|
| ACFRB | Apalachicola–Chattahoochee–Flint River basin |
| ET | evapotranspiration |
| HAWQS | Hydrologic and Water Quality System |
| PCA | principal component analysis |
| PET | potential evapotranspiration |
| PFA | principal factor analysis |
| TS | tropical storm |
| WYLD | water yield |

## References


1. Conner, W.H.; Duberstein, J.A.; Day, J.W.; Hutchinson, S. Impacts of Changing Hydrology and Hurricanes on Forest Structure and Growth Along a Flooding/Elevation Gradient in a South Louisiana Forested Wetland from 1986 to 2009. *Wetlands* **2014**, *34*, 803–814. [CrossRef]
2. Emanuel, K. Increasing destructiveness of tropical cyclones over the past 30 years. *Nature* **2005**, *46*, 686–688. [CrossRef]
3. Swiss Re Institute. News Release. 2018. Available online: https://www.swissre.com/media/news-releases/2018/nr20180410_sigma_global_insured_loses_highest_ever.html (accessed on 8 September 2021).
4. Vieten, R.; Warken, S.; Winter, A.; Schroder-Ritzrau, A.; Scholz, D.; Spotl, C. Hurricane Impact on Seepage Water in Larga Cave, Puerto Rico. *J. Geophys. Res. Biogeosci.* **2018**, *123*, 879–888. [CrossRef]
5. Vidon, P.; Marchese, S.; Rook, S. Impact of Hurricane Irene and Tropical Storm Lee on riparian zone hydrology and biogeochemistry. *Hydrol. Processes* **2017**, *31*, 476–488. [CrossRef]
6. Zhang, Y.; Ye, F.; Yu, H.C.; Sun, W.L.; Moghimi, S.; Myers, E.; Nunez, K.; Zhang, R.Y.; Wang, H.; Roland, A.; et al. Simulating compound flooding events in a hurricane. *Ocean. Dyn.* **2020**, *70*, 621–640. [CrossRef]
7. Jayakaran, A.D.; Williams, T.M.; Ssegane, H.; Amatya, D.M.; Song, B.; Trettin, C.C. Hurricane impacts on a pair of coastal forested watersheds: Implications of selective hurricane damage to forest structure and streamflow dynamics. *Hydrol. Earth Syst. Sci.* **2014**, *18*, 1151–1164. [CrossRef]
8. Torres, J.M.; Bass, B.; Irza, N.; Fang, Z.; Proft, J.; Dawson, C.; Kiani, M.; Bedient, P. Characterizing the hydraulic interactions of hurricane storm surge and rainfall-runoff for the Houston-Galveston region. *Coast. Eng.* **2015**, *106*, 7–19. [CrossRef]
9. Brena-Naranjo, J.A.; Pedrozo-Acuna, A.; Pozos-Estrada, O.; Jimenez-Lopez, S.A.; Lopez-Lopez, M.R. The contribution of tropical cyclones to rainfall in Mexico. *Phys. Chem. Earth* **2015**, *83–84*, 111–122. [CrossRef]
10. Hou, Y.P.; Zhang, M.F.; Liu, S.R.; Sun, P.S.; Yin, L.H.; Yang, T.L.; Li, Y.D.; Li, Q.; Wei, X.H. The Hydrological Impact of Extreme Weather-Induced Forest Disturbances in a Tropical Experimental Watershed in South China. *Forests* **2018**, *9*, 734. [CrossRef]
11. Vidon, P.; Karwan, D.L.; Andres, A.S.; Inamdar, S.; Kaushal, S.; Morrison, J.; Mullaney, J.; Ross, D.S.; Schroth, A.W.; Shanley, J.B.; et al. In the path of the Hurricane: Impact of Hurricane Irene and Tropical Storm Lee on watershed hydrology and biogeochemistry from North Carolina to Maine, USA. *Biogeochemistry* **2018**, *141*, 351–364. [CrossRef]
12. USGS (US Geological Survey). As One of Several National Focus Area Studies within the USGS National Water Census (NWC) the USGS Completed a 3-Year Study of Water Availability and Use in the Apalachicola-Chattahoochee-Flint (ACF) River Basin. 2021. Available online: https://www.usgs.gov/mission-areas/water-resources/science/apalachicola-chattahoochee-flint-river-basin-focus-area-study?qt-science_center_objects=0#qt-science_center_objects (accessed on 5 December 2021).
13. Stallins, J.A.; Nesius, M.; Smith, M.; Watson, K. Biogeomorphic characterization of floodplain forest change in response to reduced flows along the Apalachicola River, Florida. *River Res. Appl.* **2010**, *26*, 242–260. [CrossRef]
14. Wharton, C.H. *The Natural Environments of Georgia*; Georgia Department of Natural Resources: Atlanta, GA, USA, 1978; p. 227.
15. Ouyang, Y. New insights on evapotranspiration and water yield in crop and forest lands under changing climate. *J. Hydrol.* **2021**, *603*, 127192. [CrossRef]
16. Manly, B.F.; Alberto, J.A.N. *Multivariate Statistical Methods: A Primer*; CRC Press: Boca Raton, FL, USA, 2016.
17. Ouyang, Y. Evaluation of river water quality monitoring stations by principal component analysis. *Water Res.* **2005**, *39*, 2621–2635. [CrossRef] [PubMed]
18. Bryant, F.B.; Yarnold, P.R. Principal-components analysis and exploratory and confirmatory factor analysis. In *Reading and Understanding Multivariate Statistics*; Grimm, L.G., Yarnold, P.R., Eds.; American Psychological Association: Washington, DC, USA, 1995; pp. 99–136.
19. Betancur, S.B.; Gastmans, D.; Vasquez, K.V.; Santarosa, L.V.; dos Santos, V.; Kirchheim, R.E. Hydrological responses in equatorial watersheds indicated by Principal Components Analysis (PCA)-study case in Atrato River Basin (Colombia). *Rev. Bras. Recur. Hídr.* **2020**, *25*, 1–12. [CrossRef]
20. Olsen, R.L.; Chappell, R.W.; Loftis, J.C. Water quality sample collection, data treatment and results presentation for principal components analysis-literature review and Illinois River watershed case study. *Water Res.* **2012**, *46*, 3110–3122. [CrossRef] [PubMed]

21. Ouyang, Y.; Nkedi-Kizza, P.; Wu, Q.T.; Shinde, D.; Huang, C.H. Assessment of seasonal variations in surface water quality. *Water Res.* **2006**, *40*, 3800–3810. [CrossRef] [PubMed]
22. Sergeant, C.J.; Starkey, E.N.; Bartz, K.K.; Wilson, M.H.; Mueter, F.J. A practitioner's guide for exploring water quality patterns using principal components analysis and Procrustes. *Environ. Monit. Assess.* **2016**, *188*, 249. [CrossRef] [PubMed]

climate

MDPI

Article

# Projection of Sediment Loading from Pearl River Basin, Mississippi into Gulf of Mexico under a Future Climate with Afforestation

Ying Ouyang [1,*], Yanbo Huang [2], Prem B. Parajuli [3], Yongshan Wan [4], Johnny M. Grace [5], Peter V. Caldwell [6] and Carl Trettin [7]

1 USDA Forest Service, Center for Bottomland Hardwoods Research, Southern Research Station, 775 Stone Blvd., Thompson Hall, Room 309, Mississippi State, MS 39762, USA

2 Genetics and Sustainable Agriculture Research Unit, Crop Science Research Laboratory, USDA-Agricultural Research Service, 810 Highway 12 East, Mississippi State, MS 39762, USA; yanbo.huang@usda.gov

3 Department of Agricultural and Biological Engineering, Mississippi State University, Mississippi State, MS 39762, USA; pparajuli@abe.msstate.edu

4 Center for Environmental Measurement and Modeling, US EPA, 1 Sabine Island Drive, Gulf Breeze, FL 32561, USA; wan.yongshan@epa.gov

5 USDA Forest Service, Center for Forest Watershed Research, Southern Research Station, 1740 S. Martin Luther King Jr. Blvd., Perry-Paige Bldg., Suite 303 North, Tallahassee, FL 32307, USA; johnny.m.grace@usda.gov

6 USDA Forest Service, Center for Integrated Forest Science, Southern Research Station, 3160 Coweeta Lab Road, Otto, NC 28763, USA; peter.v.caldwell@usda.gov

7 USDA Forest Service enter for Forest Watershed Research, 3734 Hwy 402, Cordesville, SC 29434, USA; carl.c.trettin@usda.gov

* Correspondence: ying.ouyang@usda.gov

**Abstract:** Sediment load in rivers is recognized as both a carrier and a potential source of contaminants. Sediment deposition significantly changes river flow and morphology, thereby affecting stream hydrology and aquatic life. We projected sediment load from the Pearl River basin (PRB), Mississippi into the northern Gulf of Mexico under a future climate with afforestation using the SWAT (Soil and Water Assessment Tool)-based HAWQS (Hydrologic and Water Quality System) model. Three simulation scenarios were developed in this study: (1) the past scenario for estimating the 40-year sediment load from 1981 to 2020; (2) the future scenario for projecting the 40-year sediment load from 2025 to 2064, and (3) the future afforestation scenario that was the same as the future scenario, except for converting the rangeland located in the middle section of the Pearl River watershed of the PRB into the mixed forest land cover. Simulations showed a 16% decrease in sediment load for the future scenario in comparison to the past scenario due to the decrease in future surface runoff. Over both the past and future 40 years, the monthly maximum and minimum sediment loads occurred, respectively, in April and August; whereas the seasonal sediment load followed the order: spring > winter > summer > fall. Among the four seasons, winter and spring accounted for about 86% of sediment load for both scenarios. Under the future 40-year climate conditions, a 10% reduction in annual average sediment load with afforestation was observed in comparison to without afforestation. This study provides new insights into how a future climate with afforestation would affect sediment load into the northern Gulf of Mexico.

**Keywords:** afforestation; climate change; HAWQS model; Gulf of Mexico; sediment load

## 1. Introduction

Sediment load in surface waters is a potential source of contaminants to aquatic environments as sediment adsorbs and transports toxic chemicals, excess nutrients, and pathogens, while sediment deposition can change stream flow and morphology and thereby has broad impacts on hydrologic channels, aquatic life, and recreation activity [1]. Agricultural practices, forest disturbances, and urbanization activities are the major sources of sediment contamination and deposition in rivers, streams, and lakes [2–5]. When the rates and quantities of soil erosion and sediment load are sufficient to adversely affect terrestrial communities and streams, the surface water resources are impaired by sediments [6]. Consequently, this is occurring in the lower Mississippi River basin (LMRB) which discharges to the northern Gulf of Mexico (NGOM). This basin is one of the most disturbed by human activity among the world's largest coastal and river basins [5,7]. From the 1950s to the 1970s, the LMRB underwent a widespread loss of bottomland hardwood forests due to the clearcuttings for crop production, flood control, and floodplain development [8]. Such anthropogenic activities are largely responsible for the increased sediment loads in the surface water systems of the LMRB and NGOM [9]. Milliman and Meade [10] reported that the Mississippi River delivered 210 Mt (million tons)/y of sediment to the NGOM from the 1950s to the 1980s, while Bentley et al. [11] estimated that the lower reaches of the Mississippi and Atchafalaya Rivers discharged 57 and 71.5 Mt/y of sediment, respectively, from Tarbert Landing and Simmesport into the NGOM from 2008 to 2010. Xu et al. [12] reviewed and synthesized sediment dynamics in the Mississippi River deltaic plain. These authors suggested that future efforts should be focused on enhancing river sediment delivery, increasing sediment retention, and minimizing soil erosion.

Despite numerous studies having been made to investigate the sediment load from the LMRB into the NGOM [5,13–17], limited efforts have been devoted to estimating the impacts of forest watersheds, and their restoration practices, on sediment load in the region. Ouyang et al. [5] applied the HSPF (Hydrological Simulation Program-FORTRAN) model to predict the role of afforestation on sediment load in the lower Yazoo River Watershed, Mississippi. They reported that a conversion of marginal agricultural land into forests reduced sediment load. In general, a two-fold increase in forest land area resulted in approximately a two-fold reduction in annual sediment load, which occurred because forests absorb water, reduce surface water runoff, and prevent soil erosion. However, a thorough literature review reveals that very few efforts have been undertaken to estimate the contributions of forest watersheds to sediment delivery from the LMRB into the NGOM under a changing climate. This knowledge is critical to the assessment of relations among afforestation, climatic, environmental, and economic impacts in the region.

The goal of this study was to predict sediment load from the Pearl River basin (PRB), Mississippi (within the LMRB) into the NGOM under the past climate and the future climate with afforestation, using the Hydrologic and Water Quality System (HAWQS) model. Our specific objectives were to: (1) create and calibrate the PRB model; and (2) apply the model to estimate sediment load from the PRB into the NGOM over the past 40 years (1981 to 2020), future 40 years (2025 to 2064), and future 40 years with afforestation (i.e., conversion of a rangeland into a mixed forest land).

## 2. Materials and Methods

### 2.1. Study Site

The PRB covers southeastern Louisiana and southwestern Mississippi (Figure 1) with a drainage area of 22,533 km$^2$, encompassing all or parts of the 24 counties in Mississippi and 3 parishes in Louisiana. The Pearl River in the basin originates from the east central region of Mississippi and flows into the NGOM with a length of about 790 km. Nearly one million people live in the basin and more than one-third are residents of Mississippi [18]. The basin consists of five watersheds, namely the upper Pearl River (Hydrologic Unit Code (HUC) 03180001, 6379 km$^2$), middle Pearl River (HUC 03180002, 5120 km$^2$), lower-middle Pear River (HUC 03180003, 3156 km$^2$), lower Pearl River (HUC 03180004, 4717 km$^2$), and Bogue Chitto River (HUC 03180005, 3130 km$^2$) watersheds. Average maximum temperature is 33.4 °C in July and average minimum temperature is 0.3 °C in January. Annual average precipitation is 1412 mm with a wet period from November to April and a dry period from May to October. The PRB is on the 303(d) list of impaired waterways due to excessive nutrient levels (nitrogen and phosphorus), mercury concentrations, sediment loadings, and pesticide concentrations [19]. The basin is dominated by 69% forest, followed by 27%

agricultural land, and 3% wetland and water area [20]. The predominant soil types in the basin are fine sandy loam and silt loam soils [21]. The basin is the largest area of intact bottomland hardwood forest remaining in the southeastern US with a total of more than 14,000 ha, which is composed of mixed bottomland hardwood forest dominated by various oaks (*Quercus* spp.), sweetgum (*Liquidambar stryaciflua*), hickories (*Carya* spp.), and elms (*Ulmus* spp.) [22].

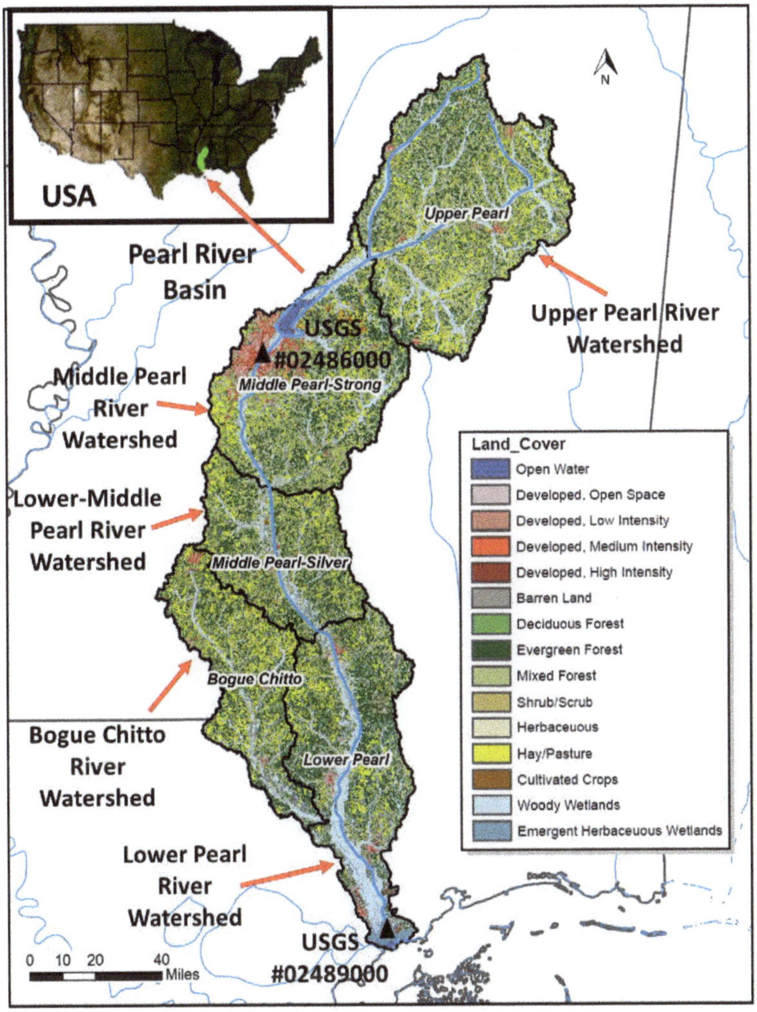


**Figure 1.** Location and land use of the Pear River basin in Mississippi, USA.

### *2.2. HAWQS Model Description*

HAWQS is a SWAT (Soil and Water Assessment Tool)-based model with a user-friendly interface and it is used to simulate hydrological processes and water quality in watersheds of continental USA, in conjunction with impacts of land use, land management, and climate change [23]. The past and future weather datasets are pre-loaded into HAWQS for users' convenience. In recent years, several modeling studies with HAWQS have been reported in the literature. Among them, Yen et al. [24] applied HAWQS to predict water quantity and quality in the Illinois River watershed; Fant et al. [25] used HAWQS to simulate future water quality conditions and economic impact in the US; and Ouyang [26] employed HAWQS to estimate ET in forest lands of Mississippi. Additional applications of HAWQS can be found elsewhere [27–30]. Overall, HAWQS is a timesaving and cost-effective modeling system for simulating water quantity and quality in complex watersheds. However, it should be noted that any limitations inherited by the SWAT model are also applied to the HAWQS model. Additionally, HAWQS has fewer input parameters that can be used for model calibration and is not as flexible as SWAT.

The PRB model was created using HAWQS with the following three major steps: (1) Create a project. The PRB model was created at the data resolution of HUC-8 for the downstream HUC # 03180004 (or lower Pearl River watershed). After entering this HUC number into the map options of HAWQS, the additional four upstream watersheds contributing to the lower Pearl River watershed were included. This created the entire PRB model (Figure 1); (2) Set the HRU (Hydrologic Response Unit). A threshold level of 1% for HRUs was used to eliminate the effects of minor land uses, soils, and slopes in each watershed; and (3) Develop scenarios. Three simulation scenarios were developed in this study: the past scenario for simulating the past 40-year sediment load from 1981 to 2020; the future scenario for projecting the future 40-year sediment load from 2025 to 2064, and the future afforestation scenario for projecting the future 40-year sediment load from 2025 to 2064 associated with conversion of rangeland into mixed forest land in the middle Pearl River watershed of the PRB. It should be noted that the general management input parameter values such as initial leaf area index, number of heat units to bring plant to maturity, and width of edge-of-field filter strip were changed when converting the rangeland to the mixed forest land. The weather data for the past scenario were from PRISM (Parameter-elevation Regressions on Independent Slopes Model), while the weather data for the future scenario were from CCSM4-RCP85 (Community Climate System Model version 4—Representative Concentration Pathway 85). Both datasets are readily available in HAWQS. Table 1 lists the major modeling input parameter values used in this study. An elaborate description on how to develop and execute the HAWQS model can be found in HAWQS [23]. A flowchart showing the HAWQS modeling procedure is given in Figure 2.

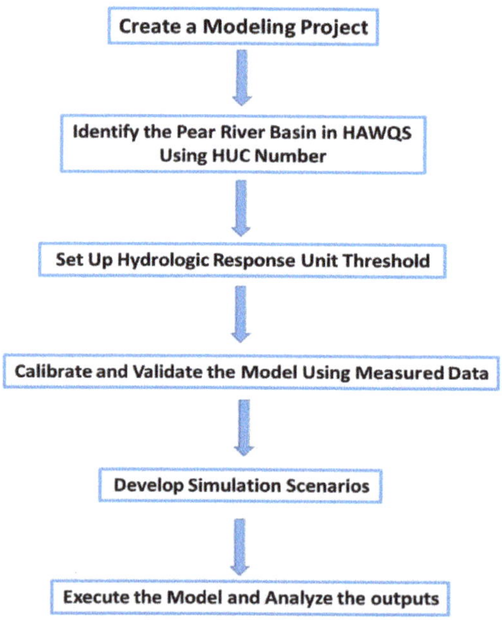


**Figure 2.** The flowchart of HAWQS modeling procedures used in this study.

**Table 1.** Major input parameter values used for the PRB model.

| Parameter | Definition | Value | Unit/method/Explanation | Reference |
| --- | --- | --- | --- | --- |
| SFTMP | Snowfall temperature | 1 | °C | Local observation |
| SMTMP | Snowmelt base temperature | 0.5 | °C | Local observation |
| SMFMX | Melt factor for snow on June 21 | 4.5 | mm $H_2O$/°C-day | Local observation |
| SMFMN | Melt factor for snow on December 21 | 4.5 | mm $H_2O$/°C-day | Local observation |
| TIMP | TIMP: Snowpack temperature lag factor | 1 | | Local observation |
| IPET | Potential evapotranspiration (PET) method | 2 | Hargreaves method | Calibrated |
| ESCO | Soil evaporation compensation factor | 0.95 | | Calibrated |
| EPCO | EPCO: Plant uptake compensation factor | 1 | | Calibrated |
| ICN | Daily curve number calculation method | 0 | Calculate daily CN value as a function of soil moisture | Calibrated |
| CNCOEF | Plant ET curve number coefficient | 1 | | Calibrated |
| ICRK | Crack flow code | 0 | Do not model crack flow in soil | Local observation |
| SURLAG | Surface runoff lag time | 4 | days | Calibrated |
| CN2 | Subbasins curve number | 0 | | Calibrated |
| IRTE | Channel water routing method | 0 | Variable Storage Method | Calibrated |
| MSK_COL1 | Calibration coefficient used to control impact of the storage time constant for normal flow | 0 | | Calibrated |
| MSK_COL2 | Calibration coefficient used to control impact of the storage time constant for low flow | 3.5 | | Calibrated |
| MSK_X | Weighting factor controlling relative importance of inflow rate and outflow rate in determining water storage in reach segment | 0.2 | | Calibrated |
| TRNSRCH | Fraction of transmission losses from main channel that enter deep aquifer | 0 | | Calibrated |
| EVRCH | Reach evaporation adjustment factor | 1 | | Calibrated |
| IDEG | Channel degradation code | 0 | Channel dimension is not updated as a result of degradation | Local observation |
| PRF | Peak rate adjustment factor for sediment routing in the main channel | 1 | | Calibrated |
| SPCON | Linear parameter for calculating the maximum amount of sediment that can be re-entrained during channel sediment routing | 0 | | Calibrated |
| SPEXP | Exponent parameter for calculating sediment re-entrained in channel sediment routing | 1 | | Calibrated |
| IWQ | In-stream water quality code | 1 | | Calibrated |
| ADJ_PKR | Peak rate adjustment factor for sediment routing in the subbasin (tributary channels) | 0.5 | | Calibrated |
| Weather dataset 1 | PRISM | Time series | Past scenario | Downloaded from HAWQS |
| Weather dataset 2 | CCSM4-RCP85 | Time series | Future scenario | Downloaded from HAWQS |

## 3. Results and Discussion

### 3.1. Model Calibration and Validation

The PRB model was calibrated and validated for the discharges and sediment loads using observed data. The observed discharges were downloaded from the US Geological Survey (USGS) Station #02489000 in the Pearl River near Columbia, MS. The observed sediment data are very limited and were obtained from both stations, #02489000 and #02486000 (near Jackson, MS) (Figure 1). The model calibration was accomplished by adjusting the input parameter values (Table 1) so that the model predictions best matched the field observations, while the model validation was performed to compare the model predictions with another independent set of observations without adjusting any input parameter values. The goodness-of-fit during the model calibration and validation was determined with coefficient of determination ($R^2$), Nash Sutcliff efficiency (NSE), and percent bias (PBIAS).

Figure 3 compared the predicted and observed discharges during the model calibration for a 10-year period from 1 January 1999 to 31 December 2008 and during the model validation for a 10-year period from 1 January 2009 to 31 December 2018. The values of $R^2$, NSE, and PBIAS for daily discharge were, respectively, 0.62, 0.47, and 9.12 for the calibration (Figure 3a), and 0.66, 0.5, and 19.38 for the validation (Figure 3b). These statistical values suggested that good agreements were attained between the predicted and observed daily discharges during the model calibration and validation [31]. Figure 3c showed the predicted and observed sediment loads during the model calibration from 12 September 1974 to 6 June 1975 as well as during the model validation from 12 March to 28 September 1981. The values of $R^2$, NSE, and PBIAS were, respectively, 0.69, 0.48, and 8.4 for the model calibration (Figure 3c), and 0.91, 0.84, and 119 for the model validation (Figure 3d). Results concluded that reasonable agreements were obtained between the predicted and observed daily sediment loads during the model calibration and validation. It should be noted that the field measured sediment data are very limited, sparse, and intermittent in the PRB. The long-term and continuous sediment data in the PRB are not attainable for rigorous model calibration and validation.

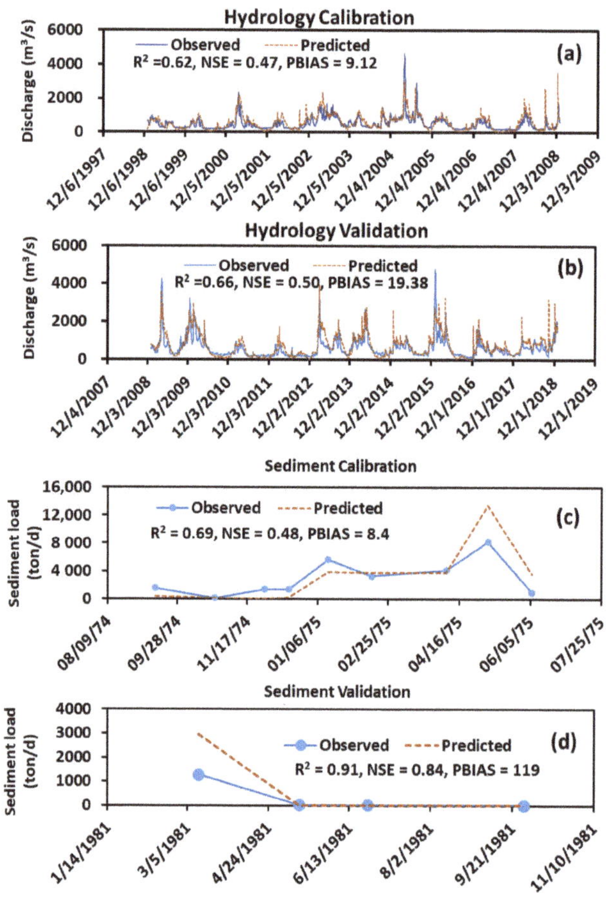


**Figure 3.** Comparisons of the observed and predicted daily discharges (**a**,**b**) and sediment loads (**c**,**d**) during the model calibration and validation.

*3.2. Daily, Monthly, and Annual Sediment Load*

The changes in the daily, monthly, and annual sediment load at the PRB outlet into the NGOM for the past and future scenarios are, respectively, shown in Figures 4 and 5. All the changes are significant at $\alpha = 0.01$ between the past and future scenarios among the daily, monthly, and annual values based on the Kolmogorov–Smirnov test. In general, the daily sediment load varied from date to date. The average daily sediment loads at the basin outlet were 872 and 753 ton/d, respectively, for the past and future scenarios. In other words, there was a 16% decrease in the sediment load for the future scenario in comparison to that of the past scenario. We attributed the future sediment load reduction to the decrease in the future daily average maximum precipitation, although the average annual precipitations between the past and future scenarios were very close (Table 2). The basin average daily maximum precipitation was 38 mm for the past scenario but was 29 mm for the future scenario. The latter was about 31% less than the former. The lower future daily average maximum precipitation had resulted in the lower annual average surface runoff (Table 2). The lower surface water runoff would result in the lower soil erosion and thus the lower sediment load in the streams. The results suggested that the higher precipitation rate is an important factor for sediment load in the streams.

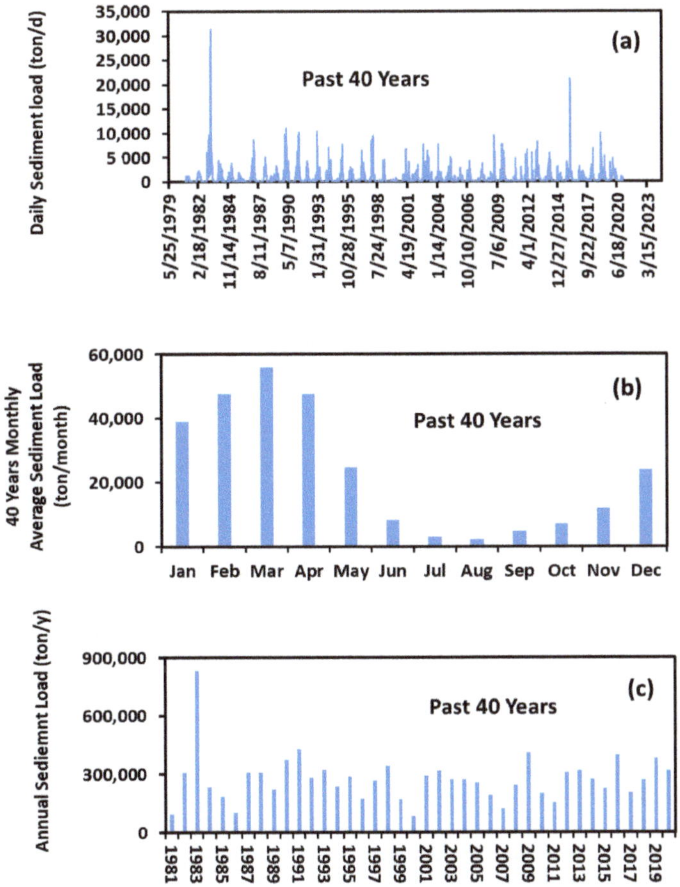


**Figure 4.** Daily (**a**), monthly (**b**), and annual (**c**) sediment loads at the PRB outlet over the past 40 years from 1981 to 2020.

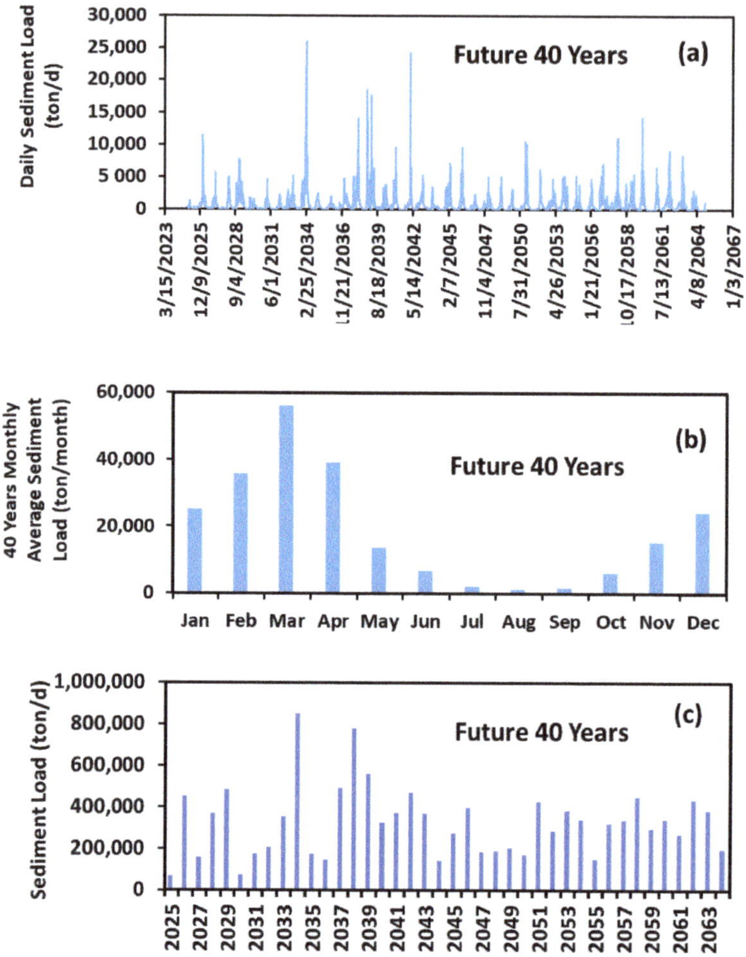


**Figure 5.** Daily (**a**), monthly (**b**), and annual (**c**) sediment loads at the PRB outlet over the future 40 years from 2025 to 2064.

Variations of the 40-year monthly average sediment load in the past scenario are shown in Figure 4b. The maximum was 55,776 ton/month in March, while the minimum was 2257 ton/month in August. Thorne et al. [32] estimated the historical sediment load in the lower Mississippi River and reported that the largest monthly sediment load was in April and the smallest one was in September, at the Tarbert Landing watershed that is adjacent to the PRB, which were somewhat comparable to our findings. Analogous to the case of the past scenario, the same temporal monthly trend was observed for the future scenario (Figure 5b). The results indicated that the monthly variation in sediment load did not change for the past and future scenarios.

A seasonal pattern of the past and future sediment loads can be deducted from Figures 4b and 5b. That is, the seasonal sediment load for both scenarios increased from winter to spring and decreased from spring through summer to fall with the following order: spring > winter > summer > fall. This order was consistent with that reported by Thorne et al. [32]. Among the four seasons, winter and spring accounted for 87% of the sediment load for the past scenario and 86% of the sediment load for the future scenario. Winter and spring are wet seasons while summer and fall are dry seasons in the PRB.

Heavy and frequent precipitation in the wet seasons would produce more sediment load in streams.

**Table 2.** Annual average precipitation, surface water runoff, and sediment load for each watershed within the PRB.

| Watershed Name | HUC Number | Past Annual Average Precipitation (mm) | Future Annual Average Precipitation (mm) | Past Daily Average Maximum Rainfall (mm) | Future Daily Average Maximum Rainfall (mm) | Past Annual Average Surface Water Runoff (mm) | Future Annual Average Surface Water Runoff (mm) |
|---|---|---|---|---|---|---|---|
| Upper Pearl River Watershed | 3180001 | 1459 | 1461 | 36 | 26 | 250 | 178 |
| Middle Pearl River Watershed | 3180002 | 1467 | 1469 | 35 | 28 | 290 | 220 |
| Lower-Middle Pearl River Watershed | 3180003 | 1535 | 1514 | 39 | 29 | 108 | 75 |
| Lower Pearl River Watershed | 3180004 | 1617 | 1629 | 40 | 30 | 242 | 181 |
| Bogue Chitto River Watershed | 3180005 | 1612 | 1608 | 41 | 30 | 270 | 194 |
| **Basin Average** | | 1538 | 1536 | 38 | 29 | 232 | 170 |

Analogous to the case of the daily sediment loads, the annual sediment loads for the past and future scenarios varied from year to year (Figures 4c and 5c). The largest annual sediment loads were found in 1983 for the past scenario (Figure 4c) and in 2034 for the future scenario (Figure 5c). These occurred because of the largest daily sediment loads during those years (Figures 4a and 5a). Overall, the 40-year sediment load was 15,926,342 tons for the past scenario and was 11,010,298 tons for the future scenario. The latter was 30% less than the former. The results indicated a reduction of the sediment load in the next 40 years (from 2025 to 2064) from the PRB into the NGOM due to the reduction of surface water runoff because of a decrease in the average daily maximum precipitation in the next 40 years.

A Mann–Kendall analysis was performed for annual precipitation, discharge, and sediment load over both the past and future 40 years. The Mann–Kendall statistic $\tau$ ranges from $-1$ to 1 and measures the relationships between variables and times. If $\tau = 0$, no relationship exists, while $\tau = 1$ indicates a perfect increasing trend, and $-1$ a perfect decreasing trend. The $p$-value is a statistical measure of a trend and, if $p \leq 0.05$, there is a monotonic trend [33]. While there were some decreasing ($-\tau$) and increasing ($+\tau$) trends in some hydrological variables, our Mann–Kendall statistical test revealed no significant trends for the annual precipitation, discharge, and sediment load over both the past and future 40 years (Figure 6).

The plots of annual sediment load with annual discharge showed the following good linear correlations: $Y = 2.53X - 100{,}612$ with $R^2 = 0.701$ for the past 40 years and $Y = 3.24X - 117{,}356$ with $R^2 = 0.84$ for the future 40 years. Similar correlations were obtained for watersheds at various locations in the USA by Ouyang [1].

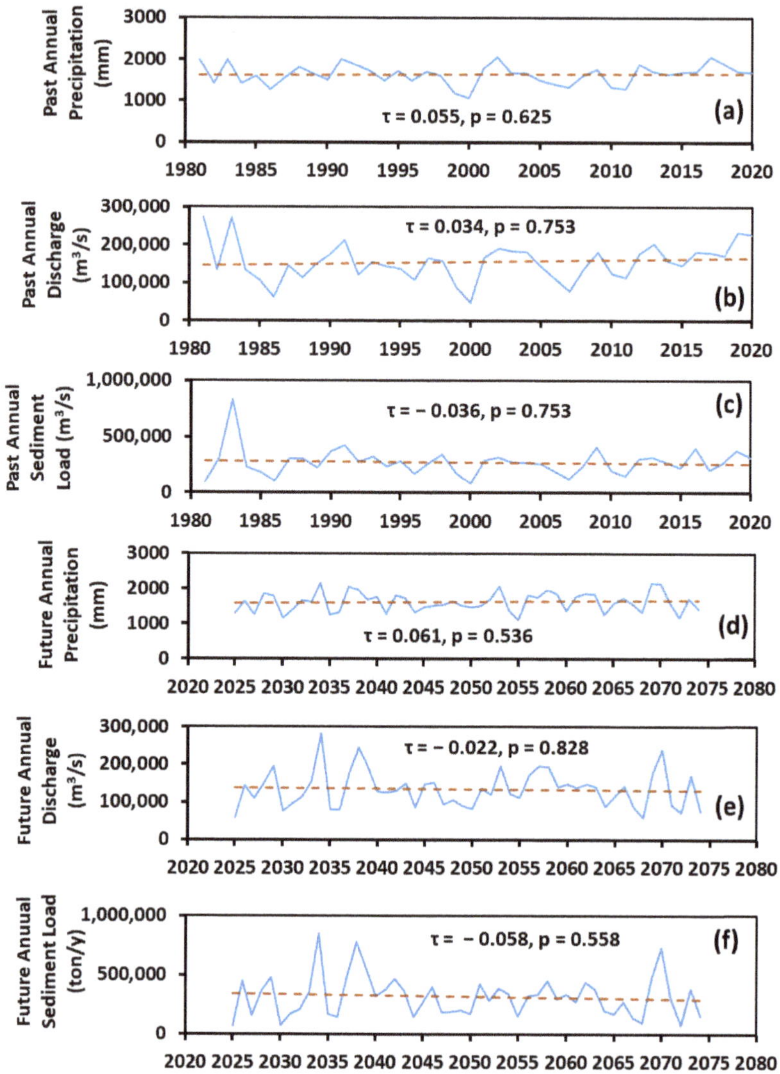


**Figure 6.** Trend analysis (Mann–Kendall) for the past annual precipitation (**a**), discharge (**b**), and sediment load (**c**) as well as for the future annual precipitation (**d**), discharge (**e**), and sediment load (**f**).

### 3.3. Effects of Afforestation under a Future Climate

Afforestation is a field process to grow trees in the non-forest lands to create forest plantations. Afforestation conserves rainwater, diffuses surface runoff, and absorbs pollutants, which mitigates river flooding, reduces stream sediment load, and generates a higher quality water (i.e., clean water) [5]. The differences in the annual average sediment yield between the future and future afforestation scenarios among the five watersheds within the PRB are shown in Figure 7. The differences were calculated by subtracting the values of the future scenario from the values of the future afforestation scenario. In the future afforestation scenario, the rangeland of the middle Pearl River watershed was converted to the mixed forest land.

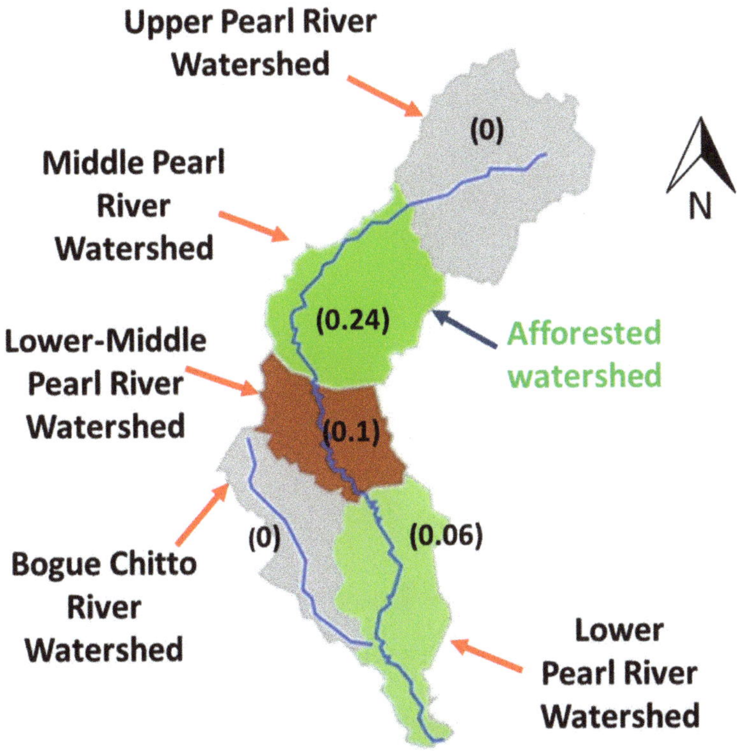


**Figure 7.** Differences (values in paratheses at ton/ha/y) in annual average sediment yield between the future and future afforestation scenarios among the watersheds within the Pearl River basin.

Little to no difference in the annual average sediment yield was observed for the upper Pearl River watershed as afforestation in the middle Pearl River watershed did not affect the hydrological processes and sediment transport at the upstream watershed. No difference in the annual average sediment yield was observed at the Bogue Chitto River watershed either after afforestation because the streams in the middle Pearl River watershed did not flow through the Bogue Chitto River watershed. There was a 0.24 ton/ha/y reduction in the annual average sediment yield at the middle Pearl River watershed after afforestation. This afforested watershed further affected its downstream watersheds, i.e., the lower-middle Pearl River and lower Pearl River watersheds. There were 0.1 and 0.06 ton/ha/y reductions at the lower-middle Pearl River and lower Pearl River watersheds, respectively. Overall, a 10% reduction in the annual average sediment yield with afforestation was observed as compared to without afforestation. The results indicated that afforestation reduced the sediment load at the afforested watershed and its downstream watersheds.

## 4. Summary

The HAWQS-based PRB model was developed to simulate sediment load from the PRB into the NGOM. Good agreements were obtained between model predictions and field measurements during the model calibration. Three simulation scenarios were then created to project sediment loads for the past 40 years (from 1981 to 2020), future 40 years (from 2025 to 2064), and future 40 years with afforestation.

Over the past and future 40 years, the maximum sediment load occurred in April and the minimum sediment load in August. The pattern of the monthly variation in sediment load was the same in both the past and future scenarios.

A seasonal pattern of the past and future sediment loads was observed: increasing from winter to spring and decreasing from spring through summer to fall with the following order: spring > winter > summer > fall. Our study revealed that there was 16% less sediment load from the PRB into the NGOM for the future 40 years than for the past 40 years due to the reduction in surface water runoff that resulted from the decrease in the daily maximum precipitations.

While there were some decreasing ($-\tau$) and increasing ($+\tau$) trends in some hydrological variables, the Mann–Kendall statistical test revealed no significant trends for the annual precipitation, discharge, and sediment load over both the past and future 40 years.

Afforestation reduced the sediment load from the PRB into the NGOM at the afforested watershed and its downstream watersheds. This finding provides useful reference for forest restoration practices.

**Author Contributions:** Conceptualization, Y.O.; methodology, Y.O.; validation, Y.O.; formal analysis, Y.O.; investigation, Y.O.; resources, Y.O.; data curation, Y.O.; writing—original draft preparation, Y.O.; writing—review and editing, Y.O., Y.H., P.B.P., Y.W., J.M.G., P.V.C. and C.T. All authors have read and agreed to the published version of the manuscript.

**Funding:** This research received no external funding.

**Institutional Review Board Statement:** Not applicable.

**Informed Consent Statement:** Not applicable.

**Data Availability Statement:** Data will be available by the authors upon request.

**Acknowledgments:** The views expressed in this article are those of the authors and do not necessarily reflect the views or policies of the United States Department of Agriculture and the U.S. Environmental Protection Agency.

**Conflicts of Interest:** The authors declare no conflict of interest.

## Abbreviations

| | |
|---|---|
| PRB | Pearl River basin |
| HAWQS | Hydrologic and Water Quality System |
| HRU | Hydrologic Response Unit |
| HUC | Hydrologic Unit Code |
| NGOM | Northern Gulf of Mexico |

## References


1. Ouyang, Y. A Gap-Filling Tool: Predicting Daily Sediment Loads Based on Sparse Measurements. *Hydrology* **2022**, *9*, 181. [CrossRef]
2. Almasalmeh, O.; Saleh, A.A.; Mourad, K.A. Soil erosion and sediment transport modelling using hydrological models and remote sensing techniques in Wadi Billi, Egypt. *Model. Earth Syst. Environ.* **2022**, *8*, 1215–1226. [CrossRef]
3. Klimas, K.; Hiesl, P.; Hagan, D.; Park, D. Prescribed fire effects on sediment and nutrient exports in forested environments: A review. *J. Environ. Qual.* **2020**, *49*, 793–811. [CrossRef] [PubMed]
4. Noe, G.B.; Cashman, M.J.; Skalak, K.; Gellis, A.; Hopkins, K.G.; Moyer, D.; Webber, J.; Benthem, A.; Maloney, K.; Brakebill, J.; et al. Sediment dynamics and implications for management: State of the science from long-term research in the Chesapeake Bay watershed, USA. *WIREs Water* **2020**, *7*, e1454. [CrossRef]
5. Ouyang, Y.; Leininger, T.D.; Moran, M. Impacts of reforestation upon sediment load and water outflow in the Lower Yazoo River Watershed, Mississippi. *Ecol. Eng.* **2013**, *61*, 394–406. [CrossRef]
6. Simon, A.; Darby, S.E. Effectiveness of grade-control structures in reducing erosion along incised river channels: The case of Hotophia Creek, Mississippi. *Geomorphology* **2002**, *42*, 229–254. [CrossRef]
7. Xu, Z.; Xu, Y.J. Assessing effectiveness of long-term forestry best management practices on stream water quality at a basin scale-a case study in Southern USA. *Environ. Monit. Assess.* **2018**, *190*, 108. [CrossRef]
8. MacDonald, P.O.; Frayer, W.E.; Clauser, J.K.; Forsythe, S.W. *Documentation, Chronology, and Future Projections of Bottomland Hardwood Habitat Loss in the Lower Mississippi Alluvial Plain*; Ecological Services, Fish and Wildlife Service, U.S. Deptartment of the Interior: Washington, DC, USA, 1979.

9. Rosen, T.; Xu, Y.J. A Hydrograph-Based Sediment Availability Assessment: Mplications for Mississippi River Sediment Diversion. *Water* **2014**, *6*, 564–583. [CrossRef]
10. Milliman, J.D.; Meade, R.H. World-Wide Delivery of River Sediment to the Oceans. *J. Geol.* **1983**, *91*, 1–21. [CrossRef]
11. Bentley, S.J.; Blum, M.D.; Maloney, J.; Pond, L.; Paulsell, R. The Mississippi River source-to-sink system: Perspectives on tectonic, climatic, and anthropogenic influences, Miocene to Anthropocene. *Earth-Sci. Rev.* **2016**, *153*, 139–174. [CrossRef]
12. Xu, K.H.; Bentley, S.J.; Day, J.W.; Freeman, A.M. A review of sediment diversion in the Mississippi River Deltaic Plain. *Estuar. Coast. Shelf Sci.* **2019**, *225*, 106241. [CrossRef]
13. Iles, R.L.; Walker, N.D.; White, J.R.; Rohli, R.V. Impacts of a Major Mississippi River Freshwater Diversion on Suspended Sediment Plume Kinematics in Lake Pontchartrain, a Semi-enclosed Gulf of Mexico Estuary. *Estuar. Coasts* **2021**, *44*, 704–721. [CrossRef]
14. Mickey, R.; Xu, K.H.; Libes, S.; Hill, J. Sediment texture, erodibility, and composition in the Northern Gulf of Mexico and their potential impacts on hypoxia formation. *Ocean Dyn.* **2015**, *65*, 269–285. [CrossRef]
15. Mitsch, W.J.; Day, J.W.; Gilliam, J.W.; Groffman, P.M.; Hey, D.L.; Randall, G.W.; Wang, N.M. Reducing nitrogen loading to the Gulf of Mexico from the Mississippi River Basin: Strategies to counter a persistent ecological problem. *Bioscience* **2001**, *51*, 373–388. [CrossRef]
16. Obelcz, J.; Xu, K.H.; Georgiou, I.Y.; Maloney, J.; Bentley, S.J.; Miner, M.D. Sub-decadal submarine landslides are important drivers of deltaic sediment flux: Insights from the Mississippi River Delta Front. *Geology* **2017**, *45*, 703–706. [CrossRef]
17. Rabalais, N.N.; Turner, R.E.; Sen Gupta, B.K.; Platon, E.; Parsons, M.L. Sediments tell the history of eutrophication and hypoxia in the Northern Gulf of Mexico. *Ecol. Appl.* **2007**, *17*, S129–S143. [CrossRef]
18. MDEQ. *Citizen's Guide to Water Quality in the Pearl River Basin*; MDEQ: Jackson, MS, USA, 2007.
19. Joyner, T.A.; Rohli, R.V. Atmospheric influences on water quality: A simulation of nutrient loading for the Pearl River Basin, USA. *Environ. Monit. Assess.* **2013**, *185*, 3467–3476. [CrossRef]
20. Cai, Y.H.; Shim, M.J.; Guo, L.D.; Shiller, A. Floodplain influence on carbon speciation and fluxes from the lower Pearl River, Mississippi. *Geochim. Cosmochim. Acta* **2016**, *186*, 189–206. [CrossRef]
21. Parajuli, P.B. Assessing sensitivity of hydrologic responses to climate change from forested watershed in Mississippi. *Hydrol. Process.* **2010**, *24*, 3785–3797. [CrossRef]
22. Chapman, E.L.; Chambers, J.Q.; Ribbeck, K.F.; Baker, D.B.; Tobler, M.A.; Zeng, H.C.; White, D.A. Hurricane Katrina impacts on forest trees of Louisiana's Pearl River basin. *For. Ecol. Manag.* **2008**, *256*, 883–889. [CrossRef]
23. HAWQS. *HAWQS System and Data to Model the Lower 48 Conterminous U.S Using the SWAT Model*; Texas Data Repository Dataverse: Austin, TX, USA, 2020; Volume 1. [CrossRef]
24. Yen, H.; Daggupati, P.; White, M.J.; Srinivasan, R.; Gossel, A.; Wells, D.; Arnold, J.G. Application of Large-Scale, Multi-Resolution Watershed Modeling Framework Using the Hydrologic and Water Quality System (HAWQS). *Water* **2016**, *8*, 164. [CrossRef]
25. Fant, C.; Srinivasan, R.; Boehlert, B.; Rennels, L.; Chapra, S.C.; Strzepek, K.M.; Corona, J.; Allen, A.; Martinich, J. Climate Change Impacts on US Water Quality Using Two Models: HAWQS and US Basins. *Water* **2017**, *9*, 118. [CrossRef]
26. Ouyang, Y. New insights on evapotranspiration and water yield in crop and forest lands under changing climate. *J. Hydrol.* **2021**, *603*, 127192. [CrossRef]
27. Brighenti, T.M.; Gassman, P.W.; Schilling, K.E.; Srinivasan, R.; Liebman, M.; Thompson, J.R. Determination of accurate baseline representation for three Central Iowa watersheds within a HAWQS-based SWAT analyses. *Sci. Total Environ.* **2022**, *839*, 156302. [CrossRef]
28. Chen, M.Y.; Cui, Y.L.; Gassman, P.W.; Srinivasan, R. Effect of Watershed Delineation and Climate Datasets Density on Runoff Predictions for the Upper Mississippi River Basin Using SWAT within HAWQS. *Water* **2021**, *13*, 422. [CrossRef]
29. Ouyang, Y.; Grace, J.M.; Parajuli, P.B.; Caldwell, P.V. Impacts of Multiple Hurricanes and Tropical Storms on Watershed Hydrological Processes in the Florida Panhandle. *Climate* **2022**, *10*, 42. [CrossRef]
30. Ouyang, Y.; Wan, Y.S.; Yang, J.; Feng, G.; Gao, F. Application of the US-EPA's HAWQS model to predict future climate impacts on hydrological processes. *Int. J. River Basin Manag.* **2022**, *2022*, 1–12. [CrossRef]
31. Moriasi, D.N.; Arnold, J.G.; Van Liew, M.W.; Bingner, R.L.; Harmel, R.D.; Veith, T.L. Model evaluation guidelines for systematic quantification of accuracy in watershed simulations. *Trans. ASABE* **2007**, *50*, 885–900. [CrossRef]
32. Thorne, C.; Harmar, O.; Watson, C.; Clifford, N.; Biedenham, D. *Current and Historical Sediment Loads in the Lower Mississippi River*; Nottingham Univ (United Kingdom) Dept of Geography: Nottingham, UK, 2008.
33. Mangiafico, S.S. *Summary and Analysis of Extension Program Evaluation in R, Version 1.11.1*; Rutgers Cooperative Extension: New Brunswick, NJ, USA, 2016.

climate

MDPI

Article

# Extreme Seasonal Droughts and Floods in the Madeira River Basin, Brazil: Diagnosis, Causes, and Trends

**Nicole Cristine Laureanti [1,2,*], Priscila da Silva Tavares [1], Matheus Tavares [1], Daniela Carneiro Rodrigues [1], Jorge Luís Gomes [1], Sin Chan Chou [1] and Francis Wagner Silva Correia [3]**

1 National Institute for Space Research (INPE), Presidente Dutra Rod. (BR-116), km 39, Cachoeira Paulista 12630-000, SP, Brazil; priscila.tavares@inpe.br (P.d.S.T.); matheus.tavares@inpe.br (M.T.); daniela.rodrigues@inpe.br (D.C.R.); jorge.gomes@inpe.br (J.L.G.); chou.chan@inpe.br (S.C.C.)

2 Department of Environmental Sciences, Rutgers, The State University of New Jersey, 14 College Farm Rd., New Brunswick, NJ 08901, USA

3 Department of Meteorology, Universidade do Estado do Amazonas (UEA), Av. Djalma Batista, 3578, Flores, Manaus 69050-010, AM, Brazil; fcorreia@uea.edu.br

* Correspondence: nicole.laureanti@inpe.br

**Abstract:** The Madeira River, a major tributary of the Amazon River, often undergoes severe flood and drought conditions. This study seeks to investigate the climate processes associated with the opposing extreme precipitation events in the Madeira River basin and to relate them to river discharge variability based on a flood awareness dataset. Despite the uncertainty in the observational datasets, the annual precipitation cycle exhibits a rainy season from November to March. A significant result is the high correlation between the rainy season variability in the Madeira River basin and the sea surface temperature (SST) anomalies in the tropical North Atlantic Ocean and the southwestern South Atlantic Ocean. This result indicates that improving the Atlantic SST representation in climate modeling allows for capturing extreme precipitation events in the region. In addition to this impact, certain Madeira River tributaries present significant climate trends. The river discharge variability reveals an increase in hydrological extremes in recent years in the upper sector, but more significantly, in the lower basin, where it has reduced by more than 400 $m^3/s$ per decade. These findings highlight the need to improve in situ data and climate and hydrological modeling, with a focus on describing the intense climate variability and trends in river discharges.

**Keywords:** extreme precipitation; Amazon rivers; climate change; hydrology

## 1. Introduction

The Madeira River basin in southwestern South America spans Brazil, Bolivia, and Peru. The upper part of the river in Bolivia has a highly dense population. In Brazil, the lower part of the river encompasses a complex of hydropower plants, several natural reserves, plantation farms, small communities, and remote indigenous communities. Among the water consumption, most of these activities depend on the river for navigation and transportation. Therefore, the Madeira River contributes significantly to the socio-economic activities in Brazil.

These valuable hydrological conditions in the region are associated with different drivers of extreme precipitation. Whereas the elevated mean annual precipitation is 1834 mm, the basin is highly vulnerable to extreme climate variability, reaching extremes higher than 5000 mm/yr [1,2]. Different studies have focused on the drought mechanisms in the Amazonian region that occurred in 2005 and 2010, as well as the floods of 2009, 2012, and 2014 [3–6]. Specifically for the Brazilian part of the Madeira River basin, ref. [7] classified 1994, 2001, 2008, and 2014 as the rainiest years and 2000, 2005, and 2010 as the driest [5,6]. The most recent and probably the most extreme drought on record was in 2023, which caused the lowest Amazon River level and increased scientific concerns about the region [8,9]. Historically, the most intense flood occurred in 2014, when the discharge from the Madeira River increased by more than 74% [10].

The discharge from the Amazon River basin has shown linear trends over the last decades [2]. From 1974 until 2004, discharge decreased in different parts of the region, particularly in the southern part, and increased in the northwestern part [2]. According to the Sixth Assessment Report of the Intergovernmental Panel on Climate Change (IPCC AR6), warmer climates induce a decrease in the total soil moisture and possibly lead to conflicts over water use in the entire Amazon basin [11].

The precipitation variability in northern South America is often related to remote changes in ocean temperature [12–15]. The variability in the Amazonian region is associated with climate indices, such as the El Niño–Southern Oscillation (ENSO) and the Pacific Decadal Oscillation (PDO), in addition to the tropical Atlantic Ocean temperature [5,6,16]. The warm ENSO phase generally reduces the precipitation in the region, and the cold ENSO phase increases it [17–19]. The 2023 drought was also related to the ENSO, but mainly to its modulation from the warm to the cold phase from 2022 to 2023, which induces a lack of moisture during the rainy season and intensive downward movement during the drought season [9]. The 2023 event's relation to the ENSO was investigated by [8]. However, the severity of the 2023 event surpassed the natural effects of the ENSO and was mainly associated with the increase in global temperatures. In the southwestern Amazon basin, where the Madeira River is located, during the 2014 flood, the high precipitation amounts were associated with abnormal conditions in regard to the tropical Atlantic SST gradient [10], as was the drought in 2005 [6].

Increasing concerns about the region, particularly over the modification in the hydrological conditions, have motivated different studies. Many have focused on different extreme precipitation events [5,6,8–10]. A few climatological diagnoses have been made about the Brazilian part of the basin [7] or the entire Amazonian basin [2,4,16]. Others have integrated hydrological models to understand the dynamics of hydrological processes [1,3,20]. However, there is a need for further exploration and comprehension of the climatological variability and its implications for the Madeira River's regional conditions.

This study aims to contribute to understanding the meteorological features and processes associated with extreme events in the Madeira River basin. The results can help to model the climate in the region more accurately. In addition to increasing the understanding of natural processes impacting seasonal and interannual precipitation variability, this study focuses on the tendencies and impacts of river discharge. In addition, as in [17], precipitation variability is identified using different datasets to provide a more robust picture of the region.

This study is organized as follows: Section 2 describes the study area, methods, and datasets. The results in Sections 3.1–3.3 illustrate the precipitation variability in the basin and how it relates to climate indices. The meteorological patterns during years of extreme precipitation are analyzed using composites to highlight their peculiarities. Lastly, Section 4 focuses on river discharge and characterizes the flow response to precipitation extremes and tendencies. The discussion and conclusions are presented in Section 4.

## 2. Materials and Methods

### 2.1. Study Area

The Madeira River basin is a transboundary basin that extends across three countries in South America: Bolivia, Brazil, and Peru (Figure 1). With a drainage area of approximately 1,420,000 km$^2$ [21], it is one of the largest sub-basins (23%) of the Amazon basin and is on the right bank of the Amazon River. Most of the basin is found in Bolivia and Brazil, and to a lesser extent in Peru, representing 51%, 42%, and 7% of the total basin area, respectively. The basin presents considerable topographic gradients, from about 25 m in the north to almost 4000 m in the Andes Mountains in the southwestern part. Although the mountain regions remain dry for most of the year, they still exhibit some mountainous features essential to maintaining river discharge.

Due to its location in lower latitudes, the basin suffers from the seasonal variations of the Intertropical Convergence Zone (ITCZ) during austral summer, which is a common feature that occurs during the South American Monsoon System (SAMS) [22]. Studies indicate a rainy period that is established between November and April and a dry period from May to October in the Brazilian portion of the basin [7]. According to [7], using observations from 1988 until 2017 for the Brazilian part of the basin, the ITCZ contributes to establishing the rainy period [7]. The precipitation regime in the region is constrained by the SAMS, whose onset period occurs in November, and its demise occurs in April in the basin [22].

According to the Köppen classification [23], the basin contains tropical forest, tropical monsoonal, and tropical wet and dry climates (types Af, Am, and Aw). These climates are characterized by super humid summers, with monsoon-type rain and a short, dry winter season. Over the Andes, in Bolivia and Peru, the climate is characterized by warmer summers, with mild to cold and relatively dry winters (types Cwb, Cfb, and Cfc).

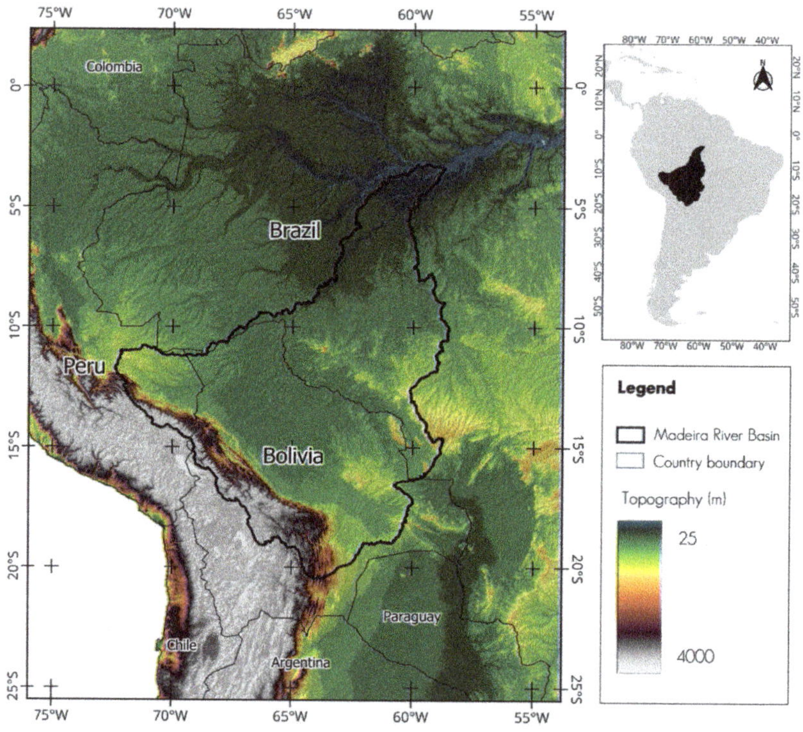


**Figure 1.** Location of the Madeira River basin. Color-shaded relief originated from a digital elevation model (DEM) derived from Shuttle Radar Topography Mission (SRTM) data, with a 3 arc-second (approximately 90 m) resolution [24].

## 2.2. *Datasets*

Different datasets explain the precipitation variability in the Madeira River basin. Initially, this study verifies the consistency between the datasets in the region due to its limited in situ gauge measurements. The two observational precipitation datasets adopted to identify the most extreme precipitation periods of the basin are the MERGE [25] and the Multi-Source Weighted-Ensemble Precipitation (MSWEP) datasets [26]. Both datasets combine gauge stations and satellite observations at a 0.1° horizontal grid resolution from

IMERG. Unlike MSWEP, the MERGE dataset is a dataset produced by the Brazilian National Institute for Space Research (in port., INPE).

Freshwater discharge from the Global Flood Awareness System (GloFAS) reanalysis dataset [27] at a 0.1° grid resolution (version 3.1) is used to diagnose the river flow in the Madeira River basin. The GloFAS derives from the daily hydrological time-series output from a model that is forced by the ECMWF Reanalysis v5 (ERA5) [28].

ERA5 [28] provides the patterns of sea surface temperature (SST), wind at lower and upper levels (850 and 200 hPa), and vertical motion at the mid-level (500 hPa) during extreme precipitation events.

### 2.3. Methodology

Figure 2 presents a detailed outline of the methodology. In the initial phase, this study compares the representation of seasonal and interannual precipitation variability in the basin for each dataset. Due to the limitations of the MERGE dataset, which includes data from June 2000, the analysis in this study considered 23 years, from 2001 until 2023. Although a more extended period would be desirable, recent studies have considered period lengths of 20 years to describe climatological features [11]. The comparison with other studies on the region over extended periods and using observational data demonstrates its sufficiency.

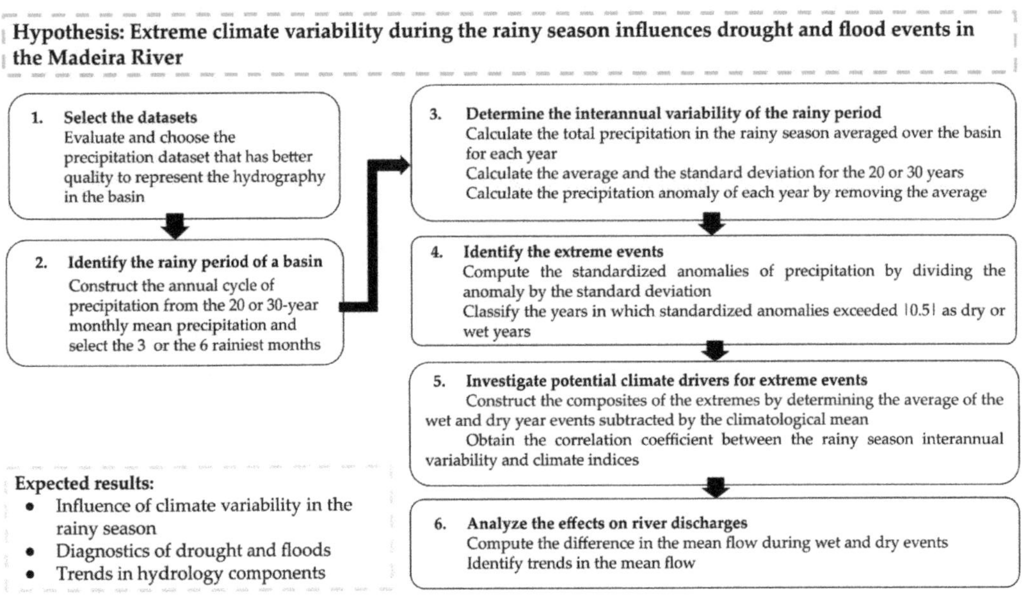


**Figure 2.** Methodology developed to analyze and identify extreme precipitation events in the Madeira River basin.

Then, the study focuses on observing the most extreme variations in the rainy season in the datasets, considering the seasonal precipitation cycle to determine the rainy season in the basin. The standardized anomalies determine the rainy season variability [29], in which rainy periods presenting deviations of $+/-0.5$ categorize extremes. The analysis of the rainy periods considers seasonal features contributing to the South American climate that infer the variability in the basin, such as the ITCZ, SACZ, and SAMS [22,30].

After identifying the most extreme events, the study investigates the climatological components that generate these events. Composite anomalies aid in characterizing meteorological variable distributions during extremes. Additionally, Pearson's correlation analysis between the precipitation variability series and climate indices, representing different variability modes, is carried out to identify their relationship to the occurrence of extremes [29]. The climate indices are a time series of the correlations/standard deviations (STDs) of the SST from the National Oceanic and Atmospheric Administration (NOAA) [31] for the determined regions. Their effects are commonly verified by an STD greater or lower than 0.5 [12]. The analysis comprises the climate indices of interannual and interdecadal variability, such as the Pacific modes Nino12, Nino34, Nino4 (representing different ENSO indices), and the PDO, as well as those for the Atlantic region, the tropical North Atlantic (TNA), the tropical South Atlantic (TSA), and the Atlantic Multidecadal Oscillation (AMO). Additionally, it considers the Quasi-Biennial Oscillation (QBO).

As indicated in Section 1, the extreme rainfall events related to the ENSO phases have different impacts on extreme precipitation in the northern parts of the continent [13]. Some studies have related the cold ENSO phase to the extreme droughts of 2005, 2010, and 2023 in the basin [5,6,9]. Along with the ENSO, different research has suggested a significant role for the TNA variability in the 2005 drought and the 2014 flood [6,10]. Other SST variability modes of interannual variability (TSA) and the interdecadal variability modes, PDO and AMO, also significantly correlate with extreme precipitation in northwestern South America [12,15]. Additionally, the Quasi-Biennial Oscillation (QBO) is considered to relate to extreme precipitation in the region, as confirmed by different models [32]. Pearson's correlation analysis investigates whether the impacts of these indices on South American precipitation influence droughts and flood periods in the Madeira River basin.

After characterizing the precipitation variability in the basin, this study links it to river discharge. In addition to anomalies and comparisons between extreme events, the trends in river discharge from 2001 until 2023 are produced and tested using the Mann–Kendall test [33,34].

## 3. Results

### 3.1. Analysis of Precipitation Variability in the Basin

A comparison of the precipitation patterns in the Madeira River basin between the two datasets (Figure 3) shows that the MSWEP dataset contains higher precipitation amounts. The elevated monthly mean precipitation amounts registered by MSWEP appear especially southwest of the basin, where the most populated cities in Bolivia are located, and northeast of the basin, in the lower Madeira basin in Brazil, where it exceeds more than 300 and 150 mm, respectively.

**Table 1.** The annual mean precipitation for six cities located in the Madeira River basin. The period is from 2001 until 2023.

| Location | Coordinates | Annual Mean Precipitation (mm) MERGE | MSWEP |
|---|---|---|---|
| Cochabamba (Bolivia) | 66.17° W, 17.41° S | 1701 | 2625 |
| La Paz (Bolivia) | 68.12° W, 16.49° S | 2321 | 2123 |
| Rurrenabaque (Bolivia) | 67.53° W, 14.44° S | 5163 | 5178 |
| Fazenda Vista Alegre (Brazil) | 60.02° W, 4.89° S | 5391 | 6276 |
| Manicoré (Brazil) | 61.20° W, 5.82° S | 6827 | 5558 |
| Porto Velho (Brazil) | 63.92° W, 8.73° S | 5442 | 7118 |

The difference in the precipitation amounts in the datasets is less concerning for some of the major cities, as indicated in Table 1. The MSWEP and MERGE estimates demonstrate consistency in the Bolivian cities of La Paz and Rurrenabaque. However, in Cochabamba, the datasets disagree by about 1000 mm, with a higher amount in the MSWEP dataset (as in Figure 3). Similarly, in the Brazilian cities of Porto Velho and Fazenda Vista Alegre, the MSWEP dataset estimates are higher than the MERGE dataset, but in Manicoré, the MERGE dataset has higher amounts.

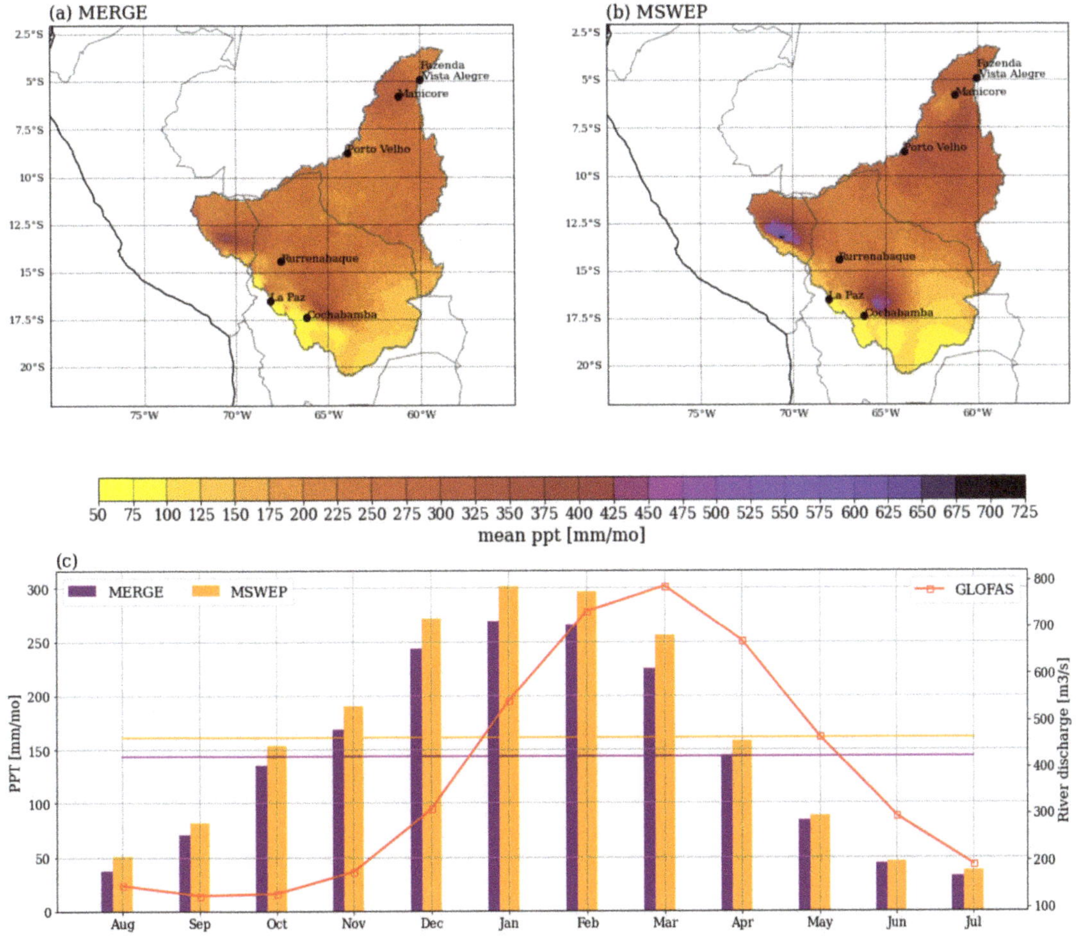


**Figure 3.** Monthly mean precipitation (mm/month) pattern from (**a**) MERGE and (**b**) MSWEP datasets. (**c**) Annual cycle of precipitation in the Madeira River basin averaged from 2001 until 2023. The GloFAS river discharge is the red curve. Black dots show the location of the cities listed in Table 1. The horizontal-colored lines are the annual mean for the corresponding precipitation dataset.

The datasets share the same horizontal resolution and consider IMERG satellite data, which does not explain the precipitation differences. As they come from different sources, their methodological approach differs. Moreover, the MSWEP dataset incorporates additional satellite observations from CHELSA (0.01° grid resolution) and GridSat (0.07°) and reanalysis data from ERA5 [26]. The assembly of different products in regions with reduced in situ data increases the discordance. Therefore, the disparities derive from the different estimation methods and/or data availability within each dataset.

The disparities in the datasets result in the mean precipitation being 20 mm higher for the MSWEP dataset, reaching 160 mm (Figure 3). Despite the differences, the estimated annual precipitation of 1717 mm for the MERGE dataset and 1926 mm for the MSWEP dataset are close to the 1834 mm in ref. [1]. The seasonal cycle of the two datasets agrees that the rainiest months in the basin occur from November to April [7].

The annual cycle of the GloFAS river discharge in the basin peaks in March and lags behind the precipitation peak by about two months. Most of the river flow in the Madeira River depends on the rainy season. The onset of the rainy season in October contributes to the flow increase in October, but with a more significant contribution in January, after three months of monsoonal precipitation. The GloFAS mean monthly discharge of 380 $m^3/s$ and an annual mean of 4565 $m^3/s$ are close to the values observed at some stations, such as the Bolivian station of Rurrenabaque and the Brazilian stations of Porto Velho and Fazenda Vista Alegre [7].

The observed precipitation in the Brazilian sector of the Madeira River basin indicates the rainiest trimester is January–February–March, and the rainy season is from November until April, characterized by precipitation above the annual average [7]. Based on the same criterion, in Figure 3, April shows precipitation below the average. Although excluding April from the rainy season disagrees with [7], the rainy season from November until March agrees with the SAMS period in the region [22]. Thus, the period from November to March defines the rainy season in the basin from now on.

The time series of the standardized anomalies of precipitation and river discharge for the rainy season reveal that the river discharge has higher interannual variability than the precipitation in the basin (Figure 4). As the river fluctuations occur four months later, the drought discharge extremes detected for the rainy season are further enhanced in the dry season (with a peak in August, Figure 3). In addition, recurrent drought events can aggravate the impact in the following seasons. For instance, the precipitation anomalies for the 2022 drought could significantly contribute to the river discharge anomalies in 2023 (Figure 4).

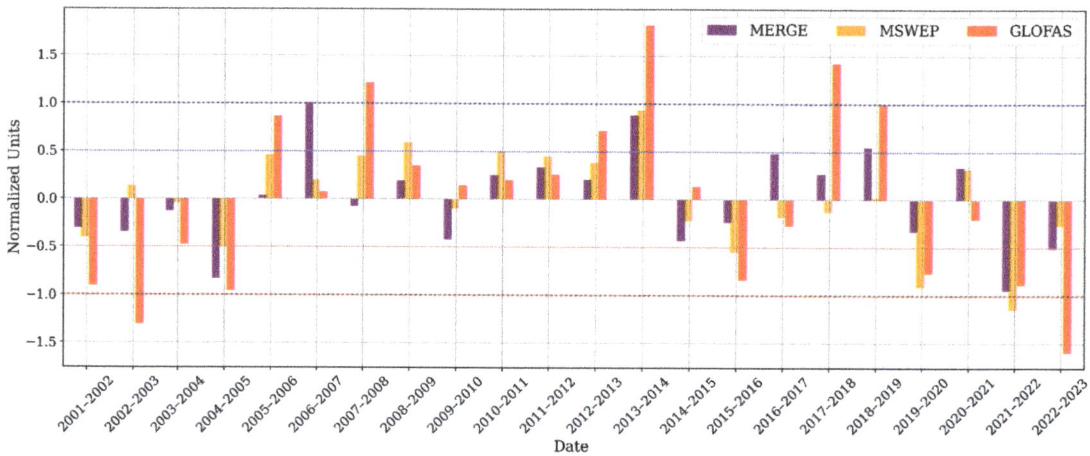


**Figure 4.** Time series of the standard deviations of precipitation and river discharge in the Madeira River basin. The rainy season starts in November of one year and ends in March of the following year.

Although GloFAS derives from a model, comparing river discharge with the precipitation datasets indicates a possible consistency in terms of characterizing droughts and floods in the Madeira River basin. In Figure 4, the MSWEP dataset shows better alignment with GloFAS. The MERGE dataset has shown more discrepancy in several years, such as 2006, 2007, 2008, 2010, 2015, 2017, and 2020, while the MSWEP dataset deviates during 2003 and 2018. Thus, the MSWEP dataset better characterizes extreme precipitation events in the Madeira River basin, confirming the superior performance of the MSWEP dataset compared to other satellite products and in situ data, as demonstrated in previous research [35,36].

### 3.2. Correlation with Climate Forcing

South American climate variability is highly related to climate indices [12–14]. Figure 4 shows the time series of the standard deviations of the MSWEP dataset precipitation in the basin, in comparison to the variability of the various climate forcing data represented by the indices: Niño12, Niño34, Niño 3, PDO, AMO, QBO, TSA, and TNA [31]. The outstanding extremes agree with the records in the literature, pointing out extreme rain events in 2009 and 2014 and extreme droughts in 2005, 2016, 2020, and 2022.

The correlation coefficients between the precipitation anomalies and the indices reveal that the basin variability is strongly modulated by the TNA SST and the AMO, with significant correlation coefficients of 0.44 and 0.27, respectively. Although earlier findings indicate a relationship between the Pacific SST and the precipitation in the Amazon region [4,9,16], the subregion covering the Madeira River basin exhibits a different relationship. The correlation coefficient with different ENSO indices, such as Nino1.2, Nino3, and Nino3.4, is low, as is the correlation with the PDO. Some drought events, such as in 2005, occurred during the ENSO warm phase (Figure 5, positive Niño indices), but the same warming occurred in the 2009 rainy event. In more recent years, the ENSO cold phase happened in the 2014 rainy event (Figure 5, negative Niño indices), but the same characteristic resulted in the 2020 drought. The absence of alignment during the extremes, as verified by the ENSO and PDO, explains the low correlation coefficients.

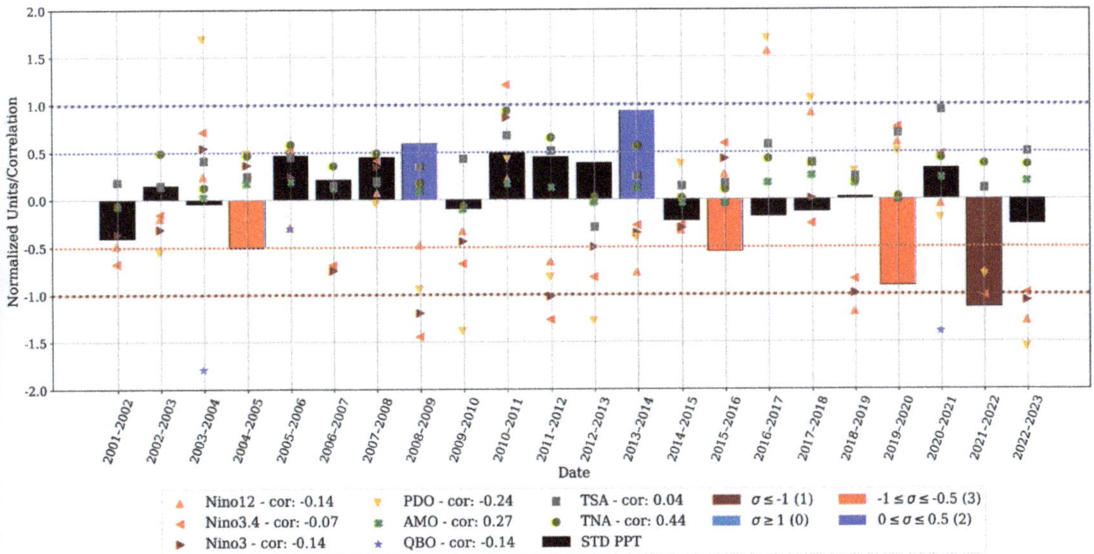


**Figure 5.** Time series of the standard deviations of the rainy season precipitation in the Madeira River basin. The bars (STD PPT) correspond to the November–March standardized anomalies from the MSWEP dataset. Red/darker red displays the most extreme drought events, and the extreme rainy events are blue/darker blue. The climate indices are monthly values derived from the NOAA [31].

### 3.3. *Composites of Extremes*

Composite anomalies of the SST confirm the relation between the precipitation variability in the basin and the TNA variability (Figure 6). Some studies have indicated this relationship during dry and rainy extremes in the Madeira River basin [5,6,10]. The SST in tropical Atlantic waters modifies the latitudinal displacement of the ITCZ [10,14]. In drier years (Figure 6b), the SST is warmer in the TNA region, indicating a northern ITCZ. At the same time, the opposite occurs in rainier years, with negative SST anomalies in the TNA (Figure 6c). The SST anomaly pattern also shows opposite signs in the southwestern South Atlantic Ocean, around 30°S. Negative SST anomalies in the southwestern South Atlantic (Figure 6b) suggest that a cooler ocean surface occurs during the drier years in the Madeira River basin. In contrast, positive SST anomalies, suggesting a warmer ocean, are associated with rainier years (Figure 6c).

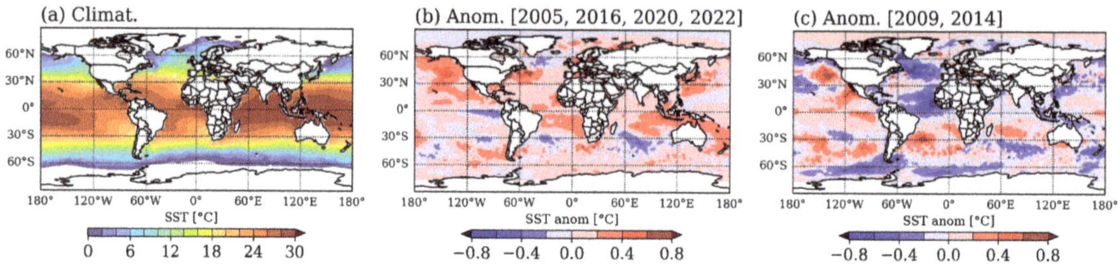


**Figure 6.** Sea surface temperature (°C): climatology (**a**) and mean anomalies during dry (**b**) and wet (**c**) years. The plots consider ERA5 data on the yearly means for the rainy season from November to March.

The lower-level winds in the 850 hPa layer (Figure 7) indicate the effects of the TNA's contribution to the precipitation variability in the basin. During drought events, the winds from the North Atlantic are weaker, revealing wind anomaly vectors in opposite directions (Figure 7c), and the southeasterly trade winds reduce their flow over northern South America (Figure 7b). The amount of moisture that the region receives from the Amazon region is related to the maintenance of this flow [10,37]. The anomalies show an average wind speed reduction of more than 1 m/s, which will contribute significantly to less moisture and precipitation in the basin. In contrast, the opposite relates to rainy events registered in the basin, and the wind patterns indicate enhanced trade winds (Figure 7e). This wind that blows out of the basin is a component of the low-level jet (LLJ) [30], and its modulation is crucial to the distribution of precipitation across the continent.

The wind anomalies observed in the ocean are consistent with the SST patterns and are related to the precipitation patterns [30]. During rainy events (Figure 7e), the anomalies intensify the northwest flow of the South Atlantic Subtropical High. Conversely, the drought events (Figure 7c) display wind anomalies from the opposite direction of the climatological flow (Figure 7a). This structure and its confluence with the LLJ are necessary to maintain the South Atlantic Convergence Zone (SACZ) activity, as observed by others [30,38]. Stronger southward winds increase precipitation over southern Brazil, while weaker southward winds increase precipitation over central–east Brazil. The relationship between precipitation and near-surface winds results in SST anomalies in the southwestern Atlantic Ocean [30,39].

The mid-level (500 hPa) mean vertical motion reveals the presence of a semi-permanent area of rising air over the Amazon basin. The negative sign indicates upward air motion as an indicator of the Walker cell circulation (Figure 8a,b,d). The extreme years modify this circulation, with interactions explained by the wind anomalies. During drought periods, the winds are weaker (Figure 7b), and, as a consequence, less moisture is transported to the region, reducing the convection and upward motion in the basin. The impact of reduced convection leads to an increase in the mean vertical velocity in the midwest of the basin (Figure 8b), and furthermore, the anomalies reveal that the entire region is affected, increasing by at least 1% of its value, with a greater contribution in the central–west and northern part of the basin (Figure 8c). This effect indicates that the wind impacts in the basin are not fully suppressing the convection activities in the region. However, it suggests a higher capability to reduce the volumes of moisture transport. Similar aspects, but opposite signs, occur during rainy periods (Figure 8d,e), and enhanced vertical motion appears in the center to the north of the area. Nevertheless, more intense vertical motion anomalies are observed northwest of the basin during extreme events. These anomalies suggest that the extremes in the Madeira River basin relate to the same system that suppresses convection in northwestern Brazil, mainly due to the modulation of SST in the TNA.

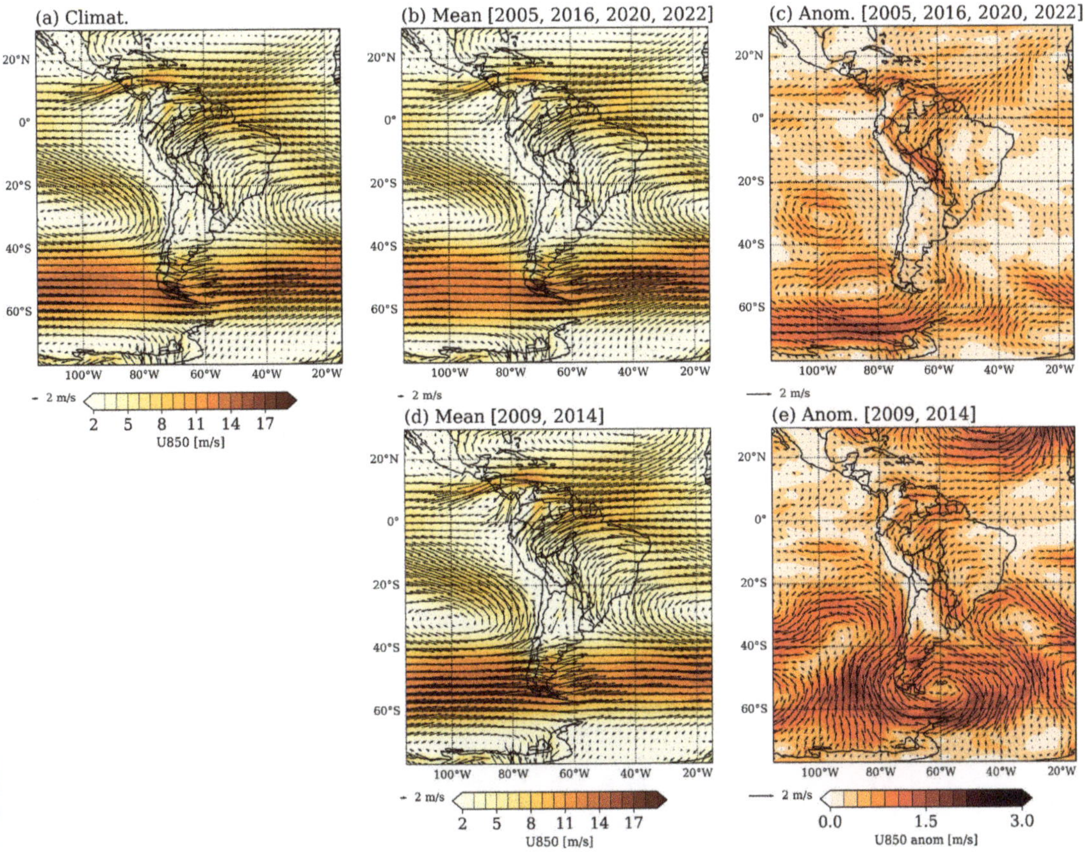


**Figure 7.** Plots of 850 hPa winds (m/s): climatology (**a**), mean (**b**), and anomalies (c) during dry years, and mean (**d**) and anomalies (**e**) during wet years. The plots consider ERA5 data on the yearly means for the rainy season from November to March.

### *3.4. Response of Madeira River Tributaries to Extremes*

The impact of extremes in the main Madeira River regions and its tributaries varies as land use and topography diverge. Figure 9 shows the differences in precipitation and river discharge between the extreme rainy and drought events in the basin. The main river course is highly susceptible to precipitation variations, and the river flow can deviate more than 20,000 $m^3/s$ from rainy to dry events. Such variations could have severe consequences for the population that depends on the river for their daily activities, such as in Manicoré and Fazenda Vista Alegre, located inside the Madeira River Sustainable Development Reserve, and the city of Porto Velho, as seen in Figure 3.

The river discharge pattern in the basin and its tributaries appears unrelated to the precipitation pattern. Considering the narrowness of the tributaries, the main Madeira River naturally shows more prominent river discharge anomalies. Nevertheless, the differences for the extremes reveal roughly homogeneous positive precipitation anomaly patterns in the southwestern Basin, consistent with a symmetrical response to the rainy and drought signals. Some northeast regions present negative anomalies, indicating the persistence of reduced precipitation, even during rainy events. Even though the precipitation varies between the different sources, their distribution is similar north of the domain, which looks unrelated to the diagnostics of the flow anomalies in Figure 9c. While the main Madeira

River path, in the northeastern region of the basin, presents a difference of more than 15,000 $m^3/s$ in terms of the river flow during extreme events, the changes nearly reach 2500 $m^3/s$ in other regions.

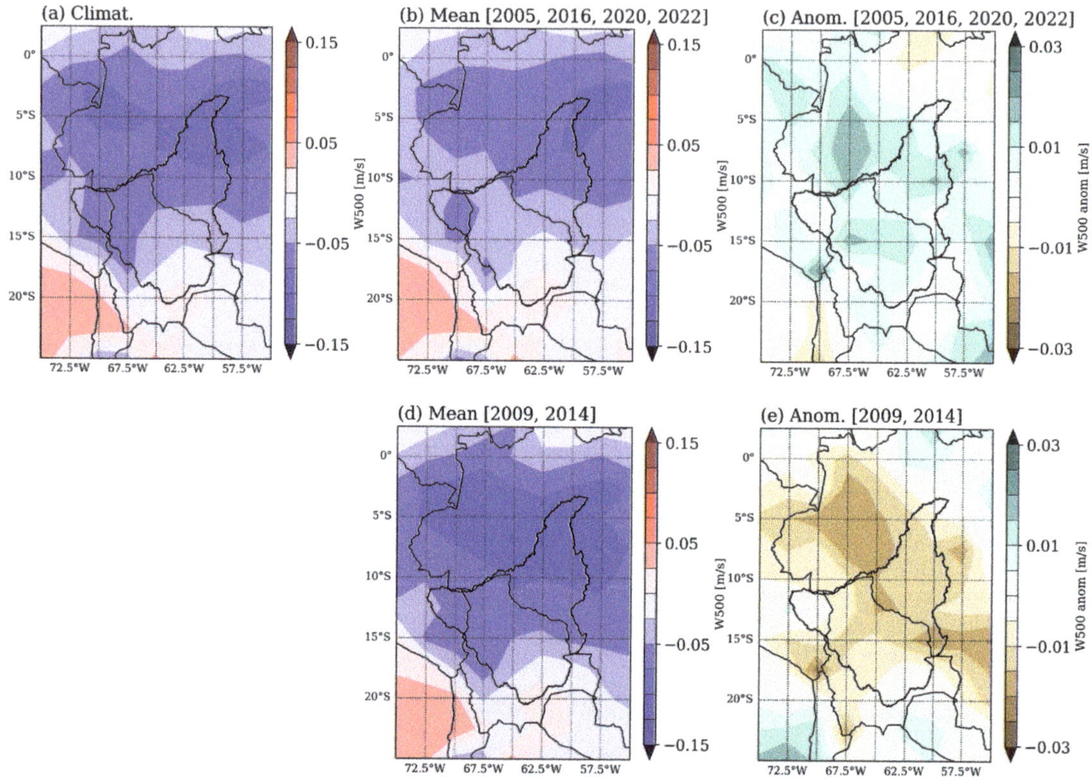


**Figure 8.** Plots of 500 hPa vertical motion: climatology (**a**), mean (**b**), and anomalies (**c**) during dry years, and mean (**d**) and anomalies (**e**) during wet years, in the Madeira River basin. The plots consider ERA5 data on the yearly means for the rainy season from November to March.

The apparent uncorrelated patterns of precipitation and river discharge during extreme events may be affected by events that present an uneven distribution of precipitation in the basin. Figure 10a,b shows the relative difference in the river discharge between more recent and older extreme events. The events are the most contrasting records and show a high variation in the precipitation amount (more than 0.5 standard deviations, as observed in Figure 4). Nevertheless, the resulting river discharge remains relatively similar in the course of the main Madeira River. Instead, the events produced different river discharge in the tributaries.

In the southern Madeira River basin, the difference between the 2014 and 2009 events is almost two (Figure 10a), which indicates that the amount of precipitation during 2014 doubled the river flow in this location. That significantly contributed to the floods registered in 2014 in northwestern Brazil and northern Bolivia [3,10,40]. Other regions experienced fewer discharges in 2014 than in 2009, with the negative values in the north, midwest, and southeast parts of the basin standing out.

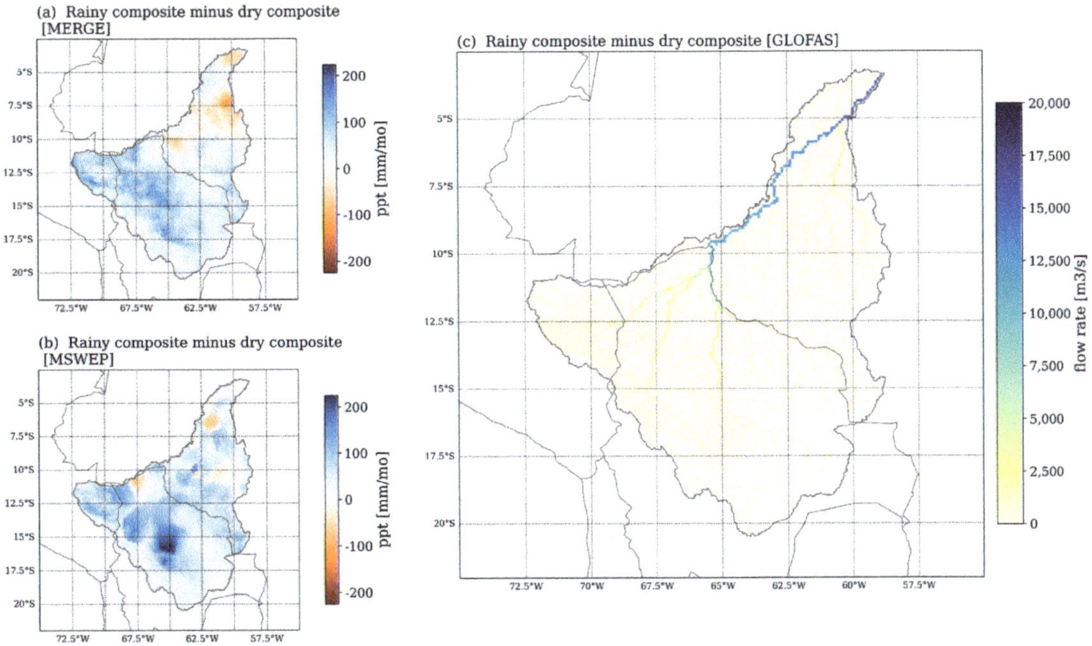


**Figure 9.** Precipitation and flow rate difference between the mean rainy (2009, 2014) and the mean drought (2005, 2016, 2020, 2022) years, using (**a**) MERGE, (**b**) MSWEP, and (**c**) GloFAS datasets.

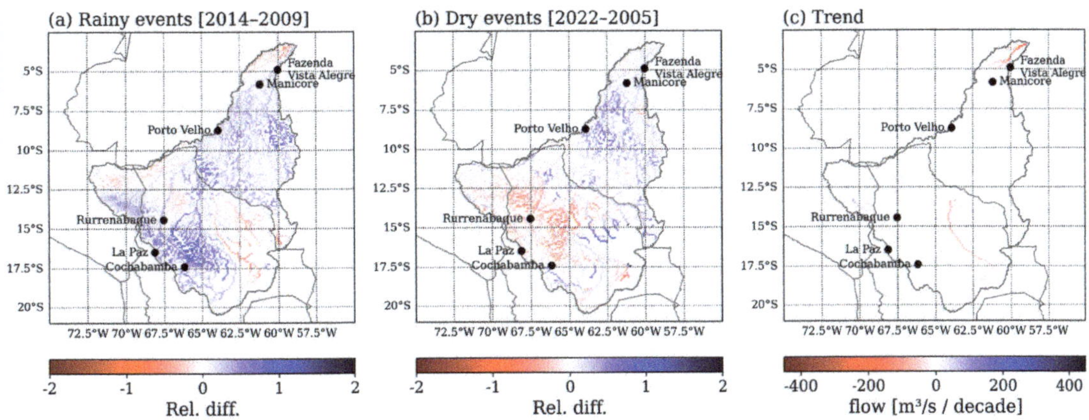


**Figure 10.** Relative difference in river discharge between the most recent and previous rainy (**a**) and dry (**b**) year events in the Madeira River basin, and (**c**) river flow trend ($m^3/s$/decade) for the period from 2001 until 2023. The trend is indicated for regions with a significance level $\alpha = 0.05$, according to the Mann–Kendall test. The GloFAS dataset disregards reservoir operations in the modeling of river discharge. The dots represent the densely populated cities in the basin.

The differences in river discharge between the drought events of 2022 and 2005 result in an interesting pattern. Although some regions had reduced discharge in 2022 compared to 2005, most areas in the central–north part of the basin had increased discharge, as revealed by the positive differences in Figure 10b. Accordingly, the 2022 drought was worse than the 2005 drought in most Peruvian and Bolivian territories, whereas higher river discharge occurred in the Brazilian portion of the basin. Other water availability, management, and monitoring aspects can be related to these differences. Regardless, the pronounced response in terms of the lower Madeira River discharge differs from that of the upper Madeira River during extreme dry events, which leads to different impacts in the tributary rivers. In addition, the river discharge observed during drought events is even lower, and, more concerning, it occurs just before the onset of the rainy season and for prolonged periods.

The relative differences in river discharge (Figure 10a,b) indicate reduced river flow in the northernmost part of the basin during both rainy and dry events. This pattern indicates tendencies confirmed by the trend evaluation in Figure 10c. Negative coefficients reveal a reduction in river flow in recent years in the eastern part of the basin. In the Peruvian part of the basin, positive coefficients indicate an increase in the river flow in recent decades. The most significant trends are found in the Brazilian part of the basin, namely a reduction trend of more than 400 $m^3/s$ per decade, which reflects the decrease in river flow in recent years. Manicoré and Fazenda Vista Alegre exhibit a reduction in river flow of 578 and 126 $m^3/s$ per decade, respectively. The uncertainty in the observational dataset in the region is obvious. Additional observational studies should increase the confidence of the results.

## 4. Conclusions

This study seeks to highlight the opposing facets of extreme dry and wet precipitation events in terms of the Madeira River basin. The temporal variability of precipitation reveals the region's susceptibility to extreme climatic events and their impacts on river discharge patterns.

The two precipitation datasets have revealed notable discrepancies, particularly in the Brazilian portion of the Madeira River basin. The MERGE and MSWEP datasets exhibit both spatial and interannual variability discrepancies. The MERGE dataset shows less precipitation in the southern part of the basin, but there are still similarities in the precipitation totals in the major cities. The results could infer a disagreement with the river discharge, which is more closely related to the MSWEP dataset values, due to their alignment with precipitation extremes in the basin. Despite the divergences, both datasets show similar seasonal precipitation cycles, revealing that the rainy season starts in November and ends in March. The variability in the extremes in the rainy season accurately identifies historical floods and drought records in the datasets, as is the case for the 2014 and 2009 enhanced floods [3,10] and the drier events in 2016, 2020, 2022, and 2005, the latter of which relates to a historical drought record [6].

An important result is the significant correlation between the anomalous pattern of the SST in the TNA and the variability of the rainy season in the basin. Composite anomalies confirm this relationship: a cooler TNA is linked with higher levels of precipitation and flooding, while a warmer TNA is linked with drought conditions. The anomalies in the TNA impact the northeasterly trade winds, consequently affecting the winds blowing over the basin and the injection of moisture toward the Amazon basin. The constructed anomalous composites agree with earlier findings [10].

In addition, this study showed that warmer waters appear in the southwestern South Atlantic Ocean during extreme rainy events in the Madeira River basin and cooler waters occur during drought events. This pattern is commonly related to the relationship between precipitation anomalies and the SACZ activity [30,39]. The results displayed anomalous vertical motion and strong wind variations in the LLJs. These results imply that extreme precipitation in the Madeira River basin can cause variations in the moisture content and convective activity in the SACZ climatological pattern [30].

The Madeira River is highly vulnerable to precipitation extremes, and river discharge can vary more than 10,000 $m^3/s$ around its mean discharge. However, the relationship between extreme precipitation events and river discharge in the basin is not straightforward. The TNA SST anomalies alone do not fully explain the river response. The comparison between extreme events driven by the same SST anomaly patterns in different decades revealed that in addition to the precipitation extremes correlated to TNA SST variability, linear trends have reduced river discharge in the Madeira River tributaries in recent years. The negative trends in river discharge are about 100 $m^3/s$ per decade in Bolivian and Peruvian territories and more than 400 $m^3/s$ per decade in Brazilian cities in the northern portion of the Madeira River.

Climate change has been inducing extreme climate events and affecting hydrological processes. This study illustrates the impacts of extreme precipitation events on river discharge across the Madeira River basin from the point of view of GloFAS, which is a dataset designed to raise awareness of flood occurrence. Future analyses should consider extended periods for climatology, as well as more advanced approaches to confirm the results. To improve the detection of drought periods, we suggest the use of the SPI (Standardized Precipitation Index) obtained with linear moments (L-moments), as in [41,42]. We believe this methodological refinement should add value beyond the studied region, potentially benefiting broader scientific applications. Furthermore, we recommend modeling approaches to river discharge that consider reservoir operations, uncertainties in the precipitation estimates, and the Amazon River discharge at the mouth of the Madeira River. Improving predictability and modeling of river discharge is crucial as extreme events impact the socio-economic activities in the Madeira River basin, such as the sustainable development of conservation units, river navigation, energy production, agriculture, fisheries, and health services.

Although every river has unique characteristics, the methodology applied in this study can be reproduced in other basins. Searching for sources of predictability for extreme events affecting rivers is helpful for properly managing and anticipating disasters. In addition, our research could expose common gaps and challenges in regions with poor scientific interest, such as the Madeira River basin.

**Author Contributions:** Conceptualization, S.C.C., N.C.L., P.d.S.T., and F.W.S.C.; methodology, S.C.C., J.L.G., P.d.S.T., and N.C.L.; software, N.C.L., M.T., and P.d.S.T.; validation, S.C.C., N.C.L., P.d.S.T., and D.C.R.; formal analysis, N.C.L., M.T., P.d.S.T., D.C.R., S.C.C., and J.L.G.; investigation, N.C.L., M.T., P.d.S.T., D.C.R., S.C.C., and J.L.G.; resources, N.C.L., M.T., and P.d.S.T.; data curation, N.C.L., M.T., and P.d.S.T.; writing—original draft preparation, N.C.L., M.T., P.d.S.T., D.C.R., S.C.C., and J.L.G.; writing—review and editing, N.C.L., M.T., P.d.S.T., D.C.R., S.C.C., J.L.G., and F.W.S.C.; visualization, N.C.L., M.T., and P.d.S.T.; supervision, S.C.C.; project administration, S.C.C., and F.W.S.C.; funding acquisition, S.C.C., and F.W.S.C. All authors have read and agreed to the published version of the manuscript.

**Funding:** The APC and Priscila Tavares were funded by FAPESP 2020/08796-2.

**Data Availability Statement:** The data presented in this study are available on request from the corresponding author.

**Acknowledgments:** The authors are thankful for the support from the Coordenação de Aperfeiçoamento de Pessoal de Nível Superior (CAPES) Finance Code 001. This work is supported jointly by Fundação de Pesquisa do Estado de São Paulo (FAPESP) 2020/08796-2 and Fundação de Pesquisa do Estado do Amazonas (FAPEAM) 01.02.016301.00268/2021. This work is also supported by Coordenação de Aperfeiçoamento de Pessoal de Nível Superior (CAPES) 88881.688972/2022-01. S.C.Chou thanks CNPq for grant 312742/2021-5. We are also thankful for the support from the Centro Nacional de Processamento de Alto Desempenho em São Paulo (CENAPAD-SP), UNICAMP/FINEP-MCTI for developing this research.

**Conflicts of Interest:** The authors declare that there are no conflicts of interest.

## References


1. Rodriguez, D.A.; Lopes, L.G.; Carriello, F.; Siqueira Júnior, J.L.; Pinto, G.L.; Tomasella, J.; Chou, S.C. Previsões de cheias extremas nos horizontes sazonais e de curto e médio prazos na bacia do rio Madeira: Estudo de caso da enchente de 30 de março de 2014 em Porto Velho. *Rev. Bras. De Cartogr.* **2017**, *69*, 827–836. [CrossRef]
2. Espinoza, J.C.E.; Guyot, J.L.; Ronchail, J.; Cochonneau, G.; Filizola, N.; Fraizy, P.; Vauchel, P. Contrasting regional discharge evolutions in the Amazon basin (1974–2004). *J. Hydrol.* **2009**, *375*, 297–311. [CrossRef]
3. Ovando, A.; Tomasella, J.; Rodriguez, D.A.; Martinez, J.M.; Siqueira Junior, J.L.; Pinto, L.N.; Passy, P.; Vauchel, P.; Noriega, L.; Von Randow, C. Extreme flood events in the Bolivian Amazon wetlands. *J. Hydrol. Reg. Stud.* **2016**, *5*, 293–308. [CrossRef]
4. Marengo, J.A.; Espinoza, J.C. Extreme seasonal droughts and floods in Amazonia: Causes, trends and impacts. *Int. J. Climatol.* **2016**, *36*, 1033–1050. [CrossRef]
5. Marengo, J.A.; Tomasella, J.; Alves, L.M.; Soares, W.R.; Rodriguez, D.A. The drought of 2010 in the context of historical droughts in the Amazon region. *Geophys. Res. Lett.* **2011**, *38*, 1–5. [CrossRef]
6. Marengo, J.A.; Nobre, C.A.; Tomasella, J.; Oyama, M.D.; Sampaio de Oliveira, G.; de Oliveira, R.; Camargo, H.; Alves, L.M.; Brown, I.F. The drought of Amazonia in 2005. *Am. Meteorol. Soc.* **2008**, *21*, 495–516. [CrossRef]
7. Gomes, D.; Lima, A.M.M.; Ferreira, N.S.; Serrão, E.A.O. Variabilidade espaço-temporal da precipitação: Bacia Hidrográfica do Rio Madeira. *Rev. Geográfica Acadêmica* **2019**, *13*, 90–104. Available online: https://revista.ufrr.br/rga/article/view/5514 (accessed on 22 July 2024).
8. Clarke, B.; Barnes, C.; Rodrigues, R.; Zachariah, M.; Stewart, S.; Raju, E.; Baumgart, N.; Heinrich, D.; Libonati, R.; Santos, D.; et al. Climate Change, not El Niño, Main Driver of Exceptional Drought in Highly Vulnerable Amazon River Basin. Grantham Institute for Climate Change. 2024. Available online: http://hdl.handle.net/10044/1/108761 (accessed on 22 July 2024).
9. Espinoza, J.C.; Jimenez, J.C.; Marengo, J.A.; Schongar, J.; Ronchail, J.; Lavado-Casimiro, W.; Ribeiro, J.V. The new record of drought and warmth in the Amazon in 2023 related to regional and global climatic features. *Sci. Rep.* **2024**, *14*, 8107. [CrossRef]
10. Espinoza, J.C.; Marengo, J.A.; Ronchail, J.; Carpio, J.M.; Flores, L.N.; Guyot, J.L. The extreme 2014 flood in southwestern Amazon basin: The role of tropical-subtropical South Atlantic SST gradient. *Environ. Res. Lett.* **2014**, *9*, 124007. [CrossRef]
11. IPCC AR6 (Intergovernmental Panel on Climate Change). Summary for Policymakers. In *Climate Change 2021: The Physical Science Basis*, 6th ed.; Contribution of Working Group I to the Sixth Assessment Report of the Intergovernmental Panel on Climate Change; Masson-Delmotte, V., Zhai, P., Pirani, A., Connors, S.L., Péan, C., Berger, S., Caud, N., Chen, Y., Goldfarb, L., Gomis, M.I., et al., Eds.; Cambridge University Press: Cambridge, UK; New York, NY, USA, 2021; pp. 3–32. [CrossRef]
12. Grimm, A.M.; Laureanti, N.C.; Rodakoviski, R.B.; Gama, C.B. Interdecadal variability and extreme precipitation events in South America during the monsoon season. *Clim. Res.* **2016**, *68*, 277–294. [CrossRef]
13. Grimm, A.M.; Tedeschi, R.G. ENSO and extreme rainfall events in South America. *J. Clim.* **2009**, *22*, 1589–1609. [CrossRef]
14. Utida, G.; Cruz, F.; Etourneau, J.; Bouloubassi, I.; Schefu, E.; Vuille, M.; Novello, V.F.; Prado, L.F.; Sifeddine, A.; Klein, V.; et al. Tropical South Atlantic influence on Northeastern Brazil precipitation and ITCZ displacement during the past 2300 years. *Sci. Rep.* **2019**, *9*, 1698. [CrossRef] [PubMed]
15. Flantua, S.G.A.; Hooghiemstra, H.; Vuille, M.; Behling, H.; Carson, J.F.; Gosling, W.D.; Hoyos, I.; Ledru, M.P.; Montoya, E.; Mayle, F.; et al. Climate variability and human impact in South America during the last 2000 years: Synthesis and perspectives from pollen records. *Clim. Past* **2016**, *12*, 483–523. [CrossRef]
16. Espinoza, J.C.; Ronchail, J.; Marengo, J.A.; Segura, H. Contrasting North–South changes in Amazon wet-day and dry-day frequency and related atmospheric features (1981–2017). *Clim. Dyn.* **2019**, *52*, 5413–5430. [CrossRef]
17. Funatsu, B.M.; Le Roux, R.; Arvor, D.; Espinoza, J.C.; Claud, C.; Ronchail, J.; Michot, V.; Dubreuil, V. Assessing precipitation extremes (1981–2018) and deep convective activity (2002–2018) in the Amazon region with CHIRPS and AMSU data. *Clim. Dyn.* **2021**, *57*, 827–849. [CrossRef]
18. Towner, J.; Cloke, H.L.; Lavado, W.; Santini, W.; Bazo, J.; Perez, E.C.; Stephens, E.M. Attribution of Amazon floods to modes of climate variability: A review. *Meteorol. Appl.* **2020**, *27*, e1949. [CrossRef]
19. Jimenez, J.C.; Libonati, R.; Peres, L.F. Droughts over Amazonia in 2005, 2010, and 2015: A cloud cover perspective. *Front. Earth Sci.* **2018**, *6*, 1–7. [CrossRef]
20. Vergasta, L.A.; Correia, F.W.S.; Chou, S.C.; Nobre, P.; Lyra, A.d.A.; Gomes, W.d.B.; Capistrano, V.; Veiga, J.A.P. Avaliação do Balanço de água na Bacia do Rio Madeira Simulado Pelo Modelo Regional Climático Eta e o Modelo Hidrológico de Grandes Bacias MGB. *Rev. Bras. De. Meteorol.* **2021**, *36*, 153–169. [CrossRef]
21. Júnior, J.S.; Tomasella, J.; Rodriguez, D.A. Impacts of future climatic and land cover changes on the hydrological regime of the Madeira River basin. *Clim. Chang.* **2015**, *129*, 117–129. [CrossRef]
22. Grimm, A.; Vera, C.; Mechoso, C. The South American monsoon system. In *The Global Monsoon System: Research and Forecast*; World Meteorological Organization: Geneva, Switzerland, 2005; pp. 219–238.
23. Peel, M.C.; Finlayson, B.L.; McMahon, T.A. Updated world map of the Köppen-Geiger climate classification. *Hydrol. Earth Syst. Sci.* **2007**, *11*, 1633–1644. [CrossRef]
24. Jarvis, A.; Reuter, H.I.; Nelson, A.; Guevara, E. Hole-Filled SRTM for the Globe Version 4. Available from the CGIAR-CSI SRTM 90m Database. 2008. Available online: https://srtm.csi.cgiar.org/ (accessed on 2 July 2024).
25. Rozante, J.R.; Moreira, D.S.; Goncalves, L.G.G.; Vila, D.A. Combining TRMM and surface observations of precipitation: Technique and validation over South America. *Am. Meteorol. Soc.* **2010**, *25*, 885–894. [CrossRef]

26. Beck, H.E.; Wood, E.F.; Pan, M.; Fisher, C.K.; Miralles, D.G.; van Dijk, A.I.J.M.; McVicar, T.R.; Adler, R.F. MSWEP V2 global 3-hourly 0.1° precipitation: Methodology and quantitative assessment. *Bull. Am. Meteorol. Soc.* **2019**, *100*, 473–500. [CrossRef]
27. Zsoter, E.; Harrigan, S.; Barnard, C.; Wetterhall, F.; Ferrario, I.; Mazzetti, C.; Alfieri, L.; Salamon, P.; Prudhomme, C. River discharge and related historical data from the Global Flood Awareness System. v3.1. Copernicus Climate Change Service (C3S) Climate Data Store (CDS). 2021. Available online: https://cds.climate.copernicus.eu/cdsapp#!/dataset/cems-glofas-historical (accessed on 22 July 2024).
28. Hersbach, H.; Bell, B.; Berrisford, P.; Hirahara, S.; Horányi, A.; Muñoz-Sabater, J.; Nicolas, J.; Peubey, C.; Radu, R.; Schepers, D.; et al. The ERA5 global reanalysis. *Q. J. R. Meteorol. Soc.* **2020**, *146*, 1999–2049. [CrossRef]
29. Wilks, D.S. *Statistical Methods in the Atmospheric Sciences*, 2nd ed.; International Geophysics Series; Academic Press: San Diego, CA, USA, 2006; Volume 91.
30. Laureanti, N.C.; Chou, S.C.; Nobre, P.; Curchitser, E. On the relationship between the South Atlantic Convergence Zone and sea surface temperature during Central-East Brazil extreme precipitation events. *Dyn. Atmos. Ocean.* **2024**, *105*, 101422. [CrossRef]
31. NOAA, National Oceanic and Atmospheric Administration; Physical Sciences Laboratory (PSL). Climate Indices: Monthly Atmospheric and Ocean Time-Series. Available online: https://psl.noaa.gov/data/climateindices/list/ (accessed on 15 May 2024).
32. Rao, J.; Garfinkel, C.I.; Ren, R.; Wu, T.; Lu, Y. Southern Hemisphere Response to the Quasi-Biennial Oscillation in the CMIP5/6 Models. *J. Clim.* **2023**, *36*, 2603–2623. [CrossRef]
33. Mann, H.B. Non-parametric tests against trend. *Econometrica* **1945**, *13*, 163–171. [CrossRef]
34. Kendall, M.G. *Rank Correlation Methods*, 4th ed.; Charles Griffin: London, UK, 1975.
35. Moreira, A.A.; Ruhoff, A.L.; Roberti, D.R.; Souza, V.A.; da Rocha, H.R.; Paiva, R.C.D. Assessment of terrestrial water balance using remote sensing data in South America. *J. Hydrol.* **2019**, *575*, 131–147. [CrossRef]
36. Sun, Q.; Miao, C.; Duan, Q.; Ashouri, H.; Sorooshian, S.; Hsu, K.-L. A review of global precipitation data sets: Data sources, estimation, and intercomparisons. *Rev. Geophys.* **2018**, *56*, 79–107. [CrossRef]
37. Satyamurty, P.; Da Costa, C.P.W.; Manzi, A.O. Moisture source for the Amazon Basin: A study of contrasting years. *Theor. Appl. Clim.* **2013**, *111*, 195–209. [CrossRef]
38. Kodama, Y. Large-scale common features of subtropical precipitation zones (the Baiu frontal zone, the SPCZ, and the SACZ) Part I: Characteristics of subtropical frontal zones. *J. Meteorol. Soc. Jpn. Ser. II* **1992**, *70*, 813–836. [CrossRef]
39. Nobre, P.; De Almeida, R.A.; Malagutti, M.; Giarolla, E. Coupled Ocean–Atmosphere Variations over the South Atlantic Ocean. *J. Clim.* **2012**, *25*, 6349–6358. [CrossRef]
40. BBC, Latin America. Record Floods in Brazil bring Chaos to Amazon Towns. Available online: https://www.bbc.com/news/world-latin-america-28123680 (accessed on 28 May 2024).
41. Ilinca, C.; Anghel, C.G. Flood-Frequency Analysis for Dams in Romania. *Water* **2022**, *14*, 2884. [CrossRef]
42. Ilinca, C.; Anghel, C.G. Flood Frequency Analysis Using the Gamma Family Probability Distributions. *Water* **2023**, *15*, 1389. [CrossRef]

climate

MDPI

*Article*

# Climate Change Impacts on Streamflow in the Krishna River Basin, India: Uncertainty and Multi-Site Analysis

**Ponguru Naga Sowjanya [1], Venkata Reddy Keesara [2], Shashi Mesapam [2], Jew Das [2] and Venkataramana Sridhar [3,*]**

[1] Department of Civil Engineering, Narasaraopeta Engineering College, Narasaraopeta 522601, India
[2] Department of Civil Engineering, National Institute of Technology Warangal, Warangal 506004, India
[3] Department of Biological Systems Engineering, Virginia Polytechnic Institute and State University, Blacksburg, VA 24061, USA
* Correspondence: vsri@vt.edu

**Abstract:** In Peninsular India, the Krishna River basin is the second largest river basin that is overutilized and more vulnerable to climate change. The main aim of this study is to determine the future projection of monthly streamflows in the Krishna River basin for Historic (1980–2004) and Future (2020–2044, 2045–2069, 2070–2094) climate scenarios (RCP 4.5 and 8.5, respectively), with the help of the Soil Water and Assessment Tool (SWAT). SWAT model parameters are optimized using SWAT-CUP during calibration (1975 to 1990) and validation (1991–2003) periods using observed discharge data at 5 gauging stations. The Cordinated Regional Downscaling EXperiment (CORDEX) provides the future projections for meteorological variables with different high-resolution Global Climate Models (GCM). Reliability Ensemble Averaging (REA) is used to analyze the uncertainty of meteorological variables associated within the multiple GCMs for simulating streamflow. REA-projected climate parameters are validated with IMD-simulated data. The results indicate that REA performs well throughout the basin, with the exception of the area near the Krishna River's headwaters. For the RCP 4.5 scenario, the simulated monsoon streamflow values at Mantralayam gauge station are 716.3 $m^3/s$ per month for the historic period (1980–2004), 615.6 $m^3/s$ per month for the future1 period (2020–2044), 658.4 $m^3/s$ per month for the future2 period (2045–2069), and 748.9 $m^3/s$ per month for the future3 period (2070–2094). Under the RCP 4.5 scenario, lower values of about 50% are simulated during the winter. Future streamflow projections at Mantralayam and Pondhugala gauge stations are lower by 30 to 50% when compared to historic streamflow under RCP 4.5. When compared to the other two future periods, trends in streamflow throughout the basin show a decreasing trend in the first future period. Water managers in developing water management can use the recommendations made in this study as preliminary information and adaptation practices for the Krishna River basin.

**Keywords:** climate change; RCM; Reliability Ensemble Averaging (REA); river basin; streamflow; SWAT

## 1. Introduction

Many studies have shown that climate change is an important factor affecting water resources [1–3]. The majority of the effects of climate change are associated with warming, shifts in precipitation patterns, and increases in annual mean temperature trends [4]. Seasonal rainfall in Southern Asia exhibits inter-decadal variability, a significant decreasing trend, and frequent deficit monsoons due to regional inhomogeneities [5]. Furthermore, most Asian countries are experiencing an increase in water demand due to increases in population, irrigated agriculture, and industry [6]. In India, there is evidence that climate extreme events will be on the rise as a result of changes in intense rainfall events and abrupt temperature changes [5,7–10]. The rate of change of extreme weather events, such as droughts or floods, will affect the quantity and quality of water resources, human health, environmental conditions, and agricultural production as a consequence [11–13]. Hence, it is important to analyze the impact of climate change on water resources at the basin level in order to understand the variability of streamflow in future periods and to provide support for regional resource management [14,15].

Representative Concentration Pathways (RCPs) are the scenarios proposed by IPCC AR5 to represent future greenhouse gas emissions. In the Indian summer monsoon, the model scenario projects an increase in both mean and extreme precipitation [5]. The use of regional climate models, as opposed to global climate models, has been shown to be more efficient for assessing the effects of climate change on hydrology at the basin level [16–18]. The reliability ensemble averaging (REA) method proposed by [19] is used to reduce model uncertainty. The nonparametric empirical bias correction method is used to correct additional bias in the ensemble climate data. Quantile–Quantile (QQ) mapping is the correction adopted to improve the efficiency of climate model data for future periods used in impact and vulnerability studies [20].

The hydrological model serves as a tool for determining the quantity and quality of hydrologic variables in combination with climate model data [21–24]. Distributed models provide detailed heterogeneity by spatially using a greater number of parameters in the watershed [25]. Several studies have emphasized the importance of streamflow variability in a watershed, both spatially and temporally, rather than the overall water budget [17,26–29]. However, the components of the water budget, including evapotranspiration fluxes and soil moisture storage, are very important for closing the water budget [1,3,30]. The SWAT model proves to be reasonably superior in predicting future streamflow under changing climate by accurately producing the observed streamflow through calibration. Yet, SWAT faces a significant challenge in identifying the appropriate model parameters during calibration. To evaluate the performance of the model's calibration and validation, Sensitivity Analysis (SA) and Uncertainty Analysis (UA) were performed [17,18,31]. SA and UA are the two processes that enable the model's uncertainty to be reduced. The SWAT-CUP combines various calibration and uncertainty analysis techniques, such as Generalized Likelihood Uncertainty Estimation (GLUE), Markov Chain Monte Carlo (MCMC), Parameter Solution (Parasol), and Sequential Uncertainty Fitting (SUFI-2) [32].The SUFI-2 calibration technique developed by [33,34] seems to be superior, as it requires fewer simulations to obtain good quality calibration and uncertainty results. The model errors are proposed to be the additional uncertainty created by comparing the simulated streamflows and RCM-simulated streamflows with the streamflow data measured at the gauging station [17,35].

In India, the impact of climate change analysis has been carried out extensively in the realm of hydrology and water resources. For instance, the impact of climate change on streamflow [36–40] and on precipitation [41–44] is analyzed over various river basins in India. In some cases, the uncertainty analysis is carried out in order to provide qualitative projections using REA [45], Bayesian analysis [20] and the possibilistic approach [46].

Decreased runoff with an increase in water consumption is seen in the Krishna River basin over the past decades due to climate change and anthropogenic activities [47]. Climate change is expected to have a significant impact on the hydrology of the Krishna River basin, with some noticeable trends in the future [42,48]. These include a decrease in water yield and average precipitation, as well as an increase in average temperature, in most areas of the Krishna River basin. According to the International Water Management Institute (IWMI) research reports [49–51], the Krishna River is categorized as an economically water scarce and food deficit region, suggesting the need for increased water-related investments. Future water supply and demand in the basin are also affected by factors such as spatial variation and future population growth, cropping intensity, and ground water use, along with industrial, domestic, and environmental water demand. Tirupathi et al. (2018) [52] performed an analysis of extreme events in the Krishna River basin using multiple models of RCP 4.5 and RCP 8.5 scenarios. They concluded that shifts in the monsoon rainfall and extreme rainfall events in the non-monsoon period must be considered to mitigate the risks in water management strategies.

Based on the availability of data, spatial and temporal variations in climate parameters and streamflow are analyzed without considering any man-made structures, such as dams, diversions, etc. The climate model data obtained from five RCMs under the representative concentration pathways (RCP) 4.5 scenario and four RCMs under the RCP 8.5 scenario are used. The RCP 4.5 is a stabilization scenario in which total radiative forcing stabilizes shortly after 2100 while not exceeding the long-run radiative forcing target level [53]. The RCP 8.5 is a high greenhouse gas emissions scenario that projects a significant increase in GHG emissions and concentrations over time, resulting in a radiative forcing of 8.5 $W/m^2$ at the end of the century [54]. Based on the uncertainties in the climate predictions of the different RCMs, as proposed by [55] and [52], the importance of developing the REA method is supported. The use of REA climate data reduces the uncertainty of climate model data. In addition to REA, precipitation must be corrected by QQ mapping in order to project future climate change.

The SWAT model was calibrated and validated using streamflow data from multiple gauge stations. The SUFI-2 algorithm was used to calibrate, validate, and analyze uncertainty [32,33]. To assess the impacts of climate change, the REA precipitation and minimum and maximum temperature data were used to simulate the streamflow in the future. From 1975 to 2100, the monsoon and monthly variations of the climate parameters and streamflow were tracked. The Mann–Kendall trend test is used to analyze the trend of streamflow in the basin at the subbasin level for all future periods. The research questions addressed in the present study are: (i) How does one minimize the uncertainty associated with the use of multiple climate models? (ii) How well does the SWAT model calibrate and validate the hydrological responses at multiple gauging points? (iii) What is the possible impact of climate change on streamflow over the Krishna River basin in India?

Following a brief introduction to the work, the study area characteristics and data used are discussed. This work also includes the overall methodology and inputs used, as well as the models used and their significance and uncertainty techniques. The results and discussions in the following section include the bias corrected climate variables, as well as the calibration, validation, and performance evaluation of the Hydrological model, followed by an analysis of the results, which associates the effects of climate change on streamflow with their trends. Finally, the conclusion section discusses the methods' strengths and weaknesses, as well as remarks and proposed elements for future research.

## 2. Study Area and Data Used

### *2.1. Study Area*

The River Krishna basin is one of the longest rivers in central southern India, with a total surface area of 258,948 sq. km. The basin is located between 13°10′ to 19°22′ North latitudes and 73°17′ to 81°9′ East longitudes, as shown in Figure 1. It covers four Indian states: Karnataka (43.8%), Andhra Pradesh and Telangana (29.81% together), and Maharashtra (26.36%). The river has several tributaries, with Ghatprabha, Malprabha, and Tungabadhra joining on the right and Bhima, Musi, and Munneru joining on the left. The Krishna basin is divided into seven subbasins: Bhima Upper, Bhima Lower, Krishna Upper, Krishna Middle, Krishna Lower, Tungabadhra Upper, and Tungabadhra Lower.

The climate of the basin is tropical, with an average annual precipitation of 960 mm and minimum and maximum temperatures of 20.73 °C and 32.2 °C, respectively. The basin experiences varying average annual precipitation, with a maximum value of 2000 mm in the Western Ghats region and values ranging from 300 mm to 1000 mm in the delta region. Streamflow has been characterized by low flows from March to May and high flows from August to November.

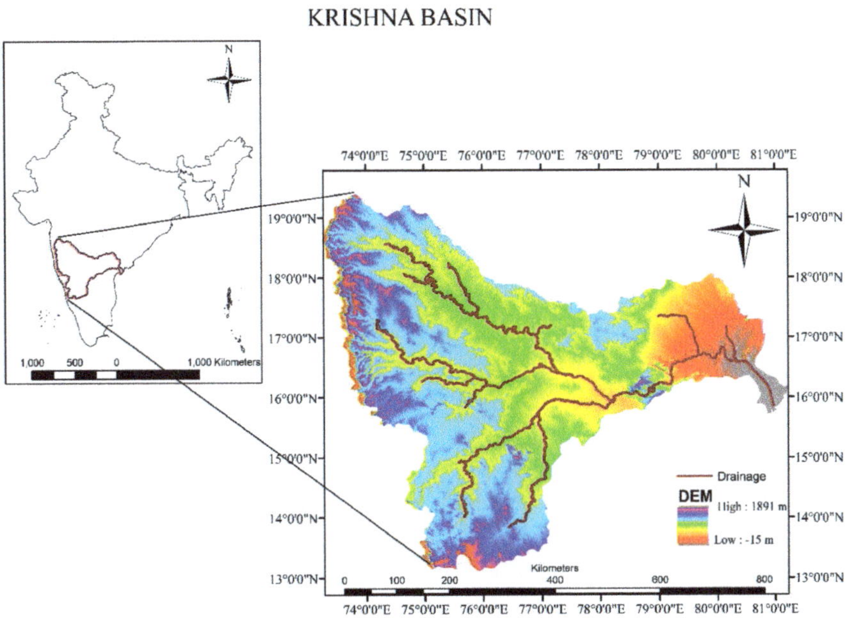


**Figure 1.** Location map of the Krishna River basin.

The total population of the Krishna River basin was 74.2 million as per the 2011 census. Approximately 68% of the population in the basin lives in rural areas, with agriculture serving as the main source of income [56]. Approximately 77% of the total geographical area of the basin is cultivable, with the main crops identified as rice, corn, cotton, sorghum, millet, and sugar cane, along with a variety of horticulture crops. Increased population necessitates increased consumption of water for domestic and industrial purposes, putting strain on the basin's water resources. The huge projects being developed in all states have caused interstate conflicts over water rights. According to [51], the basin is nearing closure because consumption exceeds the availability of water in the basin, which suggests the need to assess monthly changes in climate parameters and their impact on runoff when developing a water allocation policy.

### 2.2. Data and Methods

Table 1 lists various datasets and their resolutions. DEM estimates the basin's minimum, maximum, and mean elevations as 18 m, 1903 m, and 518 m, respectively. Approximately 50.47% of the total area falls under the elevation zone of 500 m to 750 m. The basin is made up of 132 climate grid points measuring 50 km × 50 km, as well as 47 hydrometeorological stations.

Missing values are contributed by the streamflow data obtained from hydrometeorological stations. Due to missing stream flow data at many stations, only 14 out of 47 stations' data were used for calibration and validation of the model. Land use information for the Krishna basin is divided into 14 categories, as shown in Figure 2a, with Agriculture (72.56%) the most important. Figure 2b represents a soil map dominated by fine-textured soil. Laterite and lateritic soils, red soils, alluvium, black soils, mixed soils (red and black, red and yellow, etc.), and alkaline and saline soils are the important soil types found in the basin. A slope map with a maximum of 5 classes based on the percentage rise was generated using DEM, as shown in Figure 2c.

**Table 1.** Description and source of hydro geospatial and climatological database.

| Data Type | Resolution | Source |
|---|---|---|
| Digital Elevation Model | 30 m | Advanced Space borne Thermal Emission and Reflection Radiometer (ASTER), |
| Land use/Land cover | 400 m | https://swat.tamu.edu/data/india-dataset/, accessed on 15 March 2019 |
| Soil | 1:5,000,000 | https://swat.tamu.edu/data/india-dataset/, accessed on 15 March 2019 |
| Observed Climate data | 0.5° grid | Indian Meteorological Department, Pune. |
| Climate Model data | 0.5° grid | Centre for Climate Change Research (CCCR), Indian Institute of Meteorology (IITM) Pune. ftp://cccr.tropmet.res.in/iRODS_DATA/CORDEX-Data, accessed on 23 May 2017 |
| River Discharge | 14 stations | Central Water Commission (CWC) |

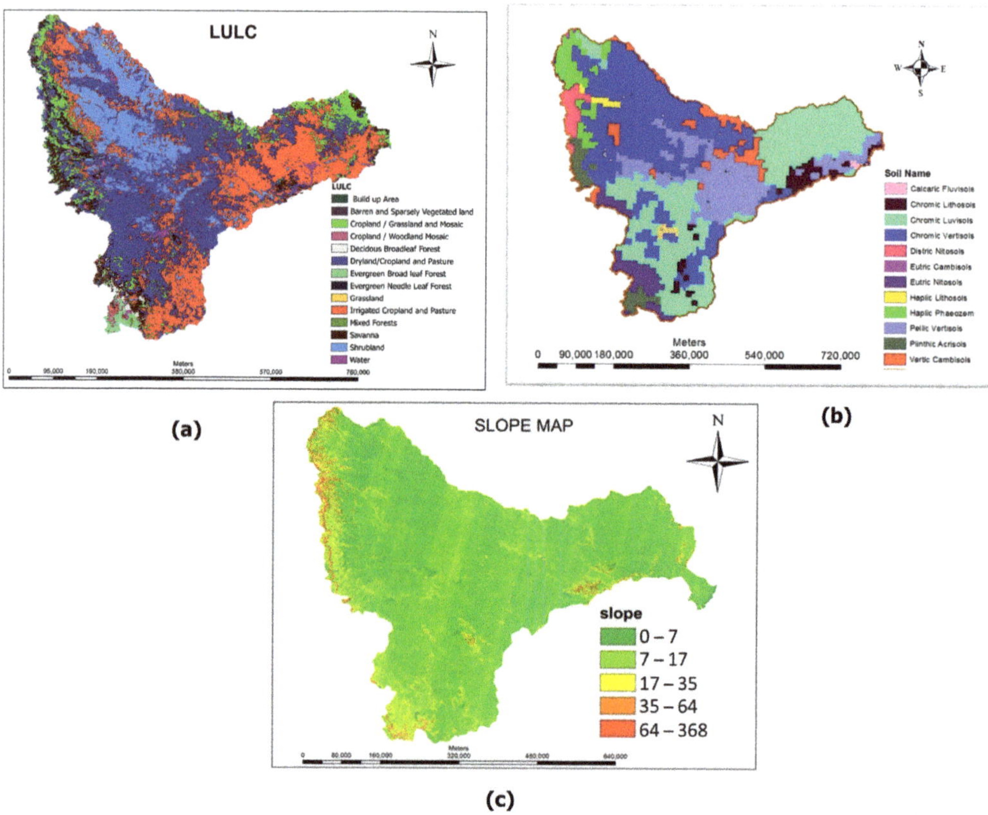


**Figure 2.** Thematic maps of the Krishna basin (**a**) Land use/Land cover, (**b**) Soil and (**c**) Slope.

Based on the availability of data, spatial and temporal variations in climate parameters and streamflow are analyzed without considering any man-made structures, such as dams, diversions, etc. The climate model data obtained from five RCMs as mentioned in Table 2 under the representative concentration pathways (RCP) 4.5 scenario and RCP 8.5 scenario are used with the assumption that emissions will be reduced by the end of the century. RCP 4.5 is a stabilization scenario in which total radiative forcing stabilizes shortly after 2100 while not exceeding the long-run radiative forcing target level [53]. RCP 8.5 is a high greenhouse gas emissions scenario that projects a significant increase in GHG emissions and concentrations over time, resulting in a radiative forcing of 8.5 W/m$^2$ at the end of the century [54].

**Table 2.** Details of RCM models.

| Acronym | RCP | | Full Name |
|---|---|---|---|
| ACCESS | 4.5 | 8.5 | Australian Community Climate and Earth System Simulator |
| CCSM4 | 4.5 | 8.5 | Community Climate System Model |
| CNRM_CM5 | 4.5 | 8.5 | Centre National de Recherché Meteorologiques |
| NorESM 1 | 4.5 | 8.5 | Norwegian Earth System Model 1 |
| MPI-ESM-LR | 4.5 | - | Max Plank Institute Earth System Model |

The SWAT model was used to simulate streamflow for the reference period using observed daily meteorological data from 1970–2005. It has a spatial resolution of 0.5° × 0.5° and includes maximum temperature ($T_{max}$), minimum temperature ($T_{min}$), and precipitation. The SWAT model was calibrated and validated using daily stream flow data collected from 14 gauge stations. To simulate the hydrologic characteristics for future periods using SWAT, an ensemble of high resolution past and future climate projections from regional scales—with mid-range and high-range concentration pathways of RCP 4.5 and 8.5 greenhouse gas (GHG) emissions scenario from the Centre for Climate Change Research (CCCR), Indian Institute of Tropical Meteorology, Pune, India—was obtained.

The inter-model uncertainty of climate models is estimated using the REA method, as discussed in the following sections.

### 2.3. *Reliability Ensemble Averaging (REA)*

Giorgi and Mearns (2003) [57] developed the REA method, which allows for the best estimate and reliable climate model data with a range of uncertainty. The calculation of REA follows a two-step approach: model performance and model convergence. For this study, REA is quantified using the algorithm developed by [19] for variables such as precipitation and minimum and maximum temperatures. Model performance is measured using the inverse of the Root Mean Square Error (RMSE), obtained from the Cumulative Distribution Functions (CDF) deviations between observed and simulated values for the period 1975–2005. Uncertainty in RCM's future data is quantified using the model convergence approach, which considers the CDF deviations between individual RCMs and the weighted mean CDF derived from the model performance approach. In REA, initial weights (Equation (2)) are calculated based on the RCM's ability to simulate historical observations in terms of root mean square error (RMSE) (Equation (1)), which indicates the model performance criteria.

$$\text{RMSE} = \left[\frac{1}{N}\sum_{i=1}^{N}(\text{Observed}_i - \text{RCM}_i)\right]^{\frac{1}{2}} \quad (1)$$

$$W_{int} = \frac{\left[\frac{1}{\text{RMSE}_i}\right]}{\left[\sum_{I=1}^{n}\frac{1}{\text{RMSE}_i}\right]} \quad (2)$$

The following are the steps used to quantify the reliability of the climate model and to obtain the reliability ensemble mean:

- Divide the total range of RCM variable data into 10 equal intervals of CDF with respect to the observed time series data and compute the RMSE. Inverse values of RMSE are considered as the proportional weights, and the sum of the weights of all RCMs is equal to one. Higher weights are assigned to models that perform better.
- Model convergence criteria are applied by considering the weights obtained from model performance criteria as the initial weight for their respective RCMs.
- The product of the initial weight ($W_{int}$) and the corresponding CDF of the future simulated with RCM ($F_{RCMi}$) is taken as the weighted mean CDF ($F_{wm}$):

$$F_{wm} = \sum_{i} W_{int(i)} \, x \, F_{RCM_i} \quad (3)$$

- The same procedure is repeated again as in Step 1, but the RMSE calculated with respect to the weighted CDF and future projection of RCM and weights obtained will be used in the next iteration for the respective RCMs, and a new weighted CDF with different weights is computed.
- Repeat Steps 2 to 4 until the same weight repeats and the model convergence criteria is met.

This procedure is adapted to all grid points and three meteorological variables, such as precipitation and minimum and maximum temperatures, under the RCP 4.5 and 8.5 scenarios in the study area due to the varied nature of RCMs for the different grids and climate variables. The ensemble average of the climate variables for a particular grid is calculated using the weighted sum of the product of corresponding final weights with the respective meteorological variables. Thus, an ensemble weighted average of each climate variable is given as input to the distributed hydrological model for each grid instead of providing each RCM input separately. The minimum and maximum temperatures of the climate model data projects values that are similar to the observed data. The REA precipitation has additional bias, which is reduced by adopting the non-parametric quantile mapping method described in the following section.

### 2.4. *Quantile Mapping Method of Bias Correction*

Quantile mapping is the most used bias correction technique due to its ease of handling large amounts of data and its computational efficiency [58,59]. The statistical bias correction used for ensemble precipitation obtained from the REA method was equal distance-based quantile-to-quantile matching. For a given variable, the CDF of REA data is first matched with the CDF of observations, resulting in the generation of a correction function based on the quantile. Then, this correction function is used to remove bias from the REA data quantile by quantile (Figure 3).

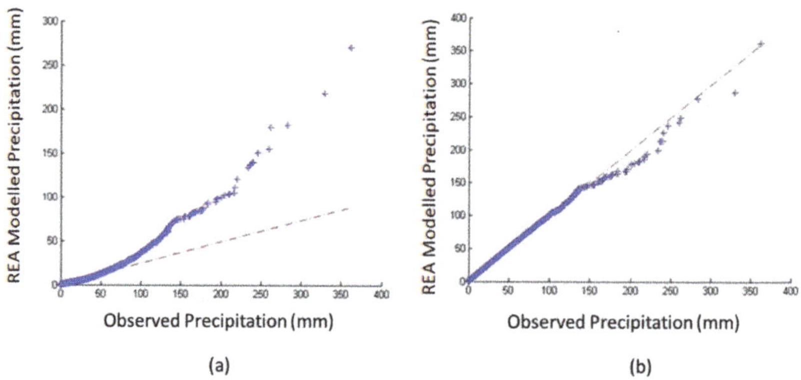


**Figure 3.** Quantile–Quantile (QQ) map of observed vs. REA modelled precipitation data (**a**) Before bias correction (**b**) After bias correction.

### 2.5. *SWAT Model Setup*

SWAT is a physically based, semi-distributed, and continuous process-based model that is used to simulate the rainfall–runoff process. It also predicts the hydrological impacts of climate change on water resources. The model was built using geospatial data (DEM, land use, soil and slope maps) and the hydrometeorological data. The watershed was divided into 50 subbasins with a threshold area of 60,000 sq km, and the subbasins were further divided into 898 hydrological response units (HRUs) based on 5, 10, and 10 threshold values for land use, soil type, and slope class of the watershed, respectively. As land use plays a major role in simulating the surface runoff, land use with less than 5% is eliminated. While determining HRUs of the basin, 10% of the soil and slope classes in the subbasin is eliminated. The increased number of subbasins and HRUs helps in projecting the variability of the watershed spatially. The observed meteorological data from 1970 to 2005 were used to simulate the streamflow, with a warm up period of 5 years. The calibration and validation of the model were carried out with SWAT-CUP.

### 2.6. Calibration and Validation of the Model Using SWAT-CUP

Uncertainties caused by input data, model structures, and parameters can be reduced using SWAT's calibration and uncertainty programs. The model was calibrated and validated using SUFI-2, which aims to cover the most measured data with the smallest uncertainty bands. It can map the uncertainties of parameter ranges and propose the 95% prediction uncertainty (95PPU), which helps in evaluating overall uncertainty in the hydrologic response output. The model performance was evaluated using the $R^2$, Nash–Sutcliff Efficiency (NSE) and Percent bias (PBias). The ranges of $R^2$ and NSE are 0 and 1, respectively, with a value close to 1 indicating high model performance and efficiency. The low-magnitude values of PBias indicate better simulations, with an optimum value as 0. Positive values indicate underestimation of the model, while negative values indicate overestimation of the model.

Monthly streamflow data from 1975 to 1990 were used to calibrate the model, and data from 1991 to 2003 were used to validate it. Streamflow is generated by the summation of variables, such as base flow, interflow, and surface runoff. The parameters responsible for these variables were used against monthly streamflow data from five gauge stations. Previous studies, as well as the minimum and maximum parameter ranges established by the SWAT user guide, are used for the calibration of the Krishna River basin. SUFI-2 was used to set up the 15-parameter combinations for each iteration, and the SWAT simulation for each combination of approximately 500 iterations generated the uncertainty measures and goodness of fit by establishing new parameter ranges. The calibration process was repeated until acceptable statistics of goodness of fit were generated. The final parameter ranges obtained were used for model validation. The uncertainty in the measured streamflow was accounted for by including the estimated error of 10%. The statistical indices used for evaluating the model simulation and predictability are the coefficient of determination ($R^2$), Nash–Sutcliff Efficiency (NSE), and Percent Bias (PBias).

### 2.7. Statistical Methods

For each basin, statistically significant trends in annual streamflow were determined using the Mann–Kendall's (M-K) tau non-parametric test. If the probability value is less than or equal to 0.10, i.e., Kendall's tau value equals zero, the trend was considered to be statistically significant. The degree of correspondence between two variables $x$ and $y$, where $x$ represents time and $y$ represents streamflow, was calculated using Kendall's tau. If $\tau = 1$, then the data shows perfect positive correlation; if $\tau = -1$, then the data exhibits perfect negative correlation; and $\tau = 0$ shows no correlation between the pairs. Thus, a positive value of $\tau$ represents an increase in trend, while a negative value of $\tau$ represents a decrease in trend. The magnitude of the trend was estimated using Sen's method [60–62].

## 3. Results and Discusion

### 3.1. Future Projections of REA Climate Data

The observed and simulated REA precipitation and maximum and minimum temperatures are compared before and after bias correction. The results show that the observed and simulated REA temperature values agreed well, and that the precipitation data projects the same after the biases are reduced using Q-Q mapping. The climate parameters obtained cover the years 1975 to 2099. The analysis divides the total period into four 25-year segments: Historic (1980–2004), Future1 (2020–2044), Future2 (2045–2069), and Future3 (2070–2094). Figure 4 depicts the annual average precipitation (in mm/year) for the historic and future periods under RCP 4.5 and 8.5 scenarios.

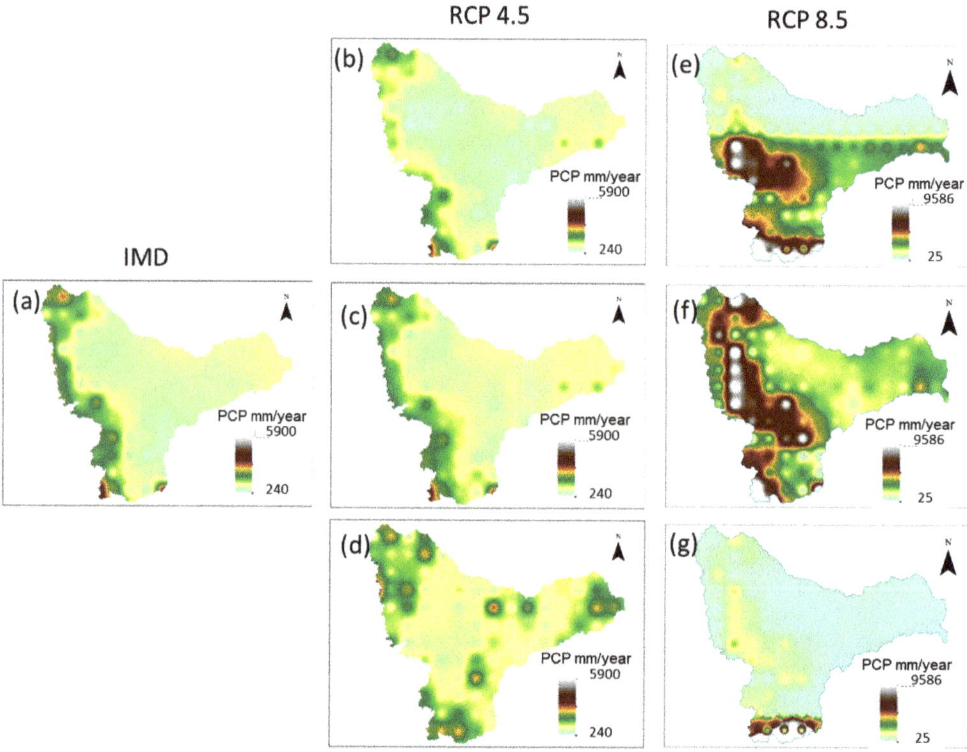


**Figure 4.** Annual average precipitation variation in the Krishna basin for (**a**) Historic, Future1 (**b**,**e**), Future2 (**c**,**f**), and Future3 (**d**,**g**) periods under RCP 4.5 and RCP 8.5 scenarios.

The annual average variations of precipitation show an increase in the western part of the basin in Future1 and a decrease in the Future2 period under RCP 4.5. Under RCP 8.5, annual average precipitation decreases by 36% in Future1, 10% in Future2, and 60% in Future3. However, Mishra and Lilhare (2016) [11] projected a continuous increasing trend of water balance components, along with rainfall and air temperature, in both the RCP 4.5 and 8.5 scenarios of CMIP5 models in the Krishna River basin. Rainfall increases approximately 8–20% under RCP 4.5 and 10–40% in the case of RCP 8.5 by the end century. Similarly, the surface runoff, streamflow, and ET were projected to increase of approximately 20–55%, 20–60%, and 4–9% under RCP 4.5 and 35–120%, 40–120%, and 2–8% in the case of RCP 8.5 by the end of the century. Tirupathi et al. (2018) [52] projected a decrease in rainfall under future scenarios for different climate models. In the present study, daily minimum and maximum temperatures within the basin show higher values in the Future periods as compared to the Historic period under both scenarios. The Future2 period has higher temperature values than the other two future periods. Figures 5 and 6 depicts seasonal variations of precipitation, i.e., monsoon (June–September), winter (October–January), and summer (February–May), for the RCP 4.5 and 8.5 scenarios.

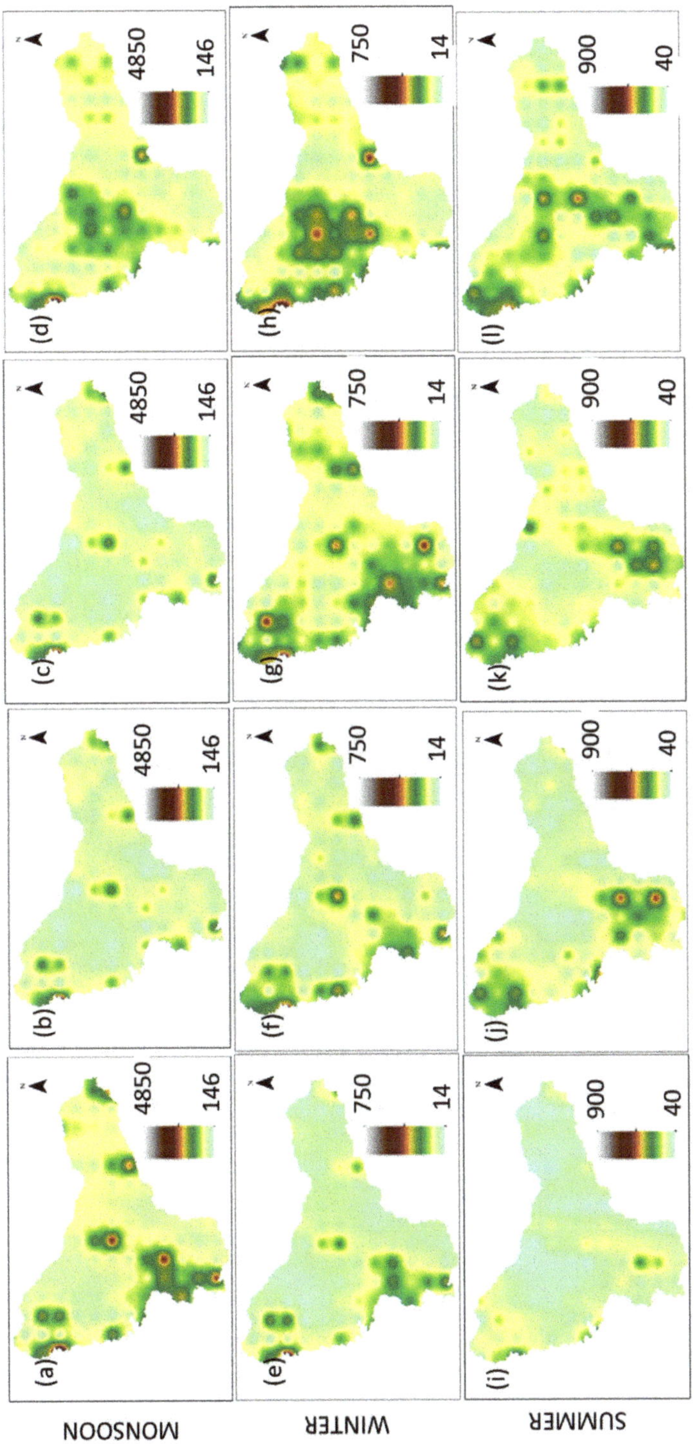


**Figure 5.** Seasonal average precipitation variation in the Krishna basin for Historic, Future1, Future2, and Future3 periods of Monsoon (**a–d**), winter (**e–h**) and summer (**i–l**) under the RCP 4.5 scenarios.

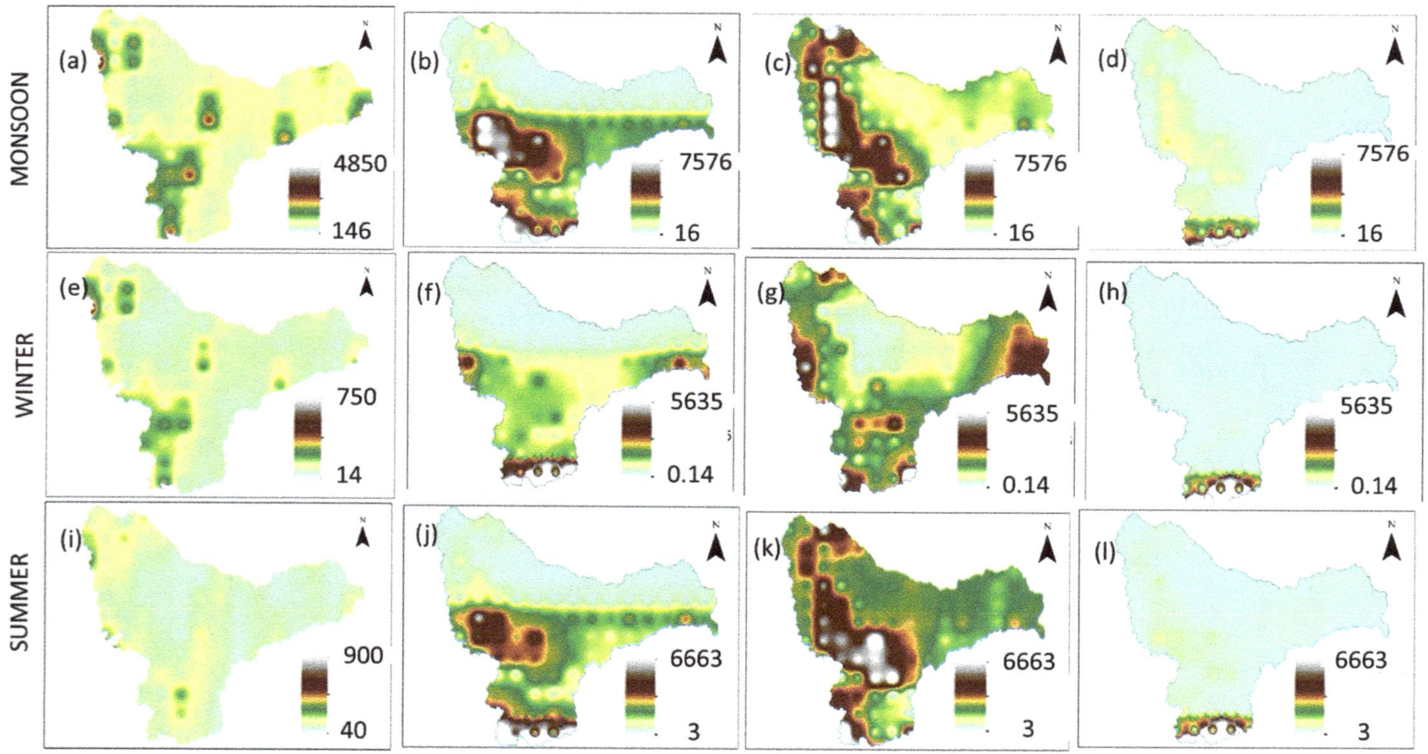


**Figure 6.** Seasonal average precipitation variation in the Krishna basin for Historic, Future1, Future2, and Future3 periods of Monsoon (**a–d**), winter (**e–h**) and summer (**i–l**) under the RCP 8.5 scenario.

From Figure 5, it is observed that the spatial distribution of precipitation is uniform throughout the basin, with extreme values projected in the Bhima and Middle Krishna subbasins of the Krishna River under RCP 4.5. A seasonal shift in precipitation is also projected in the winter and summer seasons of the Bhima and Tungabadra subbasins in the Future2 period and Bhima and Middle Krishna subbasins in the Future3 period. Extreme precipitation values are observed throughout the basin for the three future periods of the RCP 8.5 scenario, while the remaining portion of the basin projects precipitation values similar to the other scenario. The number of the extreme precipitation values is reduced in the future periods of the basin under RCP 8.5 as compared to RCP 4.5. Reduced precipitation is observed in the future periods as compared to the Historic period, with increased precipitation values in the central portion of the basin of the Future3 period. Precipitation values in the winter and summer seasons of the future periods are higher than in the Historic period. This reflects the shift in the monsoon in future periods. Even extreme values are observed in the central parts of the basin as compared to the Historic period in the winter and summer seasons. The maximum temperature represents an increase in all seasons in the future periods when compared to the Historic period. There is an overall increase in precipitation and temperatures in the Future3 period in the basin.

The temporal variations of the REA climate data of the future periods compared to the Historic period projects high precipitation values from June to October in Future2 & Future3, and they are more or less similar to the values of the Historic period. Peak values are observed in June (560 mm in Historic, 410 mm in Future1, 490 mm in Future2, and 520 mm in Future3), with a minimum value of 280 mm in Future1 and a maximum value of 485 mm in Future3 is detected in July under the RCP 4.5 scenario. Extreme precipitation values under the RCP 8.5 scenario range from 757.6 mm/day (monsoon) to 563.5 mm/day (summer). Precipitation values have increased overall in all months of the Future3 period of the RCP 4.5 scenario. Minimum and maximum temperature values rise in all months of the future period. From March to May, high temperatures are expected to range from 20 °C/day to 28 °C/day for the minimum temperature and 33 °C/day to 44 °C/day for the maximum temperature. Peak values of maximum temperatures observed in April are 34 °C/day in the Historic period, 36 °C/day in the Future1 period, 42 °C/day in the Future2 period, and 44 °C/day in the Future3 period in the RCP 4.5 scenario. Under the RCP 8.5 scenario, the basin experiences an overall temperature increase of 20 to 30 °C/day.

### 3.2. Calibration and Validation of the SWAT Model

The SUFI-2 algorithm was used to analyze the sensitivity and uncertainty of the SWAT model. For the calibration process, 15 parameters (from the SWAT-CUP manual) were chosen and used. The optimum parameter range obtained for the watershed calibration with the identical initial parameter ranges are shown in Table 3. An increase in stream flow is obtained with higher CN values and a decrease in baseflow and vice versa [63]. Whereas from Table 3, the actual CN are changed in relation to the 5% change to obtain the calibration. In this study, the rate of change of the streamflow is inconsistent with baseflow due to the change in CN values.

For example, the initial range of ESCO (Soil Evaporation Compensation Factor) is 0.4 to 0.8, but the final range is 0.94, 0.55, 0.83, 0.46, and 0.54 for the gauge stations Huvinhedgi, Narsingapur, Yadgir, Damercherla, and Keesara. The decrease in the ESCO values permits water from lower layers to upper layers to compensate the water deficit and leads to high soil evapotranspiration, thereby decreasing the surface runoff and baseflow. Thus, the changes in ESCO have a promising impact on the decreased rate in the streamflow values of the basin [64]. Table 4 represents the $R^2$, NSE, and PBias values obtained at five gauge stations during the calibration and validation periods.

**Table 3.** Calibration parameter ranges for the five-gauge stations.

| S.No | Parameter | Initial | Final | Huvinhedgi | Narsingpur | Yadgir | Damercherla | Keesara |
|---|---|---|---|---|---|---|---|---|
| 1 | R__CN2.mgt | −0.20 | 0.20 | −0.19 | 0.03 | 0.05 | −0.17 | 0.17 |
| 2 | V__ALPHA_BF.gw | 0 | 1.00 | 0.20 | 0.96 | 0.95 | 0.16 | 0.26 |
| 3 | V__GW_DELAY.gw | 0 | 500 | 275 | 45 | 427 | 414 | 352 |
| 4 | V__GWQMN.gw | 0 | 5000 | 1.24 | 1.75 | 1.70 | 1050 | 1100 |
| 5 | V__GW_REVAP.gw | 0.02 | 0.20 | 0.19 | 0.19 | 0.17 | 0.16 | 0.14 |
| 6 | R__SOL_K(..).sol | 0.14 | 0.99 | 0.99 | 0.37 | 0.66 | 0.55 | 0.72 |
| 7 | R__SOL_AWC(..).sol | −0.15 | 0.60 | 0.04 | 0.55 | 0.29 | 0.34 | 0.42 |
| 8 | R__SOL_BD(..).sol | 0.05 | 0.70 | 0.69 | 0.12 | 0.67 | 0.61 | 0.58 |
| 9 | V__ALPHA_BNK.rte | 0.00 | 1.00 | 0.60 | 0.40 | 0.43 | 0.36 | 0.65 |
| 10 | V__CH_N2.rte | −0.09 | 0.09 | 0.09 | 0.11 | 0.08 | 0.07 | 0.04 |
| 11 | V__CH_K2.rte | 18.72 | 103.96 | 88.36 | 86.75 | 47.45 | 48.42 | 53.26 |
| 12 | V__ESCO.hru | 0 | 1.00 | 0.94 | 0.55 | 0.83 | 0.46 | 0.62 |
| 13 | R__EPCO.hru | 0 | 1.00 | 0.30 | 0.59 | 0.16 | 0.51 | 0.48 |
| 14 | R__SLSUBBSN.hru | −0.12 | 0.30 | 0.13 | 0.09 | −0.05 | 0.25 | 0.06 |

Where V__ represents that existing parameter, value is to be replaced by the given value and R__ represents that existing parameter value to be multiplied by (1+ a given value). Bold values representing the sensitive param.

**Table 4.** Goodness of fit parameters for calibration and validation periods.

| Stream Gauge Station | Calibration | | | Validation | | |
|---|---|---|---|---|---|---|
| | $R^2$ | NSE | PBias | $R^2$ | NSE | PBias |
| Huvinhedgi | 0.62 | 0.62 | −13.50 | 0.42 | 0.42 | 79.80 |
| Narsingapur | 0.62 | 0.52 | −41.90 | 0.50 | 0.47 | −73.50 |
| Yadgir | 0.86 | 0.58 | −27.10 | 0.40 | 0.32 | −88.50 |
| Damercherla | 0.80 | 0.75 | −8.35 | 0.58 | 0.43 | −57.80 |
| Keesara | 0.68 | 0.62 | −12.40 | 0.52 | 0.48 | 65.40 |

The performance of the model may suffer as a result of any combination of errors, such as geospatial data, including the land use and soil map, climate data, and the difference between the observed and simulated streamflow. However, the $R^2$ and NSE values for the calibration range from 0.52 to 0.86 and 0.32 to 0.58, respectively, indicating that the performance was good and satisfactory during the calibration and validation periods.

The average thickness of the 95PPU band indicates the calibration strength and represents the width of the uncertainty interval, and it must be as small as possible. The 95PPU plot projects the uncertainty band generated during the calibration and validation of the model. The monthly time series of observed and simulated streamflows obtained with the fitted parameter value of the five gauge stations projects the minimum variations in streamflow dynamics over the calibration and validation periods.

Comparisons of the observed mean monthly streamflow and those produced by SWAT using REA data at three gauge stations (Figure 7)—i.e., the Huvinhedgi outlet of the Upper Bhima, Lower Bhima and Upper Krishna; the Mantralayam outlet of the Upper and Lower Tungabadhra; and the Pondhugala outlet of the Krishna River basin. The observed streamflow in July and August is more variable than the simulated streamflow at Huvinhedgi, whereas the model projects similar streamflow at Mantralayam in months other than June and July. The Pondhugala, downstream of the Huvinhedgi and Mantralayam, shows a similar pattern in mean monthly streamflow, with maximum flows in August, September, and October. The errors in the mean monthly streamflow may be due to the failure to account for reservoirs and changes in land use in the Krishna River basin.

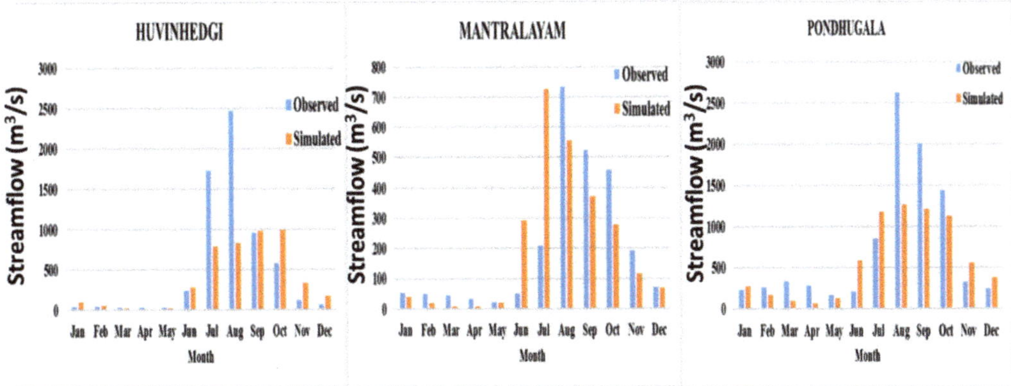


**Figure 7.** Observed and SWAT-simulated streamflow at three gauge stations for the Historic period (1975–2005).

### 3.3. Future Streamflow Projections

According to the climate projections, lower streamflows are simulated for the future periods. Flow Duration Curves (FDCs) at three gauge stations—Huvinhedgi (Figure 8), Mantralayam (Figure 9), and Pondhugala (Figure 10)—are used to explain the relationship between the magnitude and frequency of streamflow for the future periods. These curves enumerate the flow of exceedance for a given level of probability developed on an annual, monsoon, and non-monsoon basis, assisting water managers in determining water availability in the face of climate change. Figure 8 depicts the FDC of the Huvinhedgi gauge station, representing total flows upstream of the basin, which demonstrates an overall decrease in flows during the Future1 period, with flows during theFuture2 and Future3 periods comparable to the historic period. Though Future2 and Future3 project an increase in precipitation for the basin, there is a decrease in high flows (flows that exceed 10–30% of time) and low flows are (flows that exceed 80% of time), which is mainly due to a likely increase in temperatures. The median flows (flows that exceed 30–70% of time) are similar to the Historic period flows. Both scenarios project similar patterns of curves with changes in quantity, with RCP 4.5 projecting comparatively fewer quantities than RCP 8.5.

The annual FDC of the Mantralayam and Pondhugala stations shows a decrease in high and low flows, but an increase in median flows. Flows are decreasing in the Future1 period. In the monsoon period, high flows are forecast throughout the period, with lower values in future periods compared to the Historic period. However, Nikam et al. (2018) [65] projected an increase in streamflow under the RCP 4.5 and 8.5 scenarios using IITMRegCM4-4 predictions that are calibrated and validated at single gauge stations. In the present study, the streamflows are simulated using REA climate data, which is calibrated and validated at multiple gauge stations. Many studies have even reported that the Krishna River basin is becoming more vulnerable as a result of increased human consumption of surface water resources, a reduction in surface water base flows due to over abstraction of groundwater, and fewer releases to the ocean. It also implies that hydrological changes anywhere in the basin have a negative impact on total water availability and water distribution spatially during droughts. Hence, it is important to consider water availability deficits when developing water use policies in the Krishna River basin.

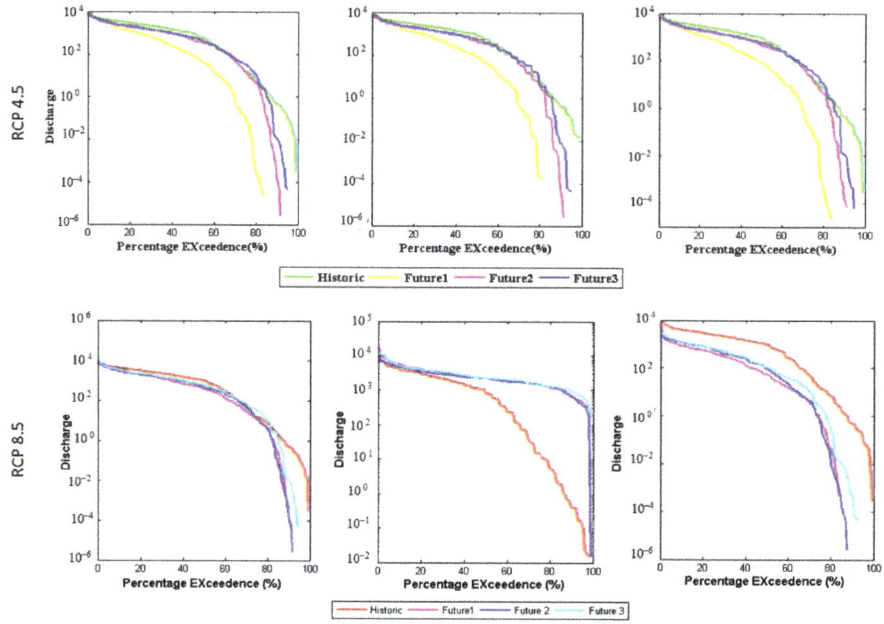

Figure 8. Annual, monsoon, and non-monsoon flow duration curves at the Huvinhedgi gauge station for historic and future periods under RCP 4.5 and RCP 8.5 scenarios.

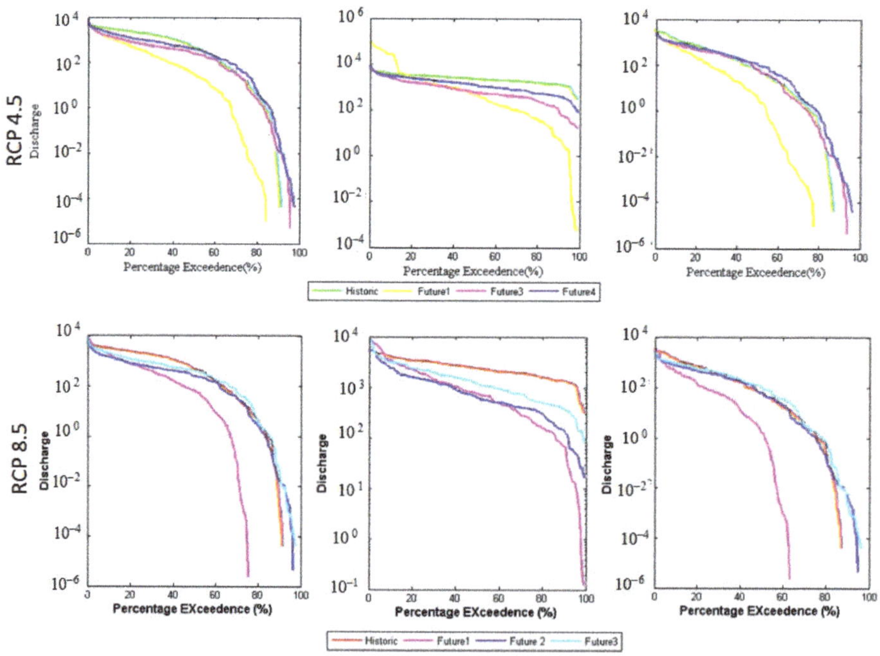

Figure 9. Annual, monsoon, and non-monsoon flow duration curves at the Mantralayam gauge station for historic and future periods under RCP 4.5 and RCP 8.5 scenarios.

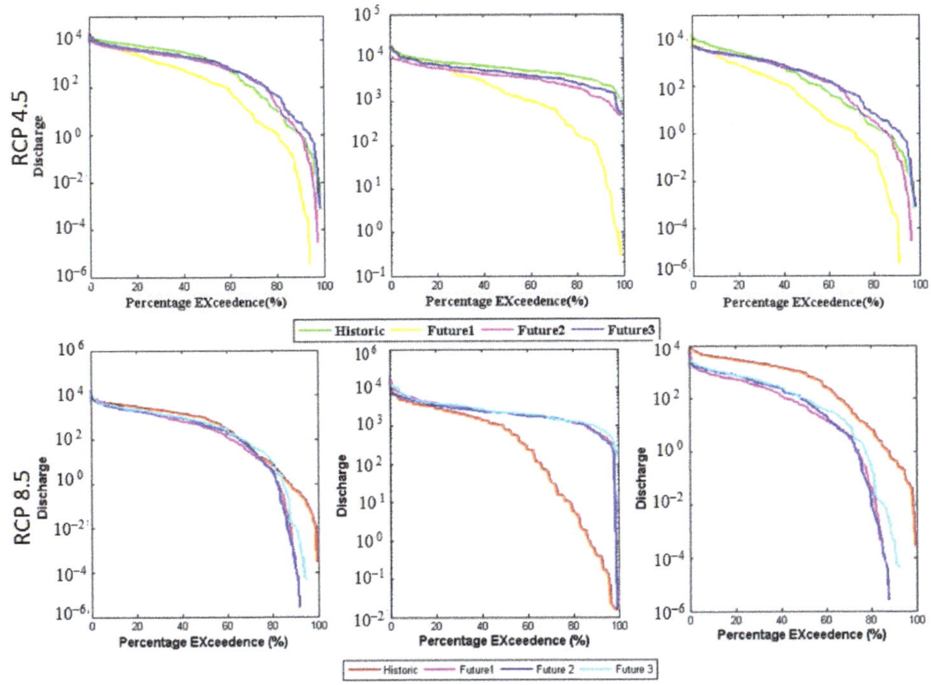


**Figure 10.** Annual, monsoon, and non-monsoon flow duration curves at the Pondhugala gauge station for historic and future periods under RCP 4.5 and RCP 8.5 scenarios.

### 3.4. Climate Change Impact Assessment

The impact of climate change on the basin's water resources was estimated by implementing the calibrated SWAT model with REA weather data obtained from an ensemble of five RCM for one historic and three future periods. The streamflow analysis was carried out on a monthly basis at three gauge stations: Huvinhedgi, Mantralayam, and Pondhugala. The streamflow from the Bhima subbasins was accumulated at Huvinhedgi, where Matralayam is the outlet of the Tungabadra. Streamflows from Huvinhedgi and Mantralayam meets the Pondhugala station, covering the remaining portions of the Krishna River basin. The REA precipitation data used in the SWAT model for the Krishna River subbasins (Figure 4) suggests a 20% decrease in the annual average values in the Future1 period and 4 to 6% decrease in the Future2 period, with Future3 projected to be the same as the Historic period. The monthly streamflow simulated by the SWAT model for the historic and future periods were analyzed, and the results are projected as normalized values of mean monthly flow as a ratio of annual flow. The absolute values of streamflow as the ratio of mean monthly flow and annual flow (Figure 11a) and relative change with respect to the Historic period (Figure 11b) show an increase in the Future2 and Future3 periods.

The absolute monthly flows simulated at Huvinhedgi during monsoon, winter, and summer are 716.3 $m^3/s$, 390.5 $m^3/s$, and 26.2 $m^3/s$ for the Historic period, whereas they are 615.65 $m^3/s$, 100.3 $m^3/s$, and 10.3 $m^3/s$ for the Future1 period of the RCP 4.5 scenario, and 2900.2 $m^3/s$, 389.8 $m^3/s$, and 259.8 $m^3/s$ are the overestimated streamflow values under the RCP 8.5 scenario. At Mantralayam and Pondhugala, the streamflows represent a decrease of about 50%, 40%, and 30% in the Future1, Future2, and Future3 periods, respectively, compared to the Historic period.

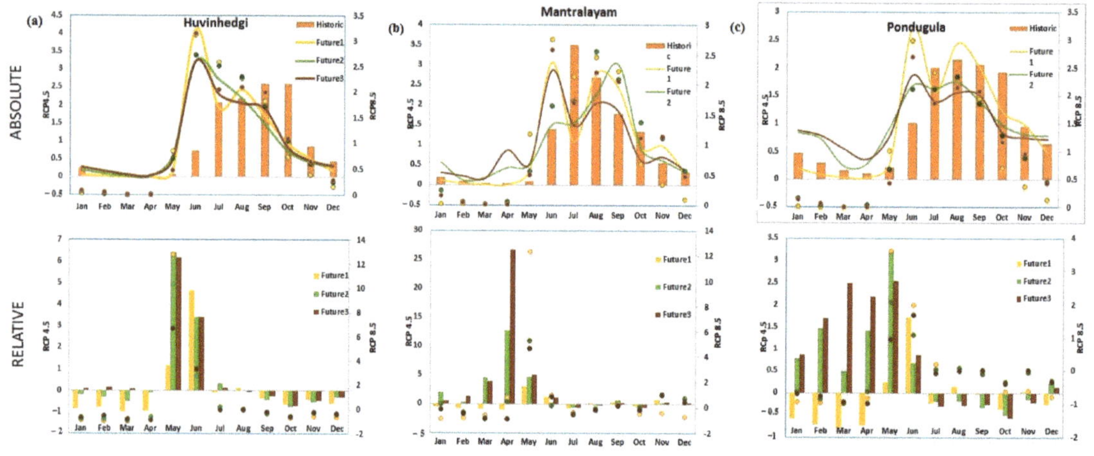


**Figure 11.** Mean monthly flow as a ratio of mean annual flow for historic and future periods of RCP 4.5 (line plot) and RCP 8.5 (dotty plot) at (**a**) Huvinhedgi, (**b**) Mantralayam, and (**c**) Pondhugula (**top**: absolute values, **bottom**: relative change).

The relative change in streamflow suggests that streamflow decreases throughout the year, with an increase in June for the three gauge stations. Figure 11 shows mean monthly flow as a ratio of the mean annual flow for the historic and future periods at Huvinhedgi, Mantralayam, and Pondhugala (top: absolute values, bottom: relative change).

The Mann–Kendall Trend test was used to determine statistically significant trends in the annual streamflow of 50 subbasins obtained from the SWAT model setup. In the analysis, the MK test results of streamflow showed no trend at 21 subbasins among 50 subbasins datasets during the Historic period. However a decreasing trend at 50 and 48 subbasins is shown in the Future1 period under the RCP 4.5 and 8.5 scenarios, respectively. Increases in trend were observed at 17 subbasins in the Future2 period under RCP 4.5 and 27 subbasins under RCP 8.5, with no trend in the remaining subbasins. Furthermore, a decreasing trend in 24 subbasins was observed under RCP 4.5, with an increasing trend in 38 subbasins under RCP 8.5 of the Future3 period. The trend analysis suggests that almost all of the subbasins have a decreasing trend in both the past and the future (Table 5). Figure 12 represents positive (increase) and negative (decrease) trends at the outlets of all subbasins spatially.

**Table 5.** Number of subbasins with increasing or decreasing trend.

| Climate Period | RCP 4.5 | | RCP 8.5 | |
|---|---|---|---|---|
| | Increasing | Decreasing | Increasing | Decreasing |
| Historic | 1 | 28 | | |
| Future I | - | 50 | - | 48 |
| Future II | 17 | 2 | 27 | 2 |
| Future III | - | 24 | 38 | 1 |

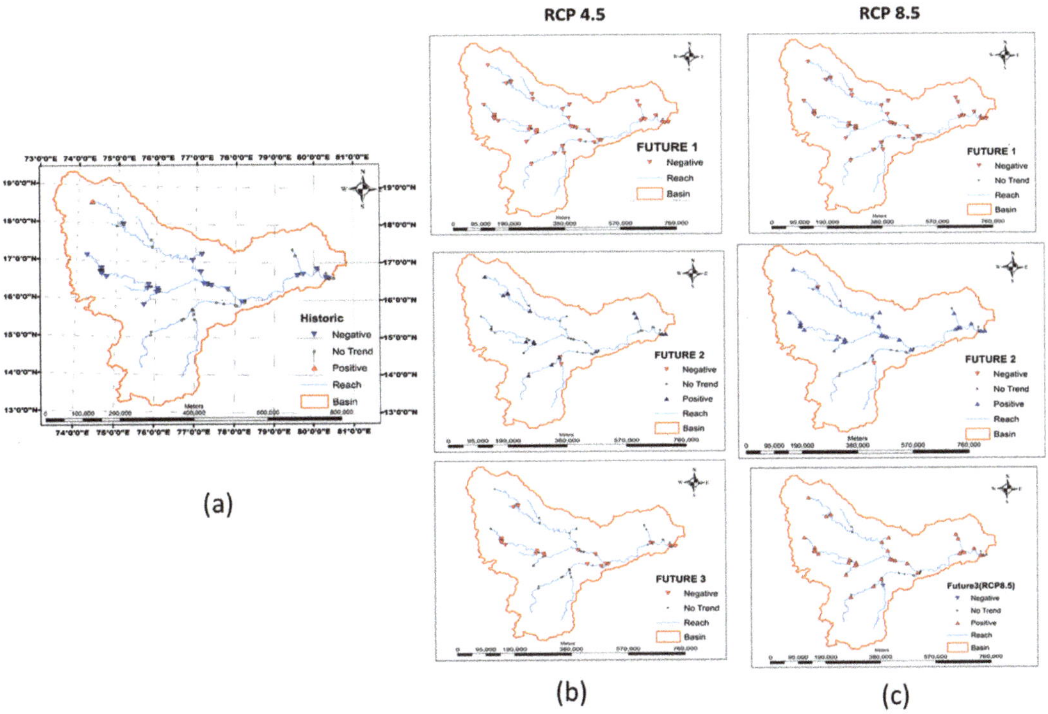


**Figure 12.** Annual streamflow trends (**a**) historic, (**b**) RCP 4.5 scenario, (**c**) RCP 8.5 scenario.

## 4. Conclusions

In this study, future streamflows are projected using REA climate data of the RCP 4.5 and 8.5 scenarios in the SWAT Model, which is calibrated and validated at multiple gauge stations using SWAT-CUP. The REA data from the five climate models have a high correlation with the observed climate data from IMD, similar results were observed for Munneru basin which is a sub-basin for Krishna river basin [66]. The ability of the climate model data to simulate climate variables near the origin of the Krishna River basin, namely, the Western Ghats regions, is limited. The mean monthly precipitation data for the Historic period obtained from REA show less variation in the Middle Krishna, Lower Krishna, and Lower Tungabhadra. Precipitation data in the Future1 period decreased by approximately 20% when compared to the Historic period. The mean monthly flows as a ratio of the annual flow suggest a decrease in the streamflow during the Future1 period and an increase during the Future2 and Future3 periods. FDCs also suggest decreased flows in the basin, emphasizing the importance of implementing necessary changes in water use policies. Annual streamflow trends in all the 50 subbasins show a statistically significant decreasing trend in the Future1 period. No significant trends are observed in most of the subbasins in the Future2 and Future3 periods. As most of the subbasins of the Krishna River basin exhibit decreasing trends over all consecutive periods, effective adaptation and water conservation strategies are required. However, in the present study, the change in LULC is not considered for simulating streamflow for future scenarios.

Specifically, the impact of climate change on water resources of the basin is high during the Future1 period, with lower simulated streamflow. The periods from April to June have lower streamflow values in the basin. Because the basin is dominated by agricultural land, it is critical to avoid water shortage conditions through controlled usage and water management during these times. Based on the projected impacts, water resource management and adaptation strategies, such as changing cropping patterns and adapting various irrigation schemes, will be developed. Future streamflow projections based on the ensemble climate variables in this study enable decision makers to be onboard, and the projections provide them with relevant insight into possible adaptation measures.

**Author Contributions:** P.N.S.: Data curation, Formal analysis, Investigation, Methodology, Software, Visualization, Writing—original draft, Writing—review and editing. V.R.K.: Conceptualization, Data curation, Investigation, Methodology, Project administration, Resources, Software, Validation, Visualization, Writing—original draft, Writing—review and editing. S.M.: Conceptualization, Data curation, Investigation, Methodology, Project administration, Resources. J.D.: Software, Validation, Visualization, Writing—original draft, Writing—review and editing. V.S.: Conceptualization, Investigation, Software, Supervision, Validation, Writing—review and editing. All authors have read and agreed to the published version of the manuscript.

**Funding:** This research received no external funding.

**Data Availability Statement:** https://swat.tamu.edu/data/india-dataset/, accessed on 10 October 2022.

**Acknowledgments:** We acknowledge Balaji Narasimhan for providing the SWAT datasets through the Workshop materials. We are thankful to CCCR, IMD Pune for providing RCM data. The corresponding author acknowledges the support of the Virginia Agricultural Experiment Station (Blacksburg) and the Hatch Program of the National Institute of Food and Agriculture, U.S. Department of Agriculture (Washington, DC, USA).

**Conflicts of Interest:** The authors declare no conflict of interest.

## References

1. Satish Kumar, K.; Venkata Rathnam, E.; Sridhar, V. Tracking seasonal fluctuations of drought indices with GRACE terrestrial water storage over major river basins in South India. *Sci. Total Environ.* **2020**, *763*, 142994. [CrossRef]
2. Sridhar, V.; Kang, H.; Ali, S.A. Human-Induced Alterations to Land Use and+ Climate and Their Responses for Hydrology and Water Management in the Mekong River Basin. *Water* **2019**, *11*, 1307. [CrossRef]
3. Sridhar, V.; Jaksa, W.; Fang, B.; Lakshmi, V.; Hubbard, K.G.; Jin, X. Evaluating Bias-Corrected AMSR-E Soil Moisture using in situ Observations and Model Estimates. *Vadose Zone J.* **2013**, *12*, 1–13. [CrossRef]
4. Field, C.B.; Barros, V.R.; Mach, K.J.; Mastrandrea, M.D.; van Aalst, M.; Adger, W.N.; Arent, D.J.; Barnett, J.; Betts, R.; Bilir, T.E.; et al. Technical summary. In *Climate Change 2014: Impacts, Adaptation, and Vulnerability. Part A: Global and Sectoral Aspects. Contribution of Working Group II to the Fifth Assessment Report of the Intergovernmental Panel on Climate Change*; IPCC: Geneva, Switzerland, 2014; pp. 35–94.
5. Noble, I.R.; Huq, S.; Anokhin, Y.A.; Carmin, J.A.; Goudou, D.; Lansigan, F.P.; Osman-Elasha, B.; Villamizar, A.; Patt, A.; Takeuchi, K.; et al. *Adaptation Needs and Options. Climate Change 2014 Impacts, Adaptation and Vulnerability: Part A: Global and Sectoral Aspects*; IPCC: Geneva, Switzerland, 2015; pp. 833–868.
6. Lal, M. Implications of climate change in sustained agricultural productivity in South Asia. *Reg. Environ. Chang.* **2011**, *11* (Suppl. S1), 79–94. [CrossRef]
7. Mukherjee, S.; Aadhar, S.; Stone, D.; Mishra, V. Increase in extreme precipitation events under anthropogenic warming in India. *Weather. Clim. Extremes* **2018**, *20*, 45–53. [CrossRef]
8. Sujatha, E.R.; Sridhar, V. Spatial Prediction of Erosion Risk of a Small Mountainous Watershed Using RUSLE: A Case-Study of the Palar Sub-Watershed in Kodaikanal, South India. *Water* **2018**, *10*, 1608. [CrossRef]
9. Sridhar, V.; Valayamkunnath, P. *Land-Atmosphere Interactions in South Asia: A Regional Earth Systems Perspective: Chapter 30, Land-Atmospheric Research Applications in South and Southeast Asia*; Vadrevu, K.P., Ohara, T., Justice, C., Eds.; Springer Remote Sensing/Photogrammetry: Berlin/Heidelberg, Germany, 2018; pp. 699–712. [CrossRef]
10. Easterling, D.R.; Meehl, G.A.; Parmesan, C.; Changnon, S.A.; Karl, T.R.; Mearns, L.O. Climate Extremes: Observations, Modeling, and Impacts. *Science* **2019**, *289*, 2068–2074. [CrossRef] [PubMed]
11. Mishra, V.; Lilhare, R. Hydrologic sensitivity of Indian sub-continental river basins to climate change. *Glob. Planet. Chang.* **2016**, *139*, 78–96. [CrossRef]
12. Leon, A.S.; Kanashiro, E.A.; Valverde, R.; Sridhar, V. Dynamic Framework for Intelligent Control of River Flooding: Case Study. *J. Water Resour. Plan. Manag.* **2014**, *140*, 258–268. [CrossRef]
13. Karl, T.R.; Meehl, G.A.; Miller, C.D.; Hassol, S.J. *Weather and Climate Extremes in a Changing Climate Regions of Focus: North America, Hawaii, Caribbean, and U.S. Pacific Islands, Synthesis and Assessment Product 3.3 Report by the U.S. Climate Change Science Program and the Subcommittee on Global Change Research*; IPCC: Geneva, Switzerland, 2008.
14. Hoekema, D.J.; Sridhar, V.; Jin, X. The Adaptability and Sustainability of Surface Water Diversions Along the Main Stem of the Snake River in Southern Idaho. In Proceedings of the 2011 World Environmental and Water Resources Congress, Palm Springs, CA, USA, 22–26 May 2011.

15. Hoekema, D.J.; Sridhar, V. Relating climatic attributes and water resources allocation: A study using surface water supply and soil moisture indices in the Snake River basin, Idaho. *Water Resour. Res.* **2011**, *47*, W07536. [CrossRef]
16. Demaria, E.M.C.; Palmer, N.R.; and Roundy, K.J. Peer review report 2 on 'Regional climate change projections of streamflow characteristics in the Northeast and Midwest. U.S.'. *J. Hydrol. Reg. Stud.* **2016**, *5*, 38–39.
17. Chien, H.; Yeh, P.J.F.; Knouft, J.H. Modeling the potential impacts of climate change on streamflow in agricultural watersheds of the Midwestern United States. *J. Hydrol.* **2013**, *491*, 73–88. [CrossRef]
18. Deshpande, K. Assessing Hydrological Response to Changing Climate in the Krishna Basin of India. *J. Earth Sci. Clim. Change* **2014**, *5*, 7. [CrossRef]
19. Chandra, R.; Saha, U.; Mujumdar, P. Model and parameter uncertainty in IDF relationships under climate change. *Adv. Water Resour.* **2015**, *79*, 127–139. [CrossRef]
20. Das, J.; Umamahesh, N.V. Uncertainty and Nonstationarity in Streamflow Extremes under Climate Change Scenarios over a River Basin. *J. Hydrol. Eng.* **2017**, *22*, 4017042. [CrossRef]
21. Bejagam, V.; Keesara, V.R.; Sridhar, V. Climate change impact on water provisional service and hydropower production of Tungabhadra basin using InVEST model. *River Res. Appl.* **2021**, *37*, 9. [CrossRef]
22. Ramabrahmam, K.; Keesara, V.R.; Srinivasan, R.; Pratap, D.; Sridhar, V. Flow Simulation and Storage Assessment in an Ungauged Irrigation Tank Cascade System Using the SWAT Model. *Sustainability* **2021**, *13*, 13158. [CrossRef]
23. Serrão, E.A.D.O.; Silva, M.T.; Ferreira, T.R.; de Ataide, L.C.P.; Wanzeler, R.T.S.; Silva, V.D.P.R.D.; de Lima, A.M.M.; Sousa, F.D.A.S.D. Large-Scale hydrological modelling of flow and hydropower production, in a Brazilian watershed. *Ecohydrol. Hydrobiol.* **2020**, *21*, 23–35, ISSN 1642-3593. [CrossRef]
24. de Andrade, C.W.; Montenegro, S.M.; Montenegro, A.A.; Lima, J.R.D.S.; Srinivasan, R.; Jones, C.A. Soil moisture and discharge modeling in a representative watershed in northeastern Brazil using SWAT. *Ecohydrol. Hydrobiol.* **2018**, *19*, 238–251, ISSN 1642-3593. [CrossRef]
25. Hillard, U. Assessing snowmelt dynamics with NASA scatterometer (NSCAT) data and a hydrologic process model. *Remote Sens. Environ.* **2003**, *86*, 52–69. [CrossRef]
26. Uniyal, B.; Jha, M.K.; Verma, A.K. Assessing Climate Change Impact on Water Balance Components of a River Basin Using SWAT Model. *Water Resour. Manag.* **2015**, *29*, 4767–4785. [CrossRef]
27. Piras, M.; Mascaro, G.; Deidda, R.; Vivoni, E.R. Quantification of hydrologic impacts of climate change in a Mediterranean basin in Sardinia, Italy, through high-resolution simulations. *Hydrol. Earth Syst. Sci.* **2014**, *18*, 5201–5217. [CrossRef]
28. Sridhar, V. Tracking the Influence of Irrigation on Land Surface Fluxes and Boundary Layer Climatology. *J. Contemp. Water Res. Educ.* **2013**, *152*, 79–93. [CrossRef]
29. Sridhar, V.; Hubbard, K.G.; Wedin, D.A. Assessment of Soil Moisture Dynamics of the Nebraska Sandhills Using Long-Term Measurements and a Hydrology Model. *J. Irrig. Drain. Eng.* **2006**, *132*, 463–473. [CrossRef]
30. Valayamkunnath, P.; Sridhar, V.; Zhao, W.; Allen, R.G. Intercomparison of surface energy fluxes, soil moisture, and evapotranspiration from eddy covariance, large-aperture scintillometer, and modeling across three ecosystems in a semiarid climate. *Agric. For. Meteorol.* **2018**, *248*, 22–47. [CrossRef]
31. Abbaspour, K.C.; Rouholahnejad, E.; Vaghefi, S.; Srinivasan, R.; Yang, H.; Kløve, B. A continental-scale hydrology and water quality model for Europe: Calibration and uncertainty of a high-resolution large-scale SWAT model. *J. Hydrol.* **2015**, *524*, 733–752. [CrossRef]
32. Abbaspour, K.C.; Yang, J.; Maximov, I.; Siber, R.; Bogner, K.; Mieleitner, J.; Zobrist, J.; Srinivasan, R. Modelling hydrology and water quality in the pre-alpine/alpine Thur watershed using SWAT. *J. Hydrol.* **2007**, *333*, 413–430. [CrossRef]
33. Abbaspour, K.C.; Johnson, C.; van Genuchten, M.T. Estimating Uncertain Flow and Transport Parameters Using a Sequential Uncertainty Fitting Procedure. *Vadose Zone J.* **2004**, *3*, 1340. [CrossRef]
34. Yang, J.; Reichert, P.; Abbaspour, K.C.; Xia, J.; Yang, H. Comparing uncertainty analysis techniques for a SWAT application to the Chaohe Basin in China. *J. Hydrol.* **2008**, *358*, 1–23. [CrossRef]
35. Jha, M.; Pan, Z.; Takle, E.S.; Gu, R. Impacts of climate change on streamflow in the Upper Mississippi River Basin: A regional climate model perspective. *J. Geophys. Res. Atmos.* **2004**, *109*, D09105. [CrossRef]
36. Bhadoriya, U.P.S.; Mishra, A.; Singh, R.; Chatterjee, C. Implications of climate change on water storage and filling time of a multipurpose reservoir in India. *J. Hydrol.* **2020**, *590*, 125542. [CrossRef]
37. Das, J.; Treesa, A.; Umamahesh, N.V. Modelling Impacts of Climate Change on a River Basin: Analysis of Uncertainty Using REA & Possibilistic Approach. *Water Resour. Manag.* **2018**, *32*, 4833–4852. [CrossRef]
38. Das, J.; Nanduri, U.V. Assessment and evaluation of potential climate change impact on monsoon flows using machine learning technique over Wainganga River basin, India. *Hydrol. Sci. J.* **2018**, *63*, 1020–1046. [CrossRef]
39. Ghosh, S.; Raje, D.; Mujumdar, P.P. Mahanadi streamflow_climate change impacts assessment and adaptive strategies.pdf. *Curr. Sci.* **2010**, *98*, 1084–1091.
40. Islam, A.; Sikka, A.K.; Saha, B.; Singh, A. Streamflow Response to Climate Change in the Brahmani River Basin, India. *Water Resour. Manag.* **2012**, *26*, 1409–1424. [CrossRef]
41. Das, J.; Nanduri, U.V. Future Projection of Precipitation and Temperature Extremes Using Change Factor Method over a River Basin: Case Study. *J. Hazard. Toxic Radioact. Waste* **2018**, *22*, 04018006. [CrossRef]
42. Gosain, A.K.; Rao, S.; Arora, A. Climate change impact assessment of water resources of {India}. *Curr. Sci.* **2011**, *101*, 356–371.

43. Anandhi, A.; Srinivas, V.V.; Nanjundiah, R.S.; Kumar, D.N. Downscaling precipitation to river basin in India for IPCC SRES scenarios using support vector machine. *Int. J. Clim.* **2007**, *28*, 401–420. [CrossRef]
44. Coulibaly, P.; Dibike, Y.; Anctil, F. Downscaling Precipitation and Temperature with Temporal Neural Networks. *J. Hydrometeorol.* **2005**, *6*, 483–496. [CrossRef]
45. Das, J.; Umamahesh, N.V. Assessment of uncertainty in estimating future flood return levels under climate change. *Nat. Hazards* **2018**, *93*, 109–124. [CrossRef]
46. Mujumdar, P.P.; Ghosh, S. Modeling GCM and scenario uncertainty using a possibilistic approach: Application to the Mahanadi River, India. *Water Resour. Res.* **2008**, *44*, W06407. [CrossRef]
47. Bouwer, L.M.; Aerts, J.C.J.H.; Droogers, P.; Dolman, A.J. Detecting the long-term impacts from climate variability and increasing water consumption on runo ff in the Krishna river basin (India). *Hydrol. Earth Syst. Sci.* **2006**, *10*, 1249–1280. [CrossRef]
48. Gosain, A.K.; Rao, S.; Basuray, D. Climate change impact assessment on hydrology of Indian river basins. *Current* **2006**, *90*, 346–353.
49. Amarasinghe, U.; Sharma, B.R.; Aloysius, N.; Scott, C.; Smakhtin, V.; De Fraiture, C. *Spatial Variation in Water Supply and Demand across River Basins of India*; International Water Management Institute: Colombo, Sri Lanka, 2004; Volume 83.
50. Amarasinghe, U.A.; Shah, T.; Turral, H.; Anand, B.K. *India's Water Future to 2025–2050: Business-as-Usual Scenario and Deviations*; IWMI: Colombo, Sri Lanka, 2007.
51. Biggs, T.W.; Gaur, A.; Scott, C.; Thenkabail, P.S.; Rao, P.G.; Gumma, M.K.; Acharya, S.; Turral, H.N. *Closing of the Krishna Basin: Irrigation, Streamflow Depletion and Macroscale Hydrology*; IWMI: Colombo, Sri Lanka, 2007.
52. Chanapathi, T.; Thatikonda, S.; Raghavan, S. Analysis of rainfall extremes and water yield of Krishna river basin under future climate scenarios. *J. Hydrol. Reg. Stud.* **2018**, *19*, 287–306. [CrossRef]
53. Thomson, A.M.; Calvin, K.V.; Smith, S.J.; Kyle, G.P.; Volke, A.; Patel, P.; Delgado-Arias, S.; Bond-Lamberty, B.; Wise, M.A.; Clarke, L.E.; et al. RCP 4.5: A pathway for stabilization of radiative forcing by 2100. *Clim. Chang.* **2011**, *109*, 77–94. [CrossRef]
54. Riahi, K.; Rao, S.; Krey, V.; Cho, C.; Chirkov, V.; Fischer, G.; Kindermann, G.; Nakicenovic, N.; Rafaj, P. RCP 8.5—A scenario of comparatively high greenhouse gas emissions. *Clim. Chang.* **2011**, *109*, 33. [CrossRef]
55. Sowjanya, P.N.; Venkata Reddy, K.; Shashi, M. Intra- and interannual streamflow variations of Wardha watershed under changing climate. *ISH J. Hydraul. Eng.* **2020**, *26*, 197–208. [CrossRef]
56. Sarma, J. *Impact of High Rinfall/Floods on Groundwater Resources in the Krishna River Basin (during 1999–2009)*; India Environment Portal: New Delhi, India, 2011.
57. Giorgi, F.; Mearns, L.O. Probability of regional climate change based on the Reliability Ensemble Averaging (REA) method. *Geophys. Res. Lett.* **2003**, *30*, 2–5. [CrossRef]
58. Gudmundsson, L.; Bremnes, J.B.; Haugen, J.E.; Engen-Skaugen, T. Technical Note: Downscaling RCM precipitation to the station scale using statistical transformations—A comparison of methods. *Hydrol. Earth Syst. Sci.* **2012**, *16*, 3383–3390. [CrossRef]
59. Piani, C.; Haerter, J.O.; Coppola, E. Statistical bias correction for daily precipitation in regional climate models over Europe. *Theor. Appl. Clim.* **2009**, *99*, 187–192. [CrossRef]
60. Mann, H.B. Nonparametric Tests Against Trend Author(s): Henry B. Mann Published by: The Econometric Society Stable URL: http://www.jstor.org/stable/1907187 REFERENCES Linked references are available on JSTOR for this article: You may need to log in to JSTOR t. *J. Econom. Soc.* **2016**, *13*, 245–259. [CrossRef]
61. Nayak, P.C.; Wardlaw, R.; Kharya, A.K. Water balance approach to study the effect of climate change on groundwater storage for Sirhind command area in India. *Int. J. River Basin Manag.* **2015**, *13*, 243–261. [CrossRef]
62. Hamed, K.H.; Rao, A.R. A modified Mann-Kendall trend test for autocorrelated data. *J. Hydrol.* **1998**, *204*, 182–196. [CrossRef]
63. Singh, L.; Saravanan, S. Simulation of monthly streamflow using the SWAT model of the Ib River watershed, India. *J. Hydro-Environ. Res.* **2020**, *3*, 95–105. [CrossRef]
64. Jha, M.K. Evaluating Hydrologic Response of an Agricultural Watershed for Watershed Analysis. *Water* **2011**, *3*, 604–617. [CrossRef]
65. Nikam, B.R.; Garg, V.; Jeyaprakash, K.; Gupta, P.K.; Srivastav, S.K.; Thakur, P.K.; Aggarwal, S.P. Analyzing future water availability and hydrological extremes in the Krishna basin under changing climatic conditions. *Arab. J. Geosci.* **2018**, *11*, 581. [CrossRef]
66. Buri, E.S.; Keesara, V.R.; Loukika, K.N.; Sridhar, V. Spatio-Temporal Analysis of Climatic Variables in the Munneru River Basin, India, Using NEX-GDDP Data and the REA Approach. *Sustainability* **2022**, *14*, 1715. [CrossRef]

climate

MDPI

*Article*

# Uganda's Hydropower System Resilience to Extreme Climate Variability

**Francis Mujjuni *, Thomas Betts and Richard Blanchard**

Centre for Renewable Energy Systems Technology (CREST), Wolfson School of Mechanical, Electrical and Manufacturing Engineering, Epinal Way, Loughborough University, Loughborough LE11 3TU, UK; t.r.betts@lboro.ac.uk (T.B.); r.e.blanchard@lboro.ac.uk (R.B.)
* Correspondence: f.mujjuni@lboro.ac.uk or francis.mujjuni@mak.ac.ug; Tel.: +44-7424-370334

**Abstract:** This study was motivated by the high reliance on hydropower plants (HPPs) developed and planned along the river Nile and the fact that drought events are the most imminent and drastic threats to Uganda's power production. The study aimed to assess HPPs' resilience and the effectiveness of selected adaptation measures. The climate, land, energy, and water system (CLEWs) framework was employed to assess resilience amidst competing water demands and stringent environmental flow requirements. Under extreme dry conditions, power generation could plummet by 91% over the next 40 years, which translates into an annual per capita consumption of 19 kWh, barely sufficient to sustain a decent socioeconomic livelihood. During arid conditions, climate models predicted an increase in streamflow with increasing radiative forcing. Restricting the ecological flow to 150 m$^3$/s could improve generation by 207%. In addition, if planned power plants were to be built 5 years ahead of schedule, the normalized mean annual plant production could increase by 23%. In contrast, increasing reservoir volumes for planned power plants will have no significant impact on generation. The path to HPP resilience could entail a combination of diversifying the generation mix, installing generators with varying capacities, and incorporating adjustable orifices on reservoirs.

**Keywords:** resilience; hydropower; climate change; extreme events; drought; CLEWs; WEAP

## 1. Introduction

It is projected that Uganda will face immense pressures arising from the increased frequency, intensity, and variability of extreme weather events [1]. Flood and drought events have been particularly cited to be the most imminent and devastating for power systems [2]. In context, 92% of Uganda's electricity production comes from hydropower plants, 64% of the installed capacity of which is developed along the river Nile (R. Nile). The share of installed capacity will grow to 76% with the expected commissioning of the 600 MW Karuma hydropower project in 2023. In addition, the R. Nile is seen as a key source of water for supporting an ambitious irrigation plan [3] and provision of water for a rapidly growing population, which is projected to double by 2050.

Since 1967, Uganda has experienced about 10 major droughts [4]. In just five of those droughts (1987, 1998, 2002, 2005, and 2008), nearly 3.5 million people were affected and average annual losses of USD 20 million were incurred, which could rise to USD 200 million in the event of a one in 200-year drought [4,5]. It is estimated that 12% of the population is exposed to droughts elevating them to be the country's most potent natural disasters [1,6]. Historically, the occurrence of droughts mainly affected food production but droughts of 2004–2005 and 2010–2011 caused massive shortages in the R. Nile flows, plunging the country into a dire electricity supply deficit and leading to prolonged daily power cuts [1]. In response, the government negotiated and commissioned emergency thermal generators and the construction of Bujagali hydropower [7]. The current high end-user energy prices can be traced back to these two hasty decisions.

Several studies [8,9] in the past have assessed the impacts of climate change on the production of hydropower amidst competing water use interests in Uganda. They mostly relied on Phase 5 Coupled Model Intercomparison Project (CMIP5) reanalysis datasets. Since then, improved datasets, i.e., CMIP6, have been produced. In addition, previous assumptions and findings on demand, energy mix, energy prices, and plant production do not correspond to current observations. Moreover, limited emissions pathways were explored and most studies did not assess the effects of different adaptative measures on stream flow and electricity production. Therefore, this study aimed to quantify the effects of projected extreme climate variability on hydropower production and assess the probable gains in resilience emerging from selected adaptation measures.

This study contributes to the existing literature in various ways. This is the first study to use the entire CMIP6 database, which is available on a monthly scale to assess projected aridity and wetness within the Nile basin. The use of groundwater nodes to control the volume of L. Victoria and L. Kyoga is a novel application to mitigate against reanalysis data limitations. In addition, this is the first study employing the CLEW framework that calibrated the R. Nile's flow using L. Victoria levels and validated it across multiple gauging stations. Moreover, no other study known to the authors modeled as many as three adaptation scenarios as demonstrated in this paper.

## 2. Materials and Methods

This section details the methods implemented in this study. First, theoretical and graphical conceptual frameworks are presented followed by a description of data sources and processing. The section then describes the main modeling and simulation process accounting for the employed tools, modeling assumptions, and limitations. Finally, the section concludes with a description of how the model was calibrated and validated.

### 2.1. Conceptual Framework of Methodology

The core of the study entailed developing a water balance model (WBM) that accounts for the main hydrological features (lakes, and rivers), water demand centers (irrigation, and domestic and municipal usage), ecological flow requirements, and hydropower elements (plants and reservoirs) within the catchments of the large hydropower stations situated or planned along the R. Nile. Within the study area, there was a general lack of sufficient observational climate and hydrological data to support a comprehensive modeling process; therefore, the study leveraged data from General Circulation Models (GCMs) archived in the ERA5-Land and CMIP6 datasets [10].

A coupled water balance and hydropower model was implemented in four major phases: (1) the reference scenario from which climate data of the past (1981–2020) were cycled into the future (2021–2060), (2) the extreme scenario, which represented both the probable wettest and driest future climatic conditions, (3) the climate change scenario, which sought to quantify the effects of different emissions pathways on R. Nile discharge and hydropower production, and (4) the adaptation scenario, which evaluated the likely impacts of various adaptation measures on both flow and plants' generation. In all the phases, the model was subjected to similar water consumption priorities, demand rates, and supply arrangements. In general, the model appropriates precipitation and downstream inflows received within a catchment to evapotranspiration, groundwater recharge, and runoff. It is the latter that forms the bulk of river discharge from which a portion is used to generate power as explained in the subsequent sections. The detailed conceptual framework of the study can be seen in Figure 1.

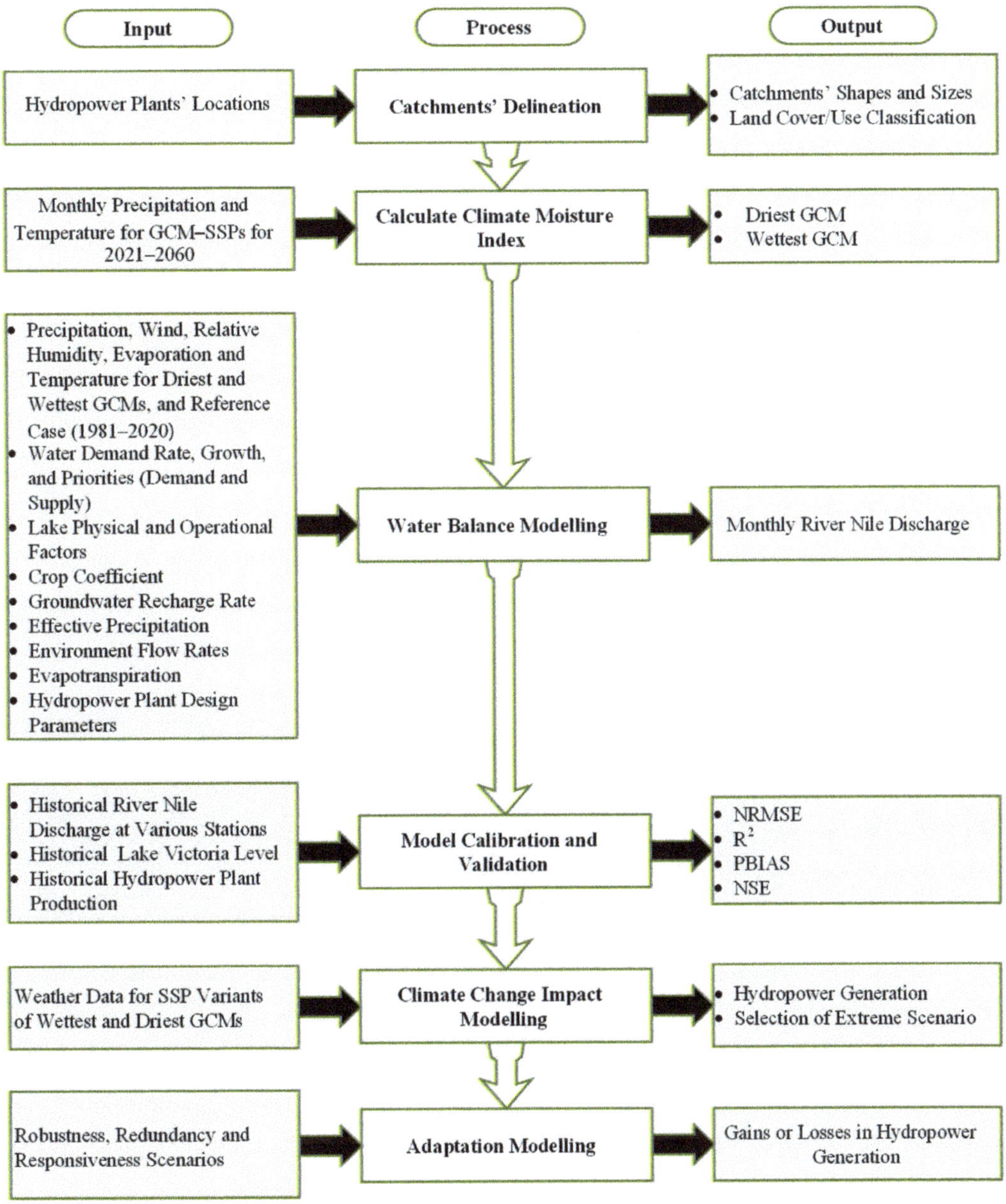


**Figure 1.** The Study Conceptual Framework.

### 2.2. Data Acquisition and Processing

Five data types were considered by this study: hydrology, hydropower, climate, land use and land cover, and demography data. The raw data files sources and relevance are listed in Table 1 and the processing of data is explained in subsequent Sections 2.2.1–2.2.5.

**Table 1.** Data Types, Sources, and Uses.

| Data Types | Parameters | Source | Use |
|---|---|---|---|
| Hydrology | • Lake Victoria levels | [11] | Calibration |
| | • River Victoria discharge | [11] | Validation |
| Large Hydropower Plants | • Digital Elevation Model | [12] | Catchments delineation |
| | • Hydropower plants design information | [13,14] | Reservoir modeling and hydropower generation |
| | • Hydropower generation | [15] | Validation |
| Climate | • Precipitation<br>• Relative humidity<br>• Radiation (shortwave and longwave)<br>• Wind<br>• Pressure<br>• Temperature<br>• Dew Point | [10] | Evaluations of run-offs<br>Reference evapotranspiration |
| Land use | • Land cover maps | [16] | Land class disaggregation |
| Demography | • Population distribution | [17] | Water consumption |
| | • Water consumption | [18] | Water consumption |
| | • Irrigation rate and plan | [3,18] | Water consumption |

2.2.1. Hydrological Data

The bulk of the hydrological data were obtained from the Directorate of Water Resource Management (DWRM) [11]. The data consisted of historical measured flows of the R. Nile at four gauging stations (i.e., Lake Victoria outflow (Jinja Pier), Mbulambuti, Masindi Port, and Paara) and Lake Victoria (L. Victoria) water levels. The flow data files were composed of single daily readings whereas lake levels were recorded twice daily (at 0800 and 1600). Table 2 shows that a sizeable span of data were obtained for each of the gauging stations and, most noticeably, the data obtained at the L. Victoria outflow (hereafter also referred to as Owen Falls Dam or headrace) were nearly complete with only 2% missing both within the raw dataset and within the span, which formed the reference case of this study (1981–2020). The data were then averaged on a monthly scale. Several suspicious observations were made in the flow data for Mbulamuti, Masindi Port, and Paara, which included sudden rises and drops and several other values, which were considerably lower than those measured at the headrace of the river (see highlighted regions in Figure 2A) but the author's attempts to reconcile these observations with DWRM went unanswered. Therefore, the data were processed as obtained. Figure 2B,C, are scatter plots demonstrating the comparison of monthly flow data for the Jinja Pier gauging station with the other three stations. Accordingly, regression equations are presented consistent with the World Meteorological Organization reference periods (1961–1990 and 1991–2020).

Given the size of L. Victoria and L. Kyoga and their influence on the observed flow of the R. Nile, the model incorporated data on the physical and operational attributes of the two lakes. The data were obtained from various sources as seen in Table 3. The lakes' physical elements were characterized by inflows, storage capacity, net evaporation, and losses to groundwater whereas their operations were defined by the top of conservation, buffer, and inactive zones as well as the buffer coefficient. These characteristic elements of lakes are described in detail in [19].

**Table 2.** Key Information of the Hydrological Data Used in this Study. L. Victoria Outflow, R. Victoria Nile, and R. Kyoga Nile are Different Names for the Specific Location of R. Nile at the Headrace, between L. Victoria and L. Kyoga, and between L. Kyoga and L. Albert Respectively.

| Name | Data Type | Lat-Lon | Elevation (Masl (Metres above Sea Level)) | Catchment Area (km$^2$) | Period | Missing Data (All) | Missing Data (1981–2020) |
|---|---|---|---|---|---|---|---|
| L. Victoria at Owen Falls Dam | Lake Level | 0.41, 33.21 | 1123 | 264,160 | 1948–2022 | 3% | 2% |
| L. Victoria at Jinja Pier | Lake Outflow | 0.41, 33.21 | 1123 | 264,160 | 1948–2022 | 3% | 2% |
| R. Victoria Nile at Mbulamuti | River Flow | 0.84, 33.03 | 1030 | 265,727 | 1956–2022 | 12% | 15% |
| R. Kyoga Nile at Masindi Port | River Flow | 1.69, 32.09 | 1021 | 338,465 | 1947–2020 | 21% | 28% |
| R. Kyoga Nile at Paraa | River Flow | 2.28, 31.56 | 641 | 349,207 | 1963–2021 | 55% | 49% |

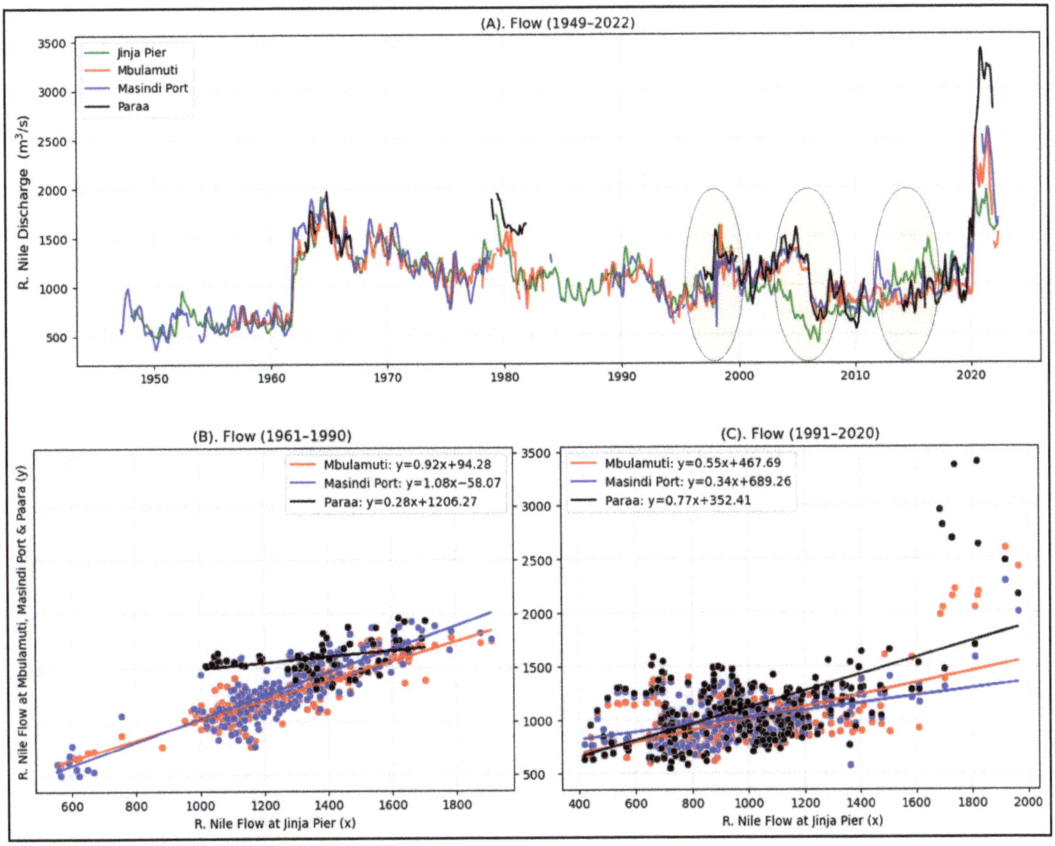


**Figure 2.** Monthly R. Nile Monthly Average Discharge at Selected Gauges: (**A**) is Plotted Raw Data with Observed Inconsistences Highlighted; (**B**,**C**) show Scatter Plots and Linear Regression Equations Comparing Data at Jinja Pier with the Other Three Gauging Sites.

**Table 3.** Physical and Operation Parameters for L. Victoria and L. Kyoga.

| Parameters | Units | L. Victoria | L. Kyoga | Data Source |
|---|---|---|---|---|
| Physical | | | | |
| • Instream flow | $m^3/s$ | Table 4 | | [9] |
| • Net Evaporation | mm | Evaporation–Precipitation | | |
| • Loss to Groundwater | % | Section 2.3.3 | | [9] |
| • Storage Capacity | MCM | 4,148,000 | 20,500 | [9,13,20] |
| • Initial Storage | MCM | 2,770,894 | 16,000 | [9] |
| • Volume Elevation Curve | MCM | $5 \times 10^{-13}H^2 + 2 \times 10^{-7}H + 7.88$ | $3 \times 10^{-3}H + 0.881$ | [20] |
| • Elevation (H) Range | m | 7.96–14 | 0–7.8 | [21–23] |
| Operational | | | | |
| • Top of Conservation | MCM | 3,128,000 | 16,000 | [9] |
| • Top of Buffer | MCM | 2,488,800 | 16,000 | [23,24] |
| • Top of Inactive | MCM | 2,000,000 | 0 | [23,24] |
| • Buffer Coefficient | | 0.5 | 1 | |

**Table 4.** Monthly Inflows in $m^3/s$ into Lakes Victoria and Kyoga.

| | Jan | Feb | Mar | Apr | May | Jun | Jul | Aug | Sep | Oct | Nov | Dec |
|---|---|---|---|---|---|---|---|---|---|---|---|---|
| L. Victoria | 638 | 531 | 744 | 1063 | 1275 | 744 | 531 | 638 | 638 | 531 | 850 | 1275 |
| L. Kyoga | 68 | 34 | 25 | 41 | 162 | 121 | 58 | 44 | 66 | 76 | 137 | 146 |

#### 2.2.2. Large Power Plant Data

Uganda's power supply system is planned to have nearly 70 hydropower plants (HPPs) installed by 2040 [13,14,25]. In the context of this study, 10 were categorized as large hydropower plants based on their capacity exceeding 100 MW as seen in Table 5. Incidentally, all operational and planned large HPPs are located along the R. Nile. Currently, 22 HPPs with a capacity of 1000 MW are operational, 82% of which comprise large HPPs, which will rise to 90% (4000 MW) if all planned power plants are developed. In other words, the R. Nile's relevance, and contribution towards Uganda's energy supply will increase in the future.

**Table 5.** Operational State of Developed and Planned HPPs in Uganda.

| Category | Name of HPPs | Commercial Operational Date (COD) | State | Capacity (MW) |
|---|---|---|---|---|
| Large Hydro | Nalubaale | 1954 | Operational | 180 |
| | Kiira | 2004 | Operational | 200 |
| | Bujagali | 2012 | Operational | 250 |
| | Isimba | 2019 | Operational | 183 |
| | Karuma | 2023 | Construction | 600 |
| | Kiba | 2030 | Feasibility study | 400 |
| | Ayago | 2033 | Feasibility study | 600 |
| | Kalagala | 2035 | Feasibility study | 330 |
| | Oriang | 2037 | Feasibility study | 392 |
| | Murchison (Uhuru) | 2040 | Feasibility study | 655 |
| Small Hydros | Several plants | | Operational | 187 |
| | | | Licensed | 157 |
| | | | Feasibility study | 145 |

The HPP catchment areas were delineated from Aster Global Digital Elevation Model (DEM) Version-3 files obtained from the National Aeronautics and Space Administration's EarthData site [12]. The DEM files were downloaded as Tag Image File Format (.Tif) images within the bounding box of 26° W, 38° E, 6° N, and −7° S. The raw images were then merged into a single raster layer and processed into catchments (see Figure 3) using ArcMap 10.8.1 software.

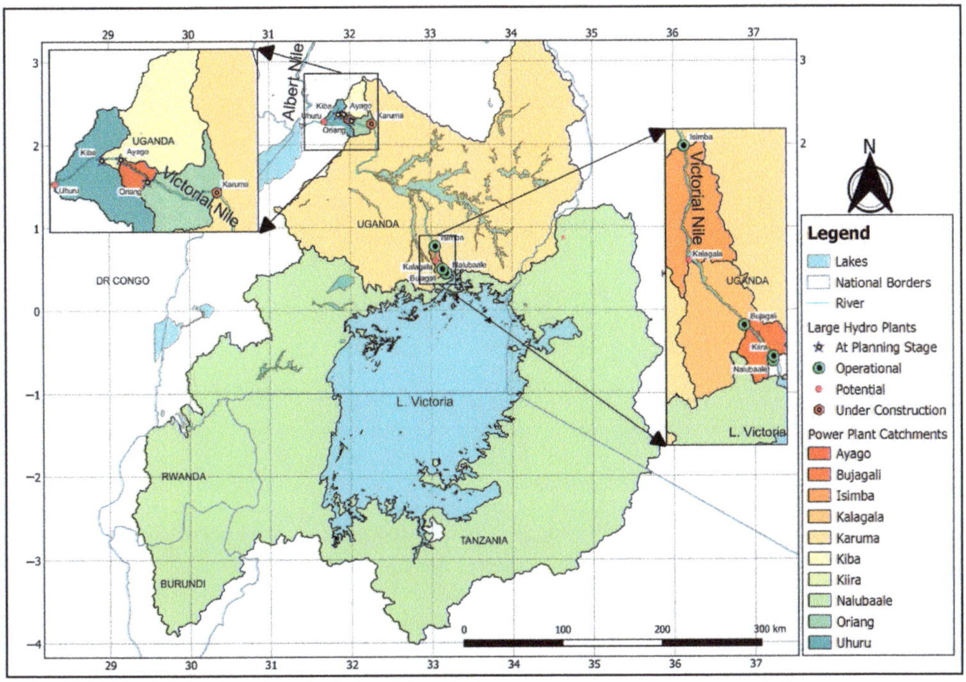


**Figure 3.** Catchments for Large Hydropower Plants.

The design data of the large HPPs as presented in Table 6 were obtained from various reports and studies [13,14,25–28]. All power plants can be classified as run-of-river given that they are essentially designed to operate with very little or no reliance on storage. Even Nalubaale and Kiira HPPs, which have L. Victoria as their reservoir, their intake flows are highly regulated by the dam operations; thus, they essentially can be considered run-of-river plants. The annual generation data recorded from 2011 to 2020 for the operational large HPPs (Nalubaale, Kiira, and Bujagali, Isimba) were obtained from the Electricity Regulatory Authority [29] This data were used to validate the output of the hydropower model.

**Table 6.** Large Hydropower Plants Salient Design Parameters.

| Parameters | Units | Nalubaale | Kiira | Bujagali | Isimba | Kalagala | Karuma | Oriang | Ayago | Kiba | Uhuru |
|---|---|---|---|---|---|---|---|---|---|---|---|
| Location | Lon, Lat | 33.2, 0.4 | 33.2, 0.5 | 33.1, 0.5 | 33.1, 0.8 | 33.1, 0.6 | 32.3, 2.3 | 32.1, 2.3 | 31.9, 2.4 | 31.9, 2.4 | 31.7, 2.3 |
| Design Head | m | 24 | 31 | 22 | 17.7 | 29 | 60 | 58 | 87 | 60 | 93 |
| Efficiency | % | 88 | 76 | 84 | 77 | 85 | 90 | 82 | 64 | 81 | 85 |
| Plant Factor | % | 62 | 38 | 100 | 65 | 64 | 65 | 81 | 81 | 82 | 41 |

**Table 6.** *Cont.*

| Parameters | Units | Nalubaale | Kiira | Bujagali | Isimba | Kalagala | Karuma | Oriang | Ayago | Kiba | Uhuru |
|---|---|---|---|---|---|---|---|---|---|---|---|
| Design Flow | $m^3/s$ | 865 | 865 | 1375 | 1375 | 1375 | 1128 | 840 | 1100 | 840 | 840 |
| Reservoir Level | masl | 1135 | 1135 | 1112 | 1059 | 1088 | 1029 | N/A (Not Applicable) | N/A | N/A | 718 |
| Tail Water Level | masl | 1132 | 1126 | 1090 | 1041 | 1059 | 969 | N/A | N/A | N/A | 625 |
| Gross Storage | MCM | Lake Victoria | | 54 | 171 | 29 | 80 | N/A | N/A | N/A | 19 |

### 2.2.3. Climate Data

The analysis was undertaken at a catchment scale. Such spatial resolution, as indicated by the catchment areas in Figure 3, could not be realistically represented by ground measurements from weather gauges sparsely spread across the basin. In any case, the authors reached out to Uganda National Meteorological Authority (UNMA) [30], the body in charge of recording, archiving, and distributing climate data, yet for several months, no response was given regarding the request of obtaining or purchasing the required data. Several studies [8,9,21,22] in the past faced similar problems, and instead, they used reanalysis data.

This study improved upon previous studies by using newer and a more expansive catalog of GCMs. Monthly precipitation, temperature, relative humidity, wind, evaporation, and dewpoint data from the fifth generation of ECMWF (European Centre for Medium-Range Weather Forecasts) atmospheric reanalysis of the global climate (ERA5) and the sixth phase of the Coupled Model Intercomparison Project (CMIP6) were obtained from [10] and processed on a catchment scale. All grids from reanalysis data files that were within or intersected a given catchment boundary were averaged at every timestep. Figure 4A,B demonstrates an example of ERA5 data grids for wind and temperature within the Karuma HPP catchment, which were averaged, and the respective values assigned to the January 1981 timestep. Similarly, monthly precipitation and evaporation data for the two large lakes, L. Victoria and L. Kyoga, were processed by obtaining the average value of all grids touching or within lake boundaries as seen in Figure 5a,b. This study acquired the ERA5-Land monthly averaged reanalysis data with a latitude–longitude resolution of $0.1° \times 0.1°$ (9 km) spanning a period of 1950–2020. These data have been demonstrated to be relatively more accurate and certainly of high spatial resolution [31,32] than the ERA interim or the Princeton Land Surface Hydrology Research Group, which was employed in previous studies [8,21,22].

In addition, monthly precipitation, wind, temperature, and relative humidity data for 58 GCMs within the CMIP6 database were accessed and processed. As seen in Table 7 below, the obtained GCM data were classified into five Shared Socioeconomic Pathway future emission scenarios (SSP1, SSP2, SSP3, SSP4, and SSP5), which indicate the different global changes and how they might affect future emissions. Different emissions scenarios are considered using Representative Concentration Pathways (RCPs), which are indicative of expected radiative forcing by the year 2100 in comparison to the 1750 baseline. This study acquired data for all RCPs (1.9, 2.6, 3.4, 3.4OS, 4.5, 7.0, and 8.5 W $m^{-2}$), which were retrievable from the ECMWF database at a monthly scale. The relationship between RCP and expected warming is presented in Table 8. For a detailed explanation of the historical context, motivation, and data structure of the CMIP6 refer to O'Neill et al. [33] and Riahi et al. [34]. This study acquired, analyzed, and evaluated data from 181 GCM-SSP combinations. For example, in Table 7, it can be observed that ACCESS-CM2 had relevant data in SSP1-2.6, SSP2-4.5, SSP3-7.0, and SSP5-8.5. That is, one GCM was considered in four different emission pathways.

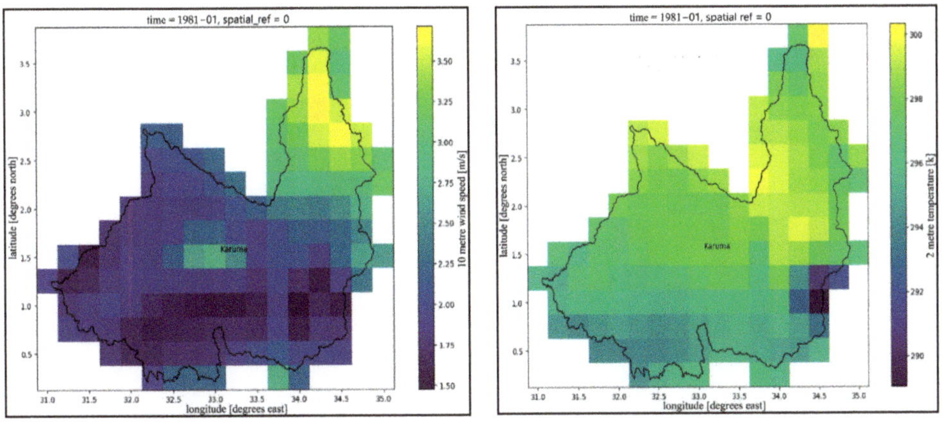

(**A**) (**B**)

**Figure 4.** ERA5 Data Grids for Karuma HPP Catchment for January 1981: (**A**) is Wind Speeds in m/s at 10 m and (**B**) is Temperature in K at 2 m.

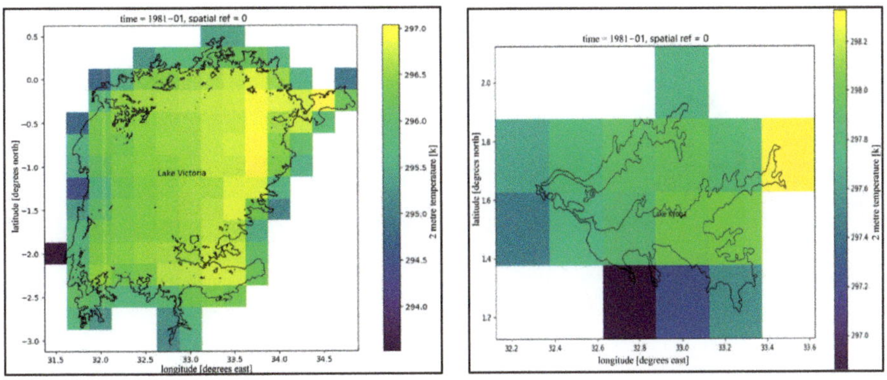

(**a**) (**b**)

**Figure 5.** ERA5 Data Grids for Lake Victoria (**a**) and Lake Kyoga (**b**) for Temperature in K for January 1981.

**Table 7.** Data Obtained from GCMs in CMIP6 and ERA5-Land Databases. In total 629 Datasets are Represented in this Table from 181 GCM-SSP Combinations (P—Precipitation, W—Wind Speed, T—Temperature, R—Relative Humidity).

| GCMs | Historical | | | | SSP1 | | | | | | | | SSP2 | | | | SSP3 | | | | SSP4 | | | | SSP5 | | | | | | | | Resolution |
|---|---|---|---|---|---|---|---|---|---|---|---|---|---|---|---|---|---|---|---|---|---|---|---|---|---|---|---|---|---|---|---|---|---|
| | | | | | 1.9 | | | | 2.6 | | | | 4.5 | | | | 7.0 | | | | 3.4 | | | | 3.4OS | | | | 8.5 | | | | |
| | P | W | R | T | P | W | R | T | P | W | R | T | P | W | R | T | P | W | R | T | P | W | R | T | P | W | R | T | P | W | R | T | |
| ERA5-Land | ✓ | ✓ | ✓ | ✓ | | | | | | | | | | | | | | | | | | | | | | | | | | | | | 9 km |
| ACCESS-CM2 | | | | | | | | | ✓ | ✓ | ✓ | ✓ | ✓ | ✓ | ✓ | ✓ | ✓ | ✓ | ✓ | ✓ | | | | | | | | | ✓ | ✓ | ✓ | ✓ | 250 km |
| ACCESS-ESM1-5 | | | | | | | | | | | | | | | | | | | | | | | | | | | | | | | | | 250 km |
| AWI-CM-1-1-MR | | | | | | | | | ✓ | ✓ | ✓ | ✓ | ✓ | ✓ | ✓ | ✓ | ✓ | ✓ | ✓ | ✓ | | | | | | | | | ✓ | ✓ | ✓ | ✓ | 100 km |

**Table 7.** *Cont.*

| GCMs | Historical | | | | SSP1 | | | | | | | | SSP2 | | | | SSP3 | | | | SSP4 | | | | SSP5 | | | | | | | | Resolution |
|---|---|---|---|---|---|---|---|---|---|---|---|---|---|---|---|---|---|---|---|---|---|---|---|---|---|---|---|---|---|---|---|---|---|
| | | | | | 1.9 | | | | 2.6 | | | | 4.5 | | | | 7.0 | | | | 3.4 | | | | 3.4OS | | | | 8.5 | | | | |
| | P | W | R | T | P | W | R | T | P | W | R | T | P | W | R | T | P | W | R | T | P | W | R | T | P | W | R | T | P | W | R | T | |
| AWI-ESM-1-1-LR | | | | | | | | | | | | | | | | | | | | | | | | | | | | | | | | | 250 km |
| BCC-CSM2-MR | | | | | | | | | ✓ | ✓ | ✓ | ✓ | ✓ | ✓ | ✓ | ✓ | ✓ | ✓ | ✓ | ✓ | | | | | | | | | | ✓ | ✓ | ✓ | 100 km |
| BCC-ESM1 | | | | | | | | | | | | | | | | | | | | | | | | | | | | | | | | | 250 km |
| CAMS-CSM1-0 | | | | | ✓ | | | ✓ | ✓ | | | ✓ | ✓ | | | ✓ | ✓ | | | ✓ | | | | | | | | | ✓ | | | ✓ | 100 km |
| CanESM5 | | | | | ✓ | ✓ | ✓ | ✓ | ✓ | ✓ | ✓ | ✓ | | | | | | | | | ✓ | ✓ | ✓ | ✓ | ✓ | ✓ | ✓ | ✓ | ✓ | | | ✓ | 500 km |
| CanESM5-CanOE | | | | | | | | | ✓ | | | ✓ | ✓ | ✓ | ✓ | ✓ | ✓ | ✓ | ✓ | ✓ | | | | | | | | | ✓ | ✓ | ✓ | ✓ | 500 km |
| CESM2 | | | | | | | | | ✓ | ✓ | ✓ | ✓ | ✓ | ✓ | ✓ | ✓ | ✓ | ✓ | ✓ | ✓ | | | | | | | | | ✓ | ✓ | ✓ | ✓ | 100 km |
| CESM2-FV2 | | | | | | | | | | | | | | | | | | | | | | | | | | | | | | | | | 250 km |
| CESM2-WACCM | | | | | | | | | | | | | | | | | ✓ | ✓ | ✓ | ✓ | | | | | ✓ | ✓ | | ✓ | | ✓ | | ✓ | 100 km |
| CESM2-WACCM-FV2 | | | | | | | | | | | | | | | | | | | | | | | | | | | | | | | | | 100 km |
| CIESM | | | | | | | | | | | ✓ | ✓ | | | ✓ | ✓ | | | | | | | | | | | | | | | ✓ | ✓ | 100 km |
| CMCC-CM2-HR4 | | | | | | | | | | | | | | | | | | | | | | | | | | | | | ✓ | | | | 100 km |
| CMCC-CM2-SR5 | | | | | | | | | ✓ | ✓ | ✓ | ✓ | ✓ | ✓ | ✓ | ✓ | ✓ | ✓ | ✓ | ✓ | | | | | | | | | ✓ | ✓ | ✓ | ✓ | 100 km |
| CMCC-ESM2 | | | | | | | | | ✓ | ✓ | ✓ | ✓ | ✓ | ✓ | ✓ | ✓ | | | | | | | | | | | | | ✓ | ✓ | ✓ | ✓ | 100 km |
| CNRM-CM6-1 | | | | | | | | | ✓ | ✓ | ✓ | ✓ | ✓ | ✓ | ✓ | ✓ | ✓ | ✓ | ✓ | ✓ | | | | | | | | | ✓ | ✓ | ✓ | ✓ | 250 km |
| CNRM-CM6-1-HR | | | | | | | | | ✓ | ✓ | ✓ | ✓ | ✓ | ✓ | ✓ | ✓ | ✓ | ✓ | ✓ | ✓ | | | | | | | | | ✓ | ✓ | ✓ | ✓ | 50 km |
| CNRM-ESM2-1 | | | | | ✓ | ✓ | ✓ | ✓ | ✓ | ✓ | ✓ | ✓ | ✓ | ✓ | ✓ | ✓ | ✓ | ✓ | ✓ | ✓ | ✓ | ✓ | ✓ | ✓ | ✓ | ✓ | ✓ | ✓ | ✓ | ✓ | ✓ | ✓ | 250 km |
| E3SM-1-0 | | | | | | | | | | | | | | | | | | | | | | | | | | | | | | | | | 100 km |
| E3SM-1-1 | | | | | | | | | | | | | | | | | | | | | | | | | | | | | ✓ | ✓ | | ✓ | 100 km |
| E3SM-1-1-ECA | | | | | | | | | | | | | | | | | | | | | | | | | | | | | | | | | 100 km |
| EC-Earth3 | | | | | ✓ | ✓ | ✓ | ✓ | | | | | | | | | | | | | ✓ | ✓ | | ✓ | ✓ | ✓ | | ✓ | | | | | 100 km |
| EC-Earth3-AerChem | | | | | | | | | | | | | | | | | ✓ | | ✓ | ✓ | | | | | | | | | | | | | 100 km |
| EC-Earth3-CC | | | | | | | ✓ | | | | | | ✓ | ✓ | ✓ | ✓ | | | | | | | | | | | | | ✓ | ✓ | ✓ | ✓ | 100 km |
| EC-Earth3-Veg | | | | | ✓ | ✓ | | ✓ | | | | | | | | | | | | | | | | | | | | | | | | | 100 km |
| EC-Earth3-Veg-LR | | | | | ✓ | ✓ | ✓ | ✓ | ✓ | ✓ | ✓ | ✓ | ✓ | ✓ | ✓ | ✓ | ✓ | ✓ | ✓ | ✓ | | | | | | | | | ✓ | ✓ | ✓ | ✓ | 250 km |
| FGOALS-f3-L | | | | | | | | | ✓ | ✓ | | ✓ | ✓ | ✓ | | ✓ | ✓ | ✓ | | ✓ | | | | | | | | | ✓ | ✓ | | ✓ | 100 km |
| FGOALS-g3 | | | | | | ✓ | ✓ | | ✓ | ✓ | ✓ | ✓ | ✓ | ✓ | ✓ | ✓ | ✓ | ✓ | ✓ | ✓ | ✓ | ✓ | ✓ | ✓ | | ✓ | ✓ | | ✓ | ✓ | ✓ | ✓ | 250 km |
| FIO-ESM-2-0 | | | | | | | | | ✓ | ✓ | ✓ | ✓ | ✓ | ✓ | ✓ | ✓ | | | | | | | | | | | | | ✓ | ✓ | ✓ | ✓ | 100 km |
| GFDL-ESM4 | | | | | ✓ | ✓ | ✓ | ✓ | ✓ | ✓ | ✓ | ✓ | ✓ | ✓ | ✓ | ✓ | ✓ | ✓ | ✓ | ✓ | | | | | | | | | ✓ | ✓ | ✓ | ✓ | 100 km |
| GISS-E2-1-G | | | | | ✓ | ✓ | | ✓ | | | | | | | | | | | | | ✓ | ✓ | | ✓ | | | | | | | | | 250 km |
| GISS-E2-1-H | | | | | | | | | | | | | | | | | | | | | | | | | | | | | | | | | 250 km |
| HadGEM3-GC31-LL | | | | | | | | | ✓ | ✓ | ✓ | ✓ | ✓ | ✓ | ✓ | ✓ | | | | | | | | | | | | | ✓ | ✓ | ✓ | ✓ | 250 km |
| HadGEM3-GC31-MM | | | | | | | | | ✓ | ✓ | ✓ | ✓ | | | | | | | | | | | | | | | | | ✓ | ✓ | ✓ | ✓ | 100 km |
| IITM-ESM | | | | | | | | | ✓ | ✓ | | ✓ | ✓ | ✓ | | ✓ | ✓ | ✓ | | ✓ | | | | | | | | | ✓ | ✓ | | ✓ | 250 km |
| INM-CM4-8 | | | | | | | | | ✓ | ✓ | | ✓ | ✓ | ✓ | | ✓ | ✓ | ✓ | | ✓ | | | | | | | | | ✓ | ✓ | ✓ | ✓ | 100 km |
| INM-CM5-0 | | | | | | | | | ✓ | ✓ | ✓ | ✓ | ✓ | ✓ | ✓ | ✓ | ✓ | ✓ | ✓ | ✓ | | | | | | | | | ✓ | ✓ | ✓ | ✓ | 100 km |
| IPSL-CM5A2-INCA | | | | | | | | | ✓ | ✓ | | ✓ | | | | ✓ | ✓ | ✓ | | ✓ | | | | | | | | | ✓ | | | | 500 km |
| IPSL-CM6A-LR | | | | | ✓ | ✓ | ✓ | ✓ | ✓ | ✓ | ✓ | ✓ | ✓ | ✓ | ✓ | ✓ | ✓ | ✓ | ✓ | ✓ | ✓ | ✓ | ✓ | ✓ | ✓ | ✓ | ✓ | ✓ | ✓ | ✓ | ✓ | ✓ | 250 km |
| KACE-1-0-G | | | | | | | | | ✓ | | ✓ | ✓ | ✓ | ✓ | ✓ | ✓ | ✓ | ✓ | ✓ | ✓ | | | | | | | | | ✓ | ✓ | ✓ | ✓ | 250 km |
| KIOST-ESM | | | | | | | | | | ✓ | ✓ | ✓ | | ✓ | ✓ | ✓ | | | | | | | | | | | | | | ✓ | ✓ | ✓ | 250 km |

**Table 7.** *Cont.*

| GCMs | Historical | | | | SSP1 | | | | | | | | SSP2 | | | | SSP3 | | | | SSP4 | | | | SSP5 | | | | | | | | Resolution |
|---|---|---|---|---|---|---|---|---|---|---|---|---|---|---|---|---|---|---|---|---|---|---|---|---|---|---|---|---|---|---|---|---|---|
| | | | | | 1.9 | | | | 2.6 | | | | 4.5 | | | | 7.0 | | | | 3.4 | | | | 3.4OS | | | | 8.5 | | | | |
| | P | W | R | T | P | W | R | T | P | W | R | T | P | W | R | T | P | W | R | T | P | W | R | T | P | W | R | T | P | W | R | T | |
| MCM-UA-1-0 | | | | | | | | | ✓ | | | ✓ | ✓ | | | ✓ | ✓ | | | ✓ | | | | | | | | | ✓ | | | ✓ | 250 km |
| MIROC6 | | | | | ✓ | ✓ | | ✓ | ✓ | ✓ | | ✓ | ✓ | ✓ | | ✓ | ✓ | ✓ | | ✓ | ✓ | ✓ | | ✓ | ✓ | ✓ | | ✓ | ✓ | ✓ | | ✓ | 250 km |
| MIROC-ES2H | | | | | | | | | | | | | | | | | | | | | | | | | | | | | | | | | 250 km |
| MIROC-ES2L | | | | | ✓ | ✓ | ✓ | ✓ | ✓ | ✓ | ✓ | ✓ | ✓ | ✓ | ✓ | ✓ | ✓ | ✓ | ✓ | ✓ | | | | | ✓ | ✓ | ✓ | ✓ | ✓ | ✓ | ✓ | ✓ | 500 km |
| MPI-ESM-1-2-HAM | | | | | | | | | | | | | | | | | ✓ | ✓ | ✓ | ✓ | | | | | | | | | | | | | 250 km |
| MPI-ESM1-2-HR | | | | | | | | | | | | | | | | | | | | | | | | | | | | | | | | | 100 km |
| MPI-ESM1-2-LR | | | | | | | | | ✓ | ✓ | ✓ | ✓ | ✓ | ✓ | ✓ | | ✓ | ✓ | ✓ | ✓ | | | | | | | | | ✓ | ✓ | ✓ | ✓ | 250 km |
| MRI-ESM2-0 | | | | | ✓ | ✓ | | ✓ | ✓ | ✓ | | ✓ | ✓ | ✓ | | ✓ | ✓ | ✓ | | ✓ | ✓ | ✓ | | ✓ | ✓ | ✓ | | ✓ | ✓ | ✓ | | ✓ | 100 km |
| NESM3 | | | | | | | | | ✓ | ✓ | ✓ | ✓ | ✓ | ✓ | ✓ | ✓ | | | | | | | | | | | | | ✓ | ✓ | ✓ | ✓ | 250 km |
| NorCPM1 | | | | | | | | | | | | | | | | | | | | | | | | | | | | | | | | | 250 km |
| NorESM2-LM | | | | | | | | | ✓ | ✓ | | ✓ | ✓ | ✓ | | ✓ | ✓ | ✓ | | ✓ | | | | | | | | | ✓ | ✓ | | ✓ | 250 km |
| NorESM2-MM | | | | | | | | | ✓ | ✓ | | ✓ | ✓ | ✓ | | ✓ | ✓ | ✓ | | ✓ | | | | | | | | | ✓ | ✓ | | ✓ | 100 km |
| SAM0-UNICON | | | | | | | | | | | | | | | | | | | | | | | | | | | | | | | | | 100 km |
| TaiESM1 | | | | | | | | | ✓ | ✓ | ✓ | ✓ | ✓ | ✓ | ✓ | ✓ | ✓ | ✓ | ✓ | ✓ | | | | | | | | | ✓ | ✓ | ✓ | ✓ | 100 km |
| UKESM1-0-LL | | | | | ✓ | ✓ | ✓ | ✓ | ✓ | ✓ | ✓ | ✓ | ✓ | ✓ | ✓ | ✓ | ✓ | ✓ | ✓ | ✓ | ✓ | ✓ | ✓ | ✓ | ✓ | ✓ | ✓ | ✓ | ✓ | ✓ | ✓ | ✓ | 250 km |

**Table 8.** Relationship between Representative Concentration Pathways and Expected Temperature Increase.

| RCPs (W $m^{-2}$) | Radiative Forcing Category | Expected Warming (°C) by 2100 |
|---|---|---|
| 1.9 | Low | 1.4 |
| 2.6 | Low | 1.8 |
| 3.4 | Low | 2.2 |
| 3.4OS | Overshoot | 2.3 |
| 4.5 | Medium | 2.7 |
| 7.0 | High | 3.6 |
| 8.5 | High | 4.4 |

Source of data [33–35].

### 2.2.4. Land Cover Data

Land cover classification data were obtained from European Space Agency-Climate Change Initiative [16]. The data for 2020 were selected and used to represent all years under consideration. In its raw form, the data contain 22 land classes as defined by the United Nations Food and Agriculture Organization's Land Cover Classification System and are archived at a horizontal resolution of 300 m. In this study, the classes were condensed into 8, namely, agriculture, forest, grassland, wetland, urban, shrubland, barren or sparse vegetation, and open water as seen in Figure 6. The area of each class within the catchments (as shown in Table 9) was computed and used as input into the WBM. The catchment areas in Table 9 represent the area in between two adjacent power plants as required input of the WBM, but the actual catchment area of each power plant is cumulative from left to right. That is, Uhuru's actual catchment area is 351,988 $km^2$ (summing all catchment areas from Nalubaale to Uhuru) and not just 875 $km^2$, which is the area between Kiba and Uhuru's planned Intakes.

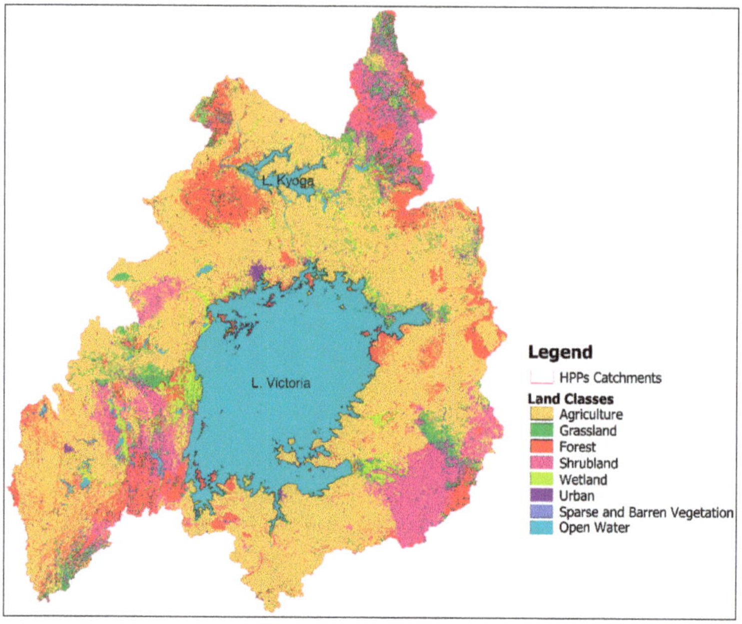


**Figure 6.** Land Cover Classes Within the Study Area.

**Table 9.** Share of Land Classes within Large Hydropower Plant Catchment Areas as Implemented in the Water Balance Model.

| | Units | Nalubaale | Kiira | Bujagali | Isimba | Kalagala | Karuma | Oriang | Ayago | Kiba | Uhuru |
|---|---|---|---|---|---|---|---|---|---|---|---|
| Catchment Area | $km^2$ | 266,078 | 0.63 | 61 | 248 | 218 | 82,431 | 679 | 135 | 1262 | 875 |
| 1. Agriculture | Share | 50.9% | 3% | 81.7% | 94.4% | 77.6% | 62.0% | 25.7% | 52.5% | 27.9% | 40.2% |
| 2. Forest | Share | 8.7% | 0% | 6.0% | 0.4% | 18.9% | 16.4% | 52.8% | 19.6% | 59.8% | 22.7% |
| 3. Grassland | Share | 2.3% | 0% | 0.0% | 0.0% | 0.0% | 3.7% | 13.3% | 14.5% | 9.1% | 29.6% |
| 4. Wetland | Share | 2.9% | 37% | 2.9% | 2.6% | 0.5% | 2.0% | 0.0% | 1.0% | 0.1% | 0.4% |
| 5. Urban | Share | 0.2% | 43% | 4.4% | 0.1% | 0.2% | 0.3% | 0.0% | 0.0% | 0.0% | 0.0% |
| 6. Shrubland | Share | 8.9% | 0% | 0.1% | 0.0% | 0.0% | 11.5% | 5.2% | 6.9% | 2.7% | 6.6% |
| 7. Barren Vegetation | Share | 0.0% | 0% | 0.0% | 0.0% | 0.0% | 0.0% | 0.0% | 0.0% | 0.0% | 0.0% |
| 8. Open Water | Share | 26.1% | 17% | 4.9% | 2.4% | 2.8% | 4.2% | 2.8% | 5.6% | 0.3% | 0.5% |

2.2.5. Demographic Data

The population within each catchment was computed from the census and household demography survey data obtained from the Uganda Bureau of Statistics (UBOS) [17]. Individual catchment populations were computed by aggregating district data bounded within a particular catchment area. In instances where a district extended to multiple catchments, the population was distributed equally amongst the relevant catchments. Given that the population data were used to account for water demand and usage within Uganda, the population outside Uganda but within the catchment areas was not considered. Figure 7 shows the population at the district scale for 2020 and illustrates its aggregation at a catchment scale. In the recent record of censuses, from 1991 to 2014, the annual population growth was 3.03%. This rate was used to estimate the annual population spanning 1981–2060.

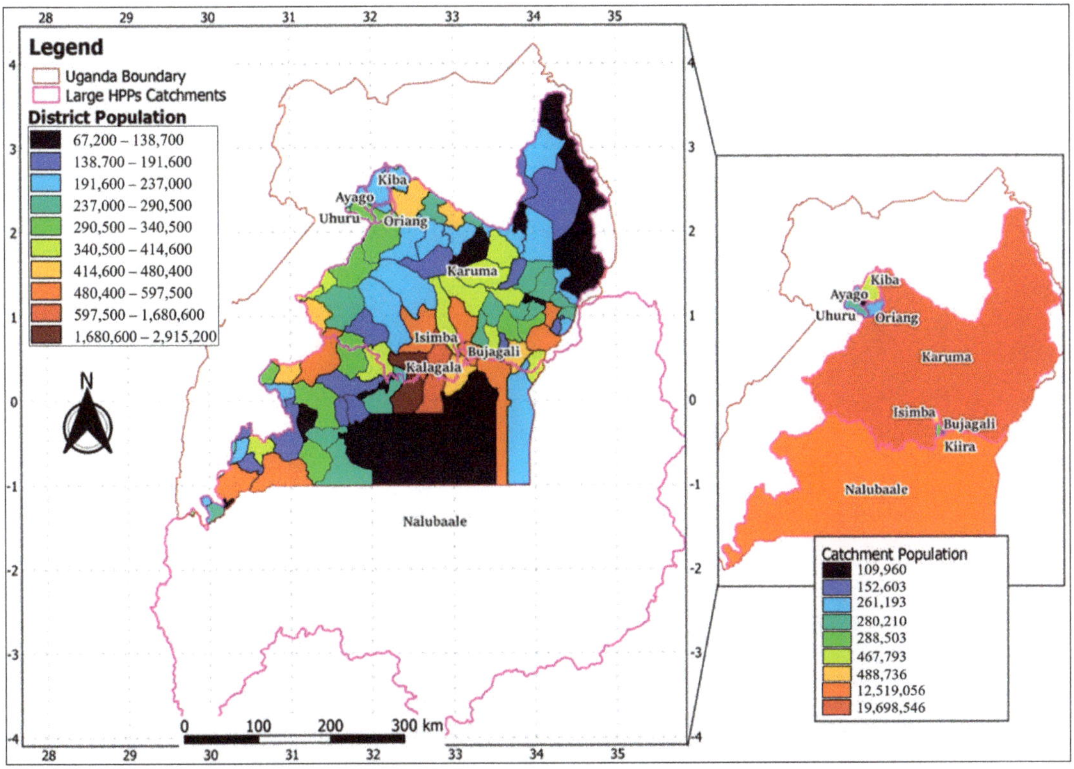


**Figure 7.** Catchment Population for 2020.

To account for the major drivers of water demand within the catchment, the study acquired and analyzed irrigation, domestic, and industrial water supply and consumption from the water accounts report published by UBOS [18]. The report covered the period from 2017 to 2020. Time series data were generated using 2020 as a baseline (captured in Table 10) with the following assumptions:

- The irrigated land area was considered to be 0.15% of the catchment agricultural area within Uganda. This was inferred from the 2020 data, in which the total irrigated area (approximately 100 km$^2$) within Uganda was 0.15% of the country's arable land area. The irrigated area for 2020 was considered to represent the period 1981–2020, whereas for 2021–2040, the study incorporated plans from the National Irrigation Policy [3], which proposed increasing irrigable land area to 15,000 km$^2$ by 2040. This plan infers an annual growth of 28.5% of irrigable land between 2020 and 2040. Given that there are no plans beyond 2040, it was considered that from 2041 to 2060, the growth rate would be 3% reflecting the projected population growth.
- The irrigation water demand was considered to be 2.6 million cubic meters per km$^2$ per year. The study considered that 50% of this was consumed or lost by the crops through evapotranspiration and the rest went to groundwater recharge.
- Municipal water demand, which aggregated both domestic and industrial water usage, was computed as 36,565 m$^3$ per person per year. The subsequent consumption was considered to be 2% of the demand and was estimated to grow at an annual rate of 2.24%.

**Table 10.** Population and Water Use Rates for 2020.

| Name | Population | Agricultural Land ($km^2$) | Irrigated Land ($km^2$) | Water Use Municipal (MCM) | Water Use Irrigation (MCM) |
|---|---|---|---|---|---|
| Ayago | 109,960 | 71 | 0.02 | 4021 | 0.044 |
| Bujagali | 152,602 | 55 | 0.05 | 5580 | 0.103 |
| Isimba | 488,735 | 244 | 0.33 | 17,870 | 0.622 |
| Kalagala | 288,502 | 211 | 0.18 | 10,549 | 0.351 |
| Karuma | 19,698,546 | 77,823 | 58.13 | 657,394 | 110.809 |
| Kiba | 397,251 | 1251 | 0.16 | 14,525 | 0.308 |
| Kiira | 152,602 | 1 | 0.00 | 5580 | 0.000 |
| Nalubaale | 12,519,056 | 30,666 | 20.33 | 367,597 | 38.754 |
| Oriang | 223,385 | 645 | 2.92 | 8168 | 0.286 |
| Uhuru | 280,210 | 855 | 0.15 | 10,246 | 0.667 |

### 2.3. Water Balance Model

This study was based on the CLEW (Climate, Land, Environment, and Water system) framework [36], which has been widely employed in recent studies [9,37,38] to holistically analyze and model water demand and usage. The framework recognizes that a nexus of these elements is not only logical (given the inherent interdependencies) but also that reality dictates that a study of any practical good towards sustainability must consider their interaction. Ramos et al. [36] cataloged 23 recent studies (2012–2020) in which the CLEW framework was implemented in (i) planning water supply systems to meet future domestic, industrial, and agricultural demand, (ii) developing decarbonization pathways, (iii) investigating the impact of climate change on electricity generation, (iv) analyzing effects of environment flow (e-flow) directives on hydropower generation, and (v) investigating the effect of national energy policies on available water sources among others.

Several tools have been employed in literature to capture the various input elements of the CLEW framework. These include Soil and Water Assessment Tool (SWAT), Water Erosion Prediction Project (WEPP), Agricultural Policy/Environnemental eXtender (APEX), (Système Hydrologique Européen (MIKE-SHE), Hydrologic Modeling System (HEC-HMS), and Water Evaluation and Planning (WEAP). Recent studies investigating the effect of projected climate change on hydrological features have increasingly preferred the WEAP tool. For example, Asghar et al. [39], Opere [40], and Abera and Ayenew [41] investigated the impact of climate change on stream flow in the central Indus basins, Narok County, and Central Rift Valley basins, respectively, and they all obtained relatively high predictions of historical stream flows. Others, like Sridharan et al. [8,37,42,43] and Cervigni et al. [44,45], developed WBMs in WEAP to analyze the R. Nile flow under extreme climate change. These too achieved high agreement between modeled and historically observed flows and were able to implement several "*what if*" scenarios.

Therefore, this study developed a WBM in WEAP given its ability to: (i) model water supply at a catchment scale using different climatic, hydrological, and socioeconomic scenarios, (ii) integrate supply with different demand types, and (iii) implement ecological flow restriction regimes whilst catering for the various demand priorities and supply preferences.

#### 2.3.1. WEAP Model Set-Up

The WEAP tool [46] was used to develop a WBM that accounts for the incoming, consumed, lost, and outgoing water within the catchments of the large HPPs. The model was built following a 5-part resilience evaluation framework, namely, threat characterization, component vulnerability analysis, system response, resilience evaluation, and evaluation of adaptation measures. The subsequent sections illustrate how these parts were modeled within WEAP, but first, this section details the general structure of the model.

The locations of HPPs were used to construct catchments in WEAP (as seen in Figure 8). The catchments were disaggregated into land classes and by selecting the rainfall–runoff (simplified coefficient) method within WEAP, three key data types were required to characterize the catchments: crop coefficient, effective precipitation, and reference evapotranspiration ($ET_{ref}$). This study considered crop coefficient (0.87) and effective precipitation (79%) for all land types across all catchments as proposed by Sundin and Lindblad [9]. The precipitation was directly derived from ERA5 and CMIP6 data files as explained in Section 2.2.3 whereas the $ET_{ref}$ was computed using the FAO Penman–Monteith Equation (1) [47] and supporting Equations (2)–(9).

$$ET_{ref} = \frac{0.408\Delta(R_n - G) + \gamma\frac{900}{T+273}u_2(e_s - e_a)}{\Delta + \gamma(1 + 0.34u_2)} \times Days \tag{1}$$

$$T = \frac{T_{max} + T_{min}}{2} \tag{2}$$

$$e^o(T) = 0.6108exp\left[\frac{17.27T}{T + 237.3}\right] \tag{3}$$

$$e_s = \frac{e^o(T_{max}) + e^o(T_{min})}{2} \tag{4}$$

$$e_a = \frac{e^o(T) \times RH}{100} \tag{5}$$

$$\Delta = 4098\frac{e^o(T)}{(T + 237.3)^2} \tag{6}$$

$$\gamma = 6.65 \times 10^{-3}P \tag{7}$$

$$R_n = (1 - 0.23)R \tag{8}$$

$$u_2 = u_z\frac{4.87}{ln(67.8z - 5.42)} \tag{9}$$

where $Days$ are the days in a month, $R_n$ is net radiation at the crop surface [MJ/m$^2$/day$^-$] at 0.23 albedo, $R$ is net incoming radiation minus net outgoing [MJ/m$^2$/day$^-$], and $G$ is soil heat flux density [MJ/m$^2$/day], which is ignored in this study. $T$ is the mean daily air temperature at 2 m height [°C], $T_{max}$ is the maximum daily air temperature at 2 m height [°C], $T_{min}$ is the minimum daily air temperature at 2 m height [°C], $u_2$ is the wind speed at 2 m height [m/s], $u_z$ is the wind speed measured at $z$ (10 m) height [m/s], $e_s$ is the saturation vapor pressure [kPa], $e^o(T)$ is the saturation vapor pressure at the air temperature T [kPa], $e_a$ is the actual vapor pressure [kPa], $RH$ is the relative humidity (%), $\Delta$ is the slope vapor pressure curve [kPa/°C], $\gamma$ is the psychrometric constant [kPa/°C] and $P$ is the atmospheric pressure [kPa].

The model has 10 power plants and subsequently 10 catchments. Each of the catchments has an irrigation and municipal water demand totaling 20 demand sites. All power plants are modeled as run-of-river with each having a diversion from the R. Nile. The model incorporates another stream element, which represents the flow into L. Kyoga. L. Kyoga and L. Victoria were modeled as reservoirs with the latter located downstream of R. Nile's headrace and linked to the R. Nile through a transmission link and the former as an instream reservoir. The transmission link mimics the Nalubaale dam complex, which is operated by an Agreed Curve [13,22] represented in Equation (10). Within WEAP, L. Victoria level is a derived parameter, therefore this study used the lake level of the 'previous month' to restrict the flow of the 'current' month. The model has 42 other transmission links connecting the river to each demand site. Given that Kiira HPP has a small catchment sandwiched between Nalubaale and Bujagali catchments, an extra transmission link was assigned to Kiira irrigation demand from the Nalubaale catchment.

$$Q = 66.3(H - 7.96)^{2.01} \tag{10}$$

where $Q$ is the outflow from Lake Victoria ($m^3/s$) and $H$ is the Lake level (m) for the 'previous' month.

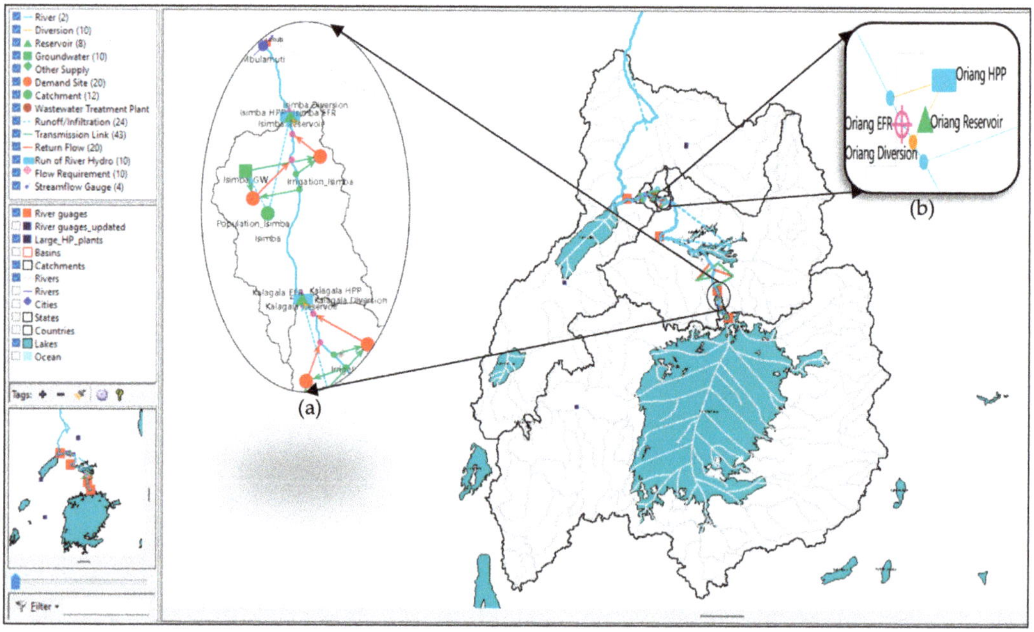


**Figure 8.** WEAP Model Showing R. Nile Basin with Insets of (**a**) Isimba Catchment Components and (**b**) a Typical arrangement of HPP Components.

Each of the catchments has a groundwater recharge station, which is supplied by a runoff/infiltration link. The model has 24 such links, with each catchment having at least two, one designated to run-offs at the catchment pour point and another to groundwater. Other runoff/infiltration links are designated to L. Victoria from the Nalubaale catchment, L. Kyoga from the Karuma catchment, the runoffs at the basin pour point, and another to a 'pseudo' catchment formed at the first gauging station near the R. Nile's headrace. The groundwater recharge was regarded to be 10% of the catchment runoff as indicated by others [8,9,48] with the rest designated to runoffs at the catchment's pour point. The exception to this allocation was made for the Nalubaale and Karuma catchments, which have large reservoirs (lakes) as explained in Section 2.3.3.

At all demand sites, return flow links were included with the unconsumed supply redirected back to the R. Nile. In addition, a flow requirement element was placed upstream of each diversion to a power plant. This was to ensure the provision of the minimum environmental flow requirements (EFRs) as required by the regulator. However, the study was not able to find data on the required EFR for individual plants. Therefore, for baseline assessment, the method proposed by Sridharan et al. [8] was adopted, which recommended an average flow requirement (AFR) derived from models by Smakhtin et al. [49], Tennant [50], Tessman [51] and two from Pastor et al. [52], as seen in Figure 9.

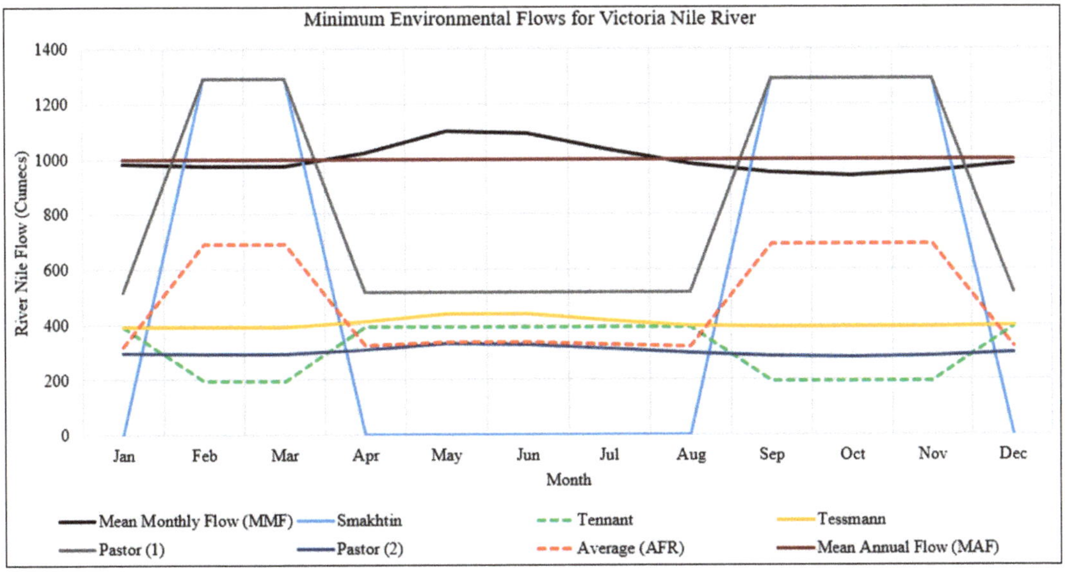


**Figure 9.** Minimum Environmental Flows Computed by Different Methods Using R. Nile Discharge at Jinja Pier.

In WEAP, water is allocated by three priorities. These are demand priority, supply preference, and distribution order. For demand priority, water allocated to competing demand sites, instream flow requirements, reservoirs, and hydropower generation can take up a priority of 1 to 99 with 1 signifying the highest and 99 the lowest priority. In essence, water will be allocated to a component with priority two if the demand for priority one is fully met. For components with the same priority, the available supply is divided equally. Supply preference criteria ensure that multiple supply sources for a given demand site are ranked. This means that a source with priority one will supply demand first and only be supplemented by the next priority source if it is not able to meet the demand. Lastly, the distribution order controls how water is internally distributed to branches within a demand site or catchment. In this study, only demand priority and supply preference criteria were considered.

Relative to all other demands, water for municipal consumption was considered priority one, priority two for irrigation and EFR, priority three for hydropower generation, and priority four for reservoir filling. For supply preference, irrigation demand was supplied first by groundwater and then by river abstraction. In contrast, municipal demand was supplied primarily by the river and then by groundwater sources. Altogether, 19 distinct '*what-if*' scenarios were created in the WEAP model. These represent the system under various conditions including the base year, reference scenario, extreme climate change, emission pathways, and adaptation, as explained in Section 2.3.2. In all scenarios, unless stated otherwise, the same assumptions were considered for consumption, growth rates, demand and supply priorities, crop coefficient, and effective precipitation.

2.3.2. Modeled Scenarios

All scenarios were considered within several categories namely, current, reference, extreme climate change, the impact of radiative forcing, and adaptation measures.

A. Current

The WEAP Model was constructed on a monthly resolution spanning the years 1981 to 2060. The current scenario, technically known as the "Current Accounts" in WEAP software (version: 2021.0.1.10_), is illustrative of the supply and demand dynamics of the base year of the model. In this study, the Current Accounts characterized the CLEWs features as they were understood to be in 1981, and as such the year served as the starting simulation year for all other scenarios.

B. Reference

The reference scenario represents the system evolution without any major intervention [19]. The study cycled climate data, from the ERA5 datasets, of the past (1981–2020) into the future (2021–2060). This scenario assumes that the future climate will be a replica of the past. The consideration of 40-year extents is informed by the recent IPCC report [35] that evaluated changes in climatic parameters by considering 20- to 40-year spans.

C. Extreme Climate Change

Given that the main objective was to ascertain the resilience of hydropower plants under extreme climatic conditions, the study utilized the Climate Moisture Index (CMI) approach, which has been employed in several studies [8,53] to select GCMs that represent the wettest and driest future outcomes. CMI is a measure of aridity that combines the effect of rainfall and temperature projections [44]. The index ranges between −1 and +1 with the latter indicative of precipitation being higher than potential evapotranspiration (PET) rates and as a result, wet conditions are expected whereas the former indicates arid conditions. Alternative indexes, such as the Palmer Drought Severity Index and Surface Water Supply Index, for characterizing rainfall deficit have been described at length by Chong et al. [54].

This study used CMI values obtained for various GCMs within CMIP6 for years spanning 2021–2060. The climate data (temperature and precipitation) over the Nalubaale catchment were used given the massive influence that L. Victoria has on the R. Nile's observed flow within the basin. CMI values were computed as the percentage difference of PET from precipitation as seen in Equation (17). PET values for individual GCMs were obtained by using Thornthwaite Models proposed by [55,56] (see Equations (11)–(16)).

$$PET = PET_c \times f \tag{11}$$

$$PET_c = 16\left[\frac{10T}{I}\right]^a \tag{12}$$

$$I = \sum_{j=1}^{12} i_j \tag{13}$$

$$i = \left[\frac{T}{5}\right]^{1.514} \tag{14}$$

$$a = 6.75 \times 10^{-7}I^3 - 7.71 \times 10^{-5}I^2 + 1.79 \times 10^{-2}I + 0.49 \tag{15}$$

$$f = \left[\left(\frac{\theta}{30}\right)\left(\frac{h}{12}\right)\right] \tag{16}$$

$$CMI = \frac{P - PET}{P} \tag{17}$$

where $PET_c$ is the normative potential evapotranspiration values (mm/month), $PET$ is the adjusted potential evapotranspiration values (mm/month), $T$ is the monthly averaged temperatures (°C), $I$ is the annual heat index, $i$ is the monthly heat index, $a$ is the constant, $f$ is the adjustment factor, $\theta$ is the length of the month (days), $h$ is the duration of daylight (in hours) on the fifteenth of the month, and $P$ is the precipitation (mm/month).

For the Nalubaale catchment, CMIs for 86 SSP-GCM combinations were valid. Others were not considered, given that their CMI values were lower than −1. The authors suspect that in such GCMs, the temperature was considerably overestimated, or precipitation was overly underestimated. Figure 10 shows that ACCESS-CM2 from SSP5-8.5 represents the probable driest conditions and INM-CM-8 from SSP3-7.0 represents the wettest conditions. Weather data from these two GCMs were used to contrast the production of hydropower plants at both extremes of the projected river discharge.

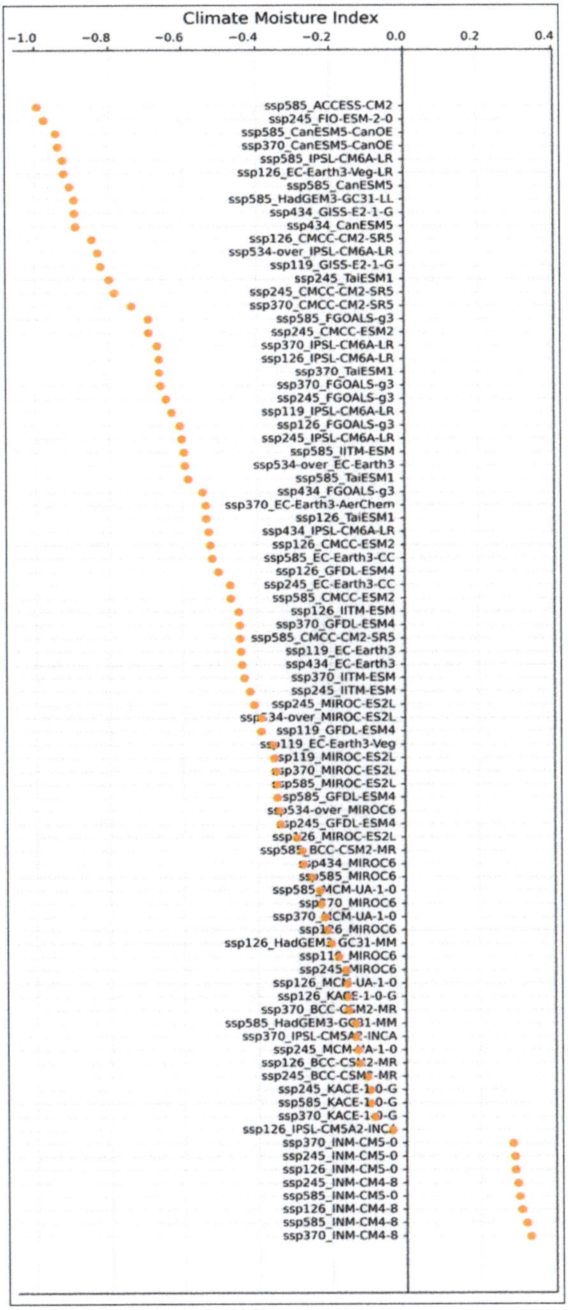


**Figure 10.** Climate Moisture Index for Various GCM-SSP Combinations.

A. Impact of Radiative Forcing

The study ascertained how changes in radiative forcing could influence the expected plant generation as projected by the 'driest' and 'wettest' GCMs. Accordingly, 4 emission pathways (SSP1-2.6, SSP2-4.5, SSP3-7.0 and SSP5-8.5) were considered for both ACCESS-CM2 and INM-CM-8.

B. Effectiveness of Adaptation Measures

The adaptation to extremely low river discharge was considered by three major interventions, namely, *robustness*, *responsiveness*, and *redundancy*. In each of these, three conditions were explored. For *robustness*, which is primarily intended to reduce exposure of the HPPs to extreme weather conditions, the study modeled this by adjusting the commercial operation date (COD) for planned power plants by 5 years ahead of schedule, and then by 5 and 10 years beyond schedule. *Responsiveness* was modeled by adjusting the EFR regime. The adjustments included the consideration of EFR at 50% and 75% of the monthly flows proposed in Section 2.3.1 as well as considering a constant EFR value of 150 $m^3/s$ for all HPPs. As for *redundancy*, the volume of each reservoir for planned HPPs was scaled by 2, 5, and 10 times. The assessment of the effectiveness of these adaptative measures can be also interpreted as a sensitivity assessment of the model.

### 2.3.3. Model Calibration and Validation

The goal of calibration is to ensure that selected values of certain independent variables can reproduce the historical values of a dependent variable. Given that such a task and its output are inevitably characterized by uncertainties, statistical metrics are used to assess the accuracy of modeled outcomes. In WEAP, the calibration can be performed manually or by automation. However, given that the flow of the R. Nile from L. Victoria is controlled by a dam as explained in Section 2.3.1, the automatic tools could not be employed. Therefore, the calibration was performed by manually adjusting (trial and error) two key parameters: crop coefficient and groundwater recharge of the catchments with lakes (Karuma and Nalubaale).

Calibration was primarily performed on L. Victoria levels at Owen Falls Dam, which accounts for the R. Nile headrace outflow. The simulated levels were compared to the observed levels for the period 1981–2020. Following what was deemed a high goodness of fit, the model was validated by simulating river flows at the 4 flow gauging stations. To assess the goodness of fit between simulated and observed values, the Nash–Sutcliffe Efficiency ($NSE$) [57], coefficient of determination ($R^2$), normalized root means square error ($nRMSE$) and percentage bias ($PBIAS$) were employed. These are expressed as seen in Equations (18)–(21), and several studies [44,58,59] have provided some guidance on the interpretation of results as summed up in Table 11. In addition to the discharge at gauging stations, another form of validation involved hydropower generation for the period 2010–2020.

$$R^2 = \frac{\left(\Sigma\left[X_i - \overline{X}\right]\left[Y_i - \overline{Y}\right]\right)^2}{\Sigma\left(X_i - \overline{X}\right)^2 \Sigma\left(Y_i - \overline{Y}\right)^2} \tag{18}$$

$$NSE = 1 - \frac{\sum_{i=1}^{n}\left(X_i - Y_i\right)^2}{\sum_{i=1}^{n}\left(X_i - \overline{X}\right)^2} \tag{19}$$

$$nRMSE\ (\%) = \frac{100}{\overline{X}}\sqrt{\frac{\sum_{i=1}^{n}\left(X_i - Y_i\right)^2}{n}} \tag{20}$$

$$PBIAS(\%) = 100\left(\frac{\overline{Y} - \overline{X}}{\overline{X}}\right) \tag{21}$$

where $X_i$ and $Y_i$ are the $i^{th}$ observed and simulated monthly discharge data, respectively; $\overline{X}$ and $\overline{Y}$ are the mean of the observed and simulated monthly discharge data, respectively, and $n$ is the total number of observations.

**Table 11.** Interpretation of Goodness-of-fit Values of Statical Indices.

| Performance Rating | *PBIAS* (%) | *nRMSE* (%) | $R^2$ | *NSE* |
|---|---|---|---|---|
| Unsatisfactory | $\lvert PBIAS \rvert > 25$ | $nRMSE > 80$ | $0 \leq R^2 \leq 0.2$ | $0 \leq NSE \leq 0.2$ |
| Satisfactory | $15 < \lvert PBIAS \rvert \leq 25$ | $60 < nRMSE \leq 80$ | $0.2 < R^2 \leq 0.6$ | $0.2 < NSE \leq 0.6$ |
| Good | $5 < \lvert PBIAS \rvert \leq 15$ | $40 < nRMSE \leq 60$ | $0.6 < R^2 \leq 0.75$ | $0.6 < NSE \leq 0.75$ |
| Very Good | $\lvert PBIAS \rvert \leq 5$ | $nRMSE \leq 40$ | $0.75 < R^2 \leq 1$ | $0.75 < NSE \leq 1$ |

Source of data: [41,44,58–60].

The crop coefficient was manually adjusted between 0 and 1.3 and it was found that the best correlation between simulated and observed flow was achieved at 0.87 as previously documented by [9]. Secondly, it was observed that a large portion of simulated flow within catchments with large reservoirs (Nalubaale and Karuma) was registered at the catchment outflow points. This meant that the two lakes were always at their maximum levels. Therefore, within the Nalubaale catchment, a proportion of the runoff was directed towards groundwater recharge using proposed models in Equations (22) and (23). The rationale of the models for the period 1981–2020 (past) was to make simulated lake level closely similar to historical observations, whereas from 2021 to 2060 (future), the models were cycled not to reintroduce new uncertainties in the study. Although this concept is novel by the proposed models and redirection of flow into a groundwater node for reservoir catchments, it builds on the work of others [13,21–23] who observed that the flow of the R. Nile can only be calibrated by directly controlling the volume and levels of L. Victoria and L. Kyoga. In addition, Hughes et al. [61] inferred that a WBM could be calibrated "with either interflows or groundwater outflow as the dominant process". Similarly, Dehghanipour et al. [62] calibrated their WBM by indirectly controlling groundwater storage using estimated pumping requirements.

$$Loss\ to\ Groundwater = \begin{cases} 0.57P_T, & 1981\text{–}1984,\ 2021\text{–}2025 \\ 0.64P_T, & 1985\text{–}1989,\ 2026\text{–}2030 \\ 0.57P_T, & 1990\text{–}1994,\ 2031\text{–}2035 \\ 0.40P_T, & 1995\text{–}1999,\ 2036\text{–}2040 \\ 0.42P_T, & 2000\text{–}2004,\ 2041\text{–}2045 \\ 0.40P_T, & 2005\text{–}2009,\ 2046\text{–}2050 \\ 0.40P_T, & 2010\text{–}2014,\ 2051\text{–}2055 \\ 0.30P_T, & 2015\text{–}2020,\ 2056\text{–}2060 \end{cases} \tag{22}$$

$$P_T = P \times A \tag{23}$$

where $P_T$ is the total rainfall over the catchment area (MCM/month) and $A$ is the catchment area ($m^2$).

The model represented in Equations (22) and (23) are not a representation of the groundwater recharge mechanism within the lake but simply a tool implemented to calibrate lake levels at the R. Nile headrace. For the Karuma catchment, the runoff fraction directed towards groundwater nodes was set to 99%. This was because, within WEAP, L. Kyoga operated at full capacity, causing the overflow, runoffs, and inflows to be part of the river discharge with no storage at all. A previous study [9] dealt with this problem by overestimating the volume of the lake to curtail it from overflowing. The method used in this study achieves similar results without changing the physical orientation of the lake, which governs the evaporation process. Nonetheless, the authors, recognize this as the most pragmatic (aim-driven), and perhaps weakest, part of the modeling exercise, which could be moderated by using downscaled and bias-corrected climate model data, or by employing observational data for lakes' precipitation, evaporation, and inflow. Alternatively, the calibration process proposed by Vanderkelen et al. [21,22] can be adopted, which could lead to a closure of the WBM without employing groundwater affixation; however, even so, one would need to restrict (fix) the operational range of the lake levels or volumes. In effect, whichever calibration method is adopted, the levels (and by extension, the volumes) of the lakes must be controlled. Once the WBM is calibrated and the historical flows reproduced with a good level of certainty, the projected flows are solely determined by input GCM data.

## 3. Results and Discussion

This section presents and discusses the findings and results of projected the R. Nile flow and subsequent hydropower generation for the modeled scenarios. Section 3.1 describes the calibration of the model and its validation under the cycled climate data scenario and Section 3.2 presents the river flow and the associated electricity production under the wettest and driest conditions. Section 3.3 discusses how different emission pathways affect river flows projected from extreme future conditions. Finally, Section 3.4 presents the effect of adaptive measures on the scenario with the lowest cumulative projected flows.

### 3.1. Calibration and Validation

In the past, different approaches were employed to calibrate and validate WEAP models. For example, Fernández-Alberti et al. [63] used data from four rain gauges; one-half of the dataset (2000–2007) was used to calibrate the model and the other half (2008–2015) for validation. Sridharan et al. [8,64] used the entire dataset to calibrate the streamflow but provided no validation results in their studies. Contrary to both approaches, this study used three data types: (i) lake levels for calibration, (ii) river flow, and (iii) hydropower generation for validation.

Table 12 and Figure 11 show the results from the calibration and validation exercise. A high degree of comparability ($NSE = 0.75$, $R^2 = 0.8$, $PBIAS = 0.4\%$, $nRMSE = 2\%$) was observed between simulated and observed monthly lake levels. Both $NSE$ and $R^2$ indicate that the simulated levels are highly correlated and are nearly similar to observed levels, whereas $PBIAS$ and $nRMSE$ indicate that the residuals between both sets of values are relatively small. It was on this basis that the flow at four gauging stations was validated. The simulated flow at the headrace was found to be highly comparable to the observed flow ($NSE = 0.77$, $R^2 = 0.8$, $PBIAS = -3.9\%$, $nRMSE = 11\%$). It should be noted that the color coding of results in Table 12 is in reference to the performance ratings in Table 11.

**Table 12.** Statistical Performance of Measured and Modelled River Nile Discharge, Lake Victoria Level, and Hydropower Generation from the Large Plants.

| Phase | Stations | *NSE* | $R^2$ | *PBIAS* (%) | *nRMSE* (%) |
|---|---|---|---|---|---|
| Calibration | L. Victoria Level (1981–2020) | 0.75 | 0.8 | 0.4 | 2 |
| Validation | 1. River Discharge (1981–2020) | | | | |
| | (i) Jinja Pier | 0.77 | 0.8 | −3.9 | 11 |
| | (ii) Mbulamuti | 0.03 | 0.3 | −5.6 | 24 |
| | (iii) Masindi Port | −0.73 | 0.1 | 6.6 | 26 |
| | (iv) Paara | 0.23 | 0.3 | 2.3 | 31 |
| | 2. Hydropower Generation (2011–2020) | | | | |
| | (i) Variable EFR | −0.26 | 0.8 | 22.8 | 32 |
| | (ii)] Constant EFR | 0.4 | 0.9 | −12.8 | 22 |

In previous related studies, Sundin and Lindblad [9] did not present any method or results for calibration and validation of the R. Nile discharge or L. Victoria levels. In Sridharan et al. [8], calibration results were presented and they cited a method from Li et al. [65]. However, there is nothing in that study [65] to suggest that the flow of L. Victoria could be calibrated. It is not well understood how Sridharan et al. [8] calibrated the flow from L. Victoria without employing a dam control model. The significance of calibrating L. Victoria levels is evident in Table 12 and Figure 11, which show that a high goodness of fit between observed and simulated lake levels translates into high comparability of simulated lake outflows.

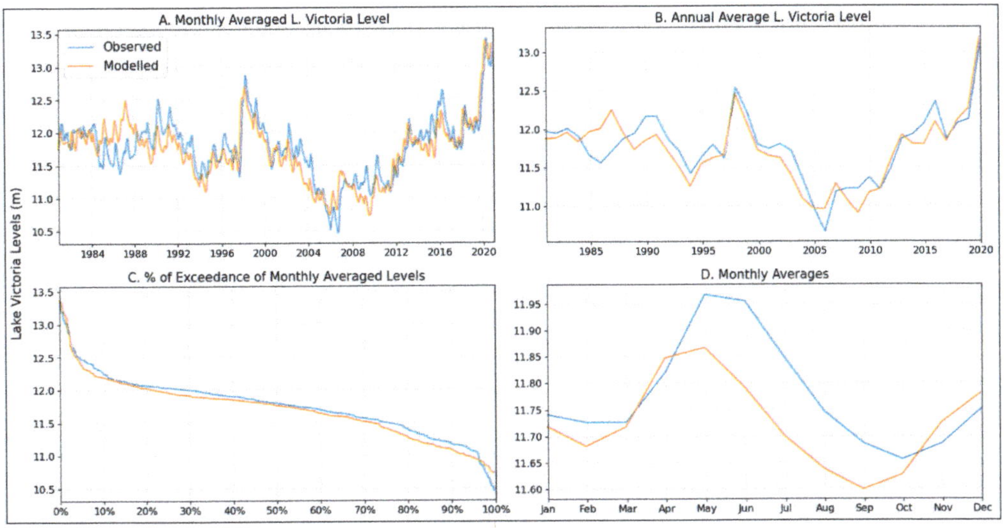


**Figure 11.** L. Victoria Simulated and Observed Levels at Owen Falls Dam for Period 1981–2020.

As seen in Table 12 and Figure 12, for gauging stations further downstream of the lake outflow, the correlation between simulated and observed flows was relatively low, although the residuals were within the acceptable range. The general trend indicates that the further away from the headrace, the higher the variance. Three reasons may be attributed to these observations. First, at the headrace, the flow is mainly governed by dam operations unlike downstream where consumption patterns and runoffs from adjacent catchments play significant roles in the observed stream flow. Second, the study used reanalysis data without downscaling or bias correcting, given that such methods rely on observed weather data, which were unavailable at the time of the study. Downstream of the lake, the simulated catchment runoffs are highly sensitive to input weather data, and the larger the biases within the data, the larger the deviation of simulated from observed discharge. In the past, several studies [21,22] observed the inability of unprocessed reanalysis data to reproduce observed flows within the basin. Third, the WEAP model is rigid in its operation of reservoirs. Water can only be drawn out of lakes to meet a particular demand or if the lake overflows. In addition, the model does not consider interyear storage such that a deficit in a dry year can be met by surplus from pervious wet years. This meant that the simulated flows largely depicted downstream demand more than the natural flow of the river.

The WEAP model was also validated by quantifying the generation from large hydropower plants. Two scenarios were used in this exercise; the reference scenario, which employed a variable EFR as explained in Section 2.3.1 and a constant EFR (150 $m^3/s$) as is likely to be in practice. Figure 13 and Table 12 show that generation from plants was better predicted by a constant EFR ($NSE = 0.4$, $R^2 = 0.9$, $PBIAS = -13\%$, $nRMSE = 22\%$) than the variable EFR ($NSE = -0.3$, $R^2 = 0.8$, $PBIAS = 23\%$, $nRMSE = 32\%$). Nonetheless, the variable EFR was adopted and applied to other scenarios, since it led to better prediction of historical (1981–2020) river discharge.

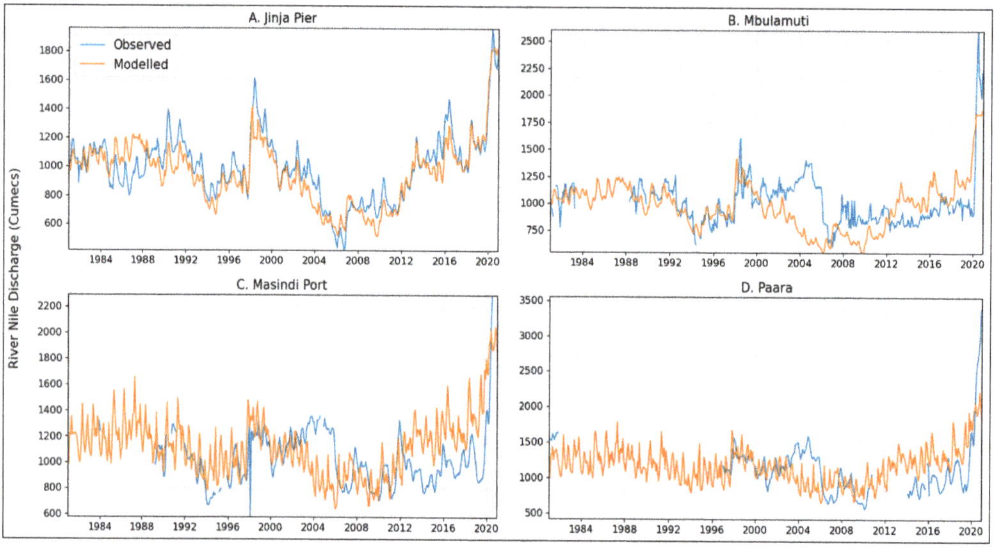


**Figure 12.** Simulated and Observed River Nile Flow at Various Gauging Stations.

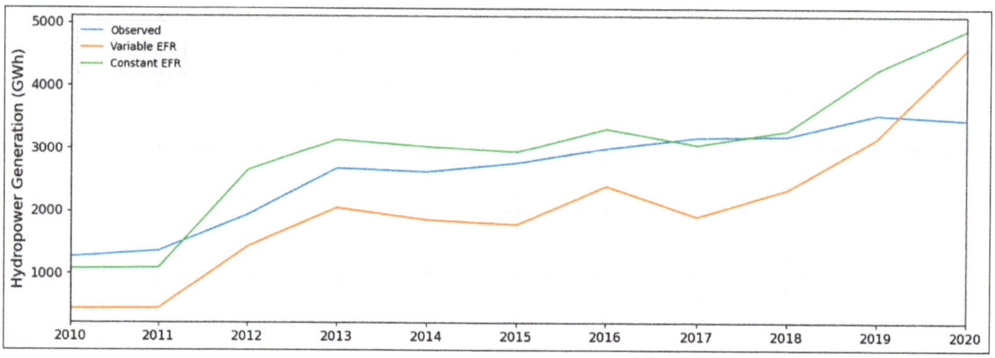


**Figure 13.** Modelled and Observed Hydropower Generation under Two EFR Conditions.

Note that $NSE = 1$, suggesting that modeled parameters are equivalent to observed values such that the variance is zero. $NSE = 0$ indicates that the modeled values are as good as the observed mean, whereas an $NSE < 0$ shows that the mean predicts better values than the model [44]. In this respect, the prediction of the flow at Masindi Port and HPP generation (under variable EFR) performed less than if the mean of the observed values were assumed across the modeled period. Nevertheless, given that the $RMSE$ of all validated parameters remained <40% infers that the model has its usefulness for long-term prediction and planning even at those stations where the goodness of fit was weak.

In essence, the model performed exceptionally well in predicting the flow at the river headrace. The residuals within the simulated lake levels, river flow at all gauging stations, and HPP generation are, at the least, within what is considered satisfactory. However, simulated flow at Mbulamuti, Masindi Port, and Paara exhibit low correlation and considerable variance, in part due to biases within the data, the inadequacy of applied assumptions to reflect reality, and the inherent flexibility limitation of the WEAP tool. Moreover, the biases seem to be most significant and sustained in the period 2003–2006. This observation was also reported by JICA [13], in which they attributed the deviation to poor calibration of lake levels and outflows. The JICA study also reported that in 2007, the agreed curve had been adjusted to deal with flooding downstream of the dam. Therefore, a significant portion of biases observed in this study is a direct result of changes made to the Owen Falls dam operations in the past. Unfortunately, some of those changes could not be outrightly depicted.

### *3.2. Hydropower Generation under Projected Extreme Climate*

This section achieves four aims: (i) it juxtaposes the two projected R. Nile flows under extreme climatic conditions with a reference scenario; (ii) it quantifies the projected changes in the flow thereof; (iii) it discusses the likelihood of occurrence of the projected extremes; and (iv) it compares flows of the extreme scenarios to short-term observed 'future' flows. As described in Section 2.3.2, the CMI values were used to select climate models, which could predict the likely wettest and driest conditions. The GCM models ACCESS-CM2 from SSP5-8.5 and INM-CM-8 from SSP3-7.0 were found to represent the projected driest and wettest conditions, respectively.

The CMI approach achieved the objective of modeling the flow of the river at extreme climatic conditions necessary for investigating the resilience of HPPs and the effectiveness of adaptation measures. However, it also brought to the fore some concerns regarding the validity of CMIP6 data. First, it can be seen in Figure 14 that from the year 2021, the flow for the two extreme conditions widely varies from each other and the reference case. That is, if future climatic conditions are to follow the projection of SSP3-7.0 INM-CM4-8 (hereafter also referred to as "B3"), the river discharge will grow exponentially until 2050, and thereafter a cyclical trend will be observed. On the other hand, if future climatic conditions follow the trend projected by SSP5-8.5 ACCESS-CM2 (hereafter also referred to as "A4"), the river discharge will drastically plummet between 2021 and 2035 and thereafter an increase will be observed, although the discharge would remain significantly lower than for the reference case.

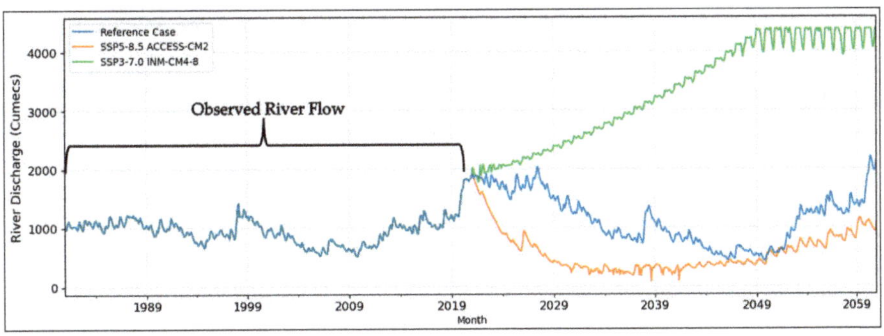


**Figure 14.** Monthly Average Discharge at L. Victoria Outflow for Reference, Driest and Wettest Scenarios.

Table 13 shows the disparity of the extreme scenarios relative to the reference case. The minus "−" sign denotes a decrease and the plus "+" sign, an increase. In the future (2021–2060), A4 predicts a decrease in the mean, minimum, maximum, 50th percentile (P50), and 10th percentile (P10) flows relative to the reference case (50%, 82%, 14%, 65% and 55%), respectively. In contrast, B3 predicts a future with extremely large flows with the mean flow of the river rising by 188% compared to the reference case. Note that the P50 and P10 flows are included in this analysis because they serve as proxies for HPP design flow and EFR, respectively. Similar results have been reported by Vanderkelen et al. [21], who modeled daily outflows using GCMs from RCPs 4.5 and 8.5 and found them to range between −85% to 229% of the historically observed flows. In addition, the phenomenon of the river flow falling between 2021 and 2050, as seen in Figure 14 (reference case), and thereafter rising is comparable with findings of other studies [8,21,66,67].

**Table 13.** Projected R. Nile Flow at the Headrace Under Extreme Climatic Conditions Relative to Reference Case (2021–2060).

| GCM/Stats | Mean | Min | Max | P50 | P10 |
|---|---|---|---|---|---|
| A4 (Driest) | −50% | −82% | −14% | −65% | −55% |
| B3 (Wettest) | +188% | +313% | +99% | +186% | +269% |
| Historical Reference case (1981–2020) | +20% | −17% | +20% | +20% | −15% |

In addition, in comparing the historical reference case (1981–2020) with the future reference case (2021–2060), the future mean, maximum and P50 flows are all projected to increase by 20% and, in contrast, a decrease in the minimum and P10 flows are projected to be 17% and 15%, respectively, as seen in Table 13. In part, the projected increase is attributed to the river flow being strongly dependent on lake levels. Leading up to 2020, the flow of the river exponentially increased following a drastic increase in the lake levels but the relatively normal cycled climate for the period 2020–2038 meant that the lake levels of the immediate 'future' were projected to remain relatively high and, as such, facilitated the high river flows. Second, the simulation of L. Victoria outflow within WEAP is informed by demand, part of the increment in river outflow indicated in Table 13 is an indication of growth in water consumption downstream of the headrace.

Regarding the likelihood that either one of the extreme climate scenarios would materialise and indeed lead to results indicated in this study, it is hard to tell for the long term but it is unlikely that the driest scenario would be as drastic in the short to medium term as observed in Figure 14. Based on the three Mann–Kendal Tests (original, seasonal, and multivariate tests), which are used to test for trends within time series data, there is no detection of any trend at 0.05 significance in the historical headrace flow spanning the period 1948–2020. This infers that the river flows for the last 70 years have been relatively stable. Moreover, moving averages of one and five years of observed river discharge (seen in Figure 15A,B) show that in the 1960s, the river had a drastic increment in its flows, which was followed by a 50-year downward trend, although every 5 years (except for 2000–2010) there was an observed spike. From 2007 onwards, the river flow increased relatively faster than it had previously decreased. The sharp rise in the 1960s was attributed to excessive rainfall and increased tributary inflows [24], although it has been suggested from another study [68] that part of the rise could have resulted from computation errors given that the lake levels rose beyond the dam gauge limit at the time. Other than the period 2000–2006, there is no precedence for the drastic plummeting of river flows, and even so, the flows did not remain low for long.

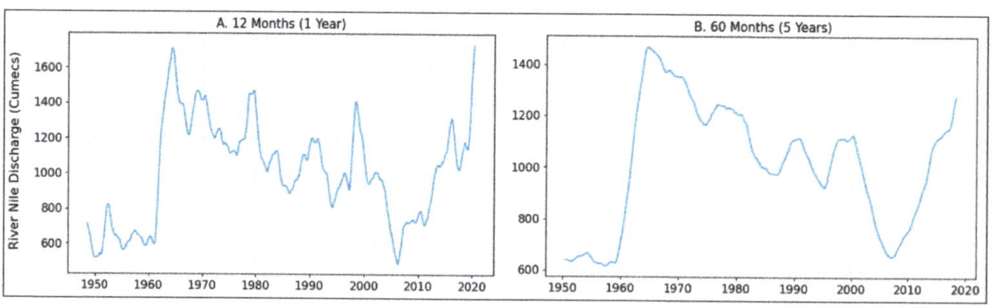


**Figure 15.** Moving Average of Monthly Flows for R. Nile at L. Victoria Outflow for Period 1948–2020.

This analysis is further complicated by the fact that when anecdotal observed data for 17 months spanning January 2021 to May 2022 are compared to both extremes and reference scenarios, the observed river flows seem to be better correlated to the simulated flows of the driest scenario as seen in Figure 16. Moreover, the mean flow for the driest condition (2021–2022) is 72% higher than the observed flow within the historical reference scenario (1981–2020). Given that only 17 months of observed data are available, there is not so much that the authors can deduce on the future river flow trend leading to 2060. It does, though, appear that the river has a 50-year cycle in which flows rise and fall given that the rise in the 2010s seems similar to what happened in the 1960s. However, this observation is inconclusive since it is only based on a single cycle, and consideration of other influential parameters, such as dating river sediments, was beyond the scope of this study.

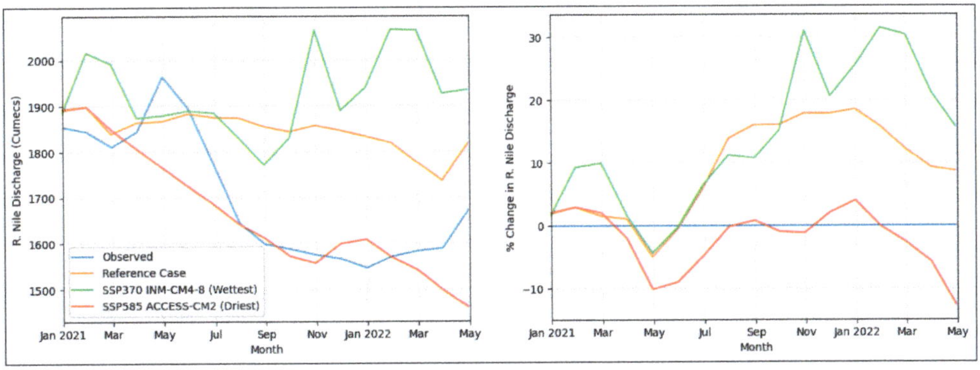


**Figure 16.** Comparison of R. Nile Observed Flow at L. Victoria Outflow for the period 2021–2022 with Selected Simulation Scenarios.

The GCMs B3 and A4 emerge from SSP3 and SSP5, respectively, which are well outside the preferred SSP1 (colloquially known as the "green road") or SSP2 (known as the "middle-of-the-road") scenarios. Essentially, GCMs developed under SSP3 and SSP5 take into consideration that progressively the objective of climate change mitigation will take a back seat as countries prioritise energy security and local development commitments ahead of global climate goals. It is, therefore, paradoxical that this study finds that GCMs developed on such common assumptions provide widely varying results. It might be that B3 and A4 are not well adapted for the study area and, in that respect, the validity of models within CMIP6 to represent global climate as remarked in several studies [35,69] was overstated. If indeed so, this detracts from the study from primarily being one of investigating HPPs' operations under extreme climate to one of investigating HPPs' operations under projections of extremely biased GCMs. The latter, although intellectually stimulating, might not be of practical benefit if the GCMs employed in the study are not able to predict future climate with a good level of certainty.

Although the results presented in this section are mixed and engender several questions all of which are not resolved, the study is useful in demonstrating the projected extreme flows reflective of the recent state-of-the-art data, tools, and methods. It then begs the question of how the simulated flows are affected by different emission pathways.

### 3.3. Impact of Projected Climate Change on Hydropower Generation

The results presented in this section show the impact of emission scenarios on projected river flows and subsequent hydropower production. GCMs from the family of the two extreme scenarios as discussed in previous sections are used such that A1, A2, A3, and A4 represent, in increasing order of radiative forcing, ACCESS-CM2 (dry scenario) from SSP1-2.6, SSP2-4.5, SSP3-7.0, and SSP5-8.5, respectively, as is similarly B1, B2, B3, and B4 for INM-CM4-8 (wet scenario).

The study shows that generally, the flow of the river and the respective generation from hydropower will increase with increasing radiative forcing. This is most pronounced with GCMs that predict a dry scenario (ACCESS-CM2) as demonstrated in Figure 17A–C. Except for A3, the flow of the river is projected to increase as demonstrated by its mean, P50, P10, and cumulative flow. The effect of radiative forcing for the wet scenario (INM-CM4-8) is such that the increase is most evident in the simulated minimum flows (Figure 17D–F). Subsequently, the wet scenario is projected to lead to an increase in hydropower generation as opposed to the dry scenario, which predicts an immense decrease. Even so, under dry conditions, the general effect of forcing is such that the higher the emissions, the more the production for hydropower plants, as seen in Figure 18A,B. This increase in precipitation with the projected increase in temperature, contrary to expectations, has previously been referred to as the "East African climate paradox" [21].

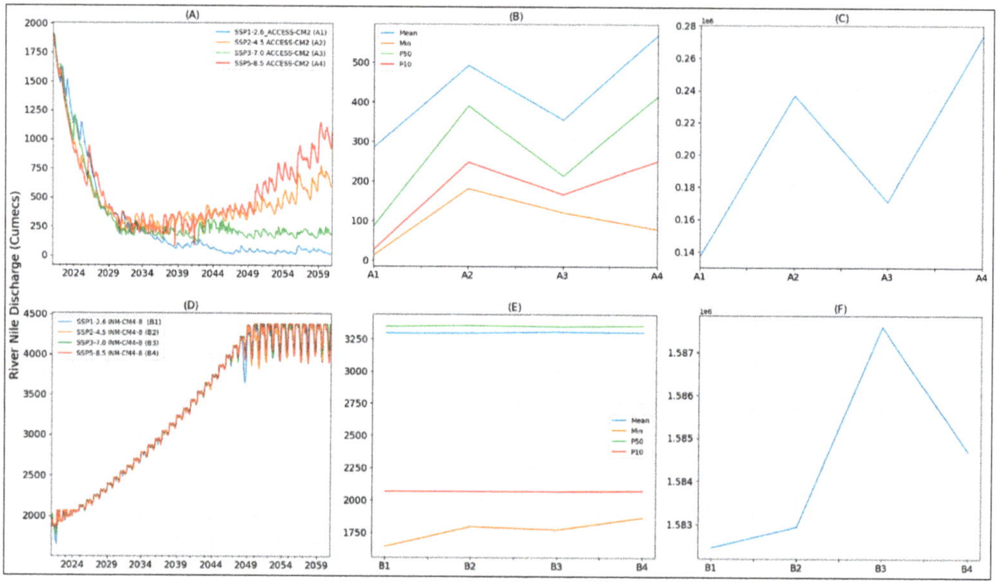


Figure 17. Comparison of Simulated Flow of R. Nile at L. Victoria Outflow Under Varying Radiative Forcing Scenarios for Period 2021–2060. (**A**,**D**) represent Monthly Flows Under Extreme Dry and Wet Conditions Respectively. (**B**,**E**) show the Trend of the Mean, Minimum, 50th, and 10th Percentile Flows of the Extreme Dry and Wet Conditions Respectively. (**C**,**F**) show the Cumulative Average Monthly Discharge for Dry and Wet Conditions.

This study estimates that under dry conditions with varying emission pathways, hydropower generation will reduce by 58 to 91% over the next 40 years, whereas under wet conditions, the increase could be about 53%, as seen in Figure 18. These results are comparable to Sewagudde [67], who projected a 50% fall in the net basin supply for the L. Victoria outflow by 2100. In addition, several studies [21,70] report a generally positive impact of increasing radiative forcing on L. Victoria outflows and a subsequent increase in generation. In contrast, other studies have reported slight variations between the driest and wettest outflows and generation. For example, Tate et at., [66] projected that outflow at the two extremes could be between −2.9% and 6.3%, comparable to Sridharan et al. [8], who projected hydropower generation to only fall by 2.6% under the driest scenario and increase by 11.6% under the wettest. Nonetheless, the underlying observation in most studies [66,70,71] estimating flows in the Nile basin is that with projected climate change, the flow will increase but so will its variability.

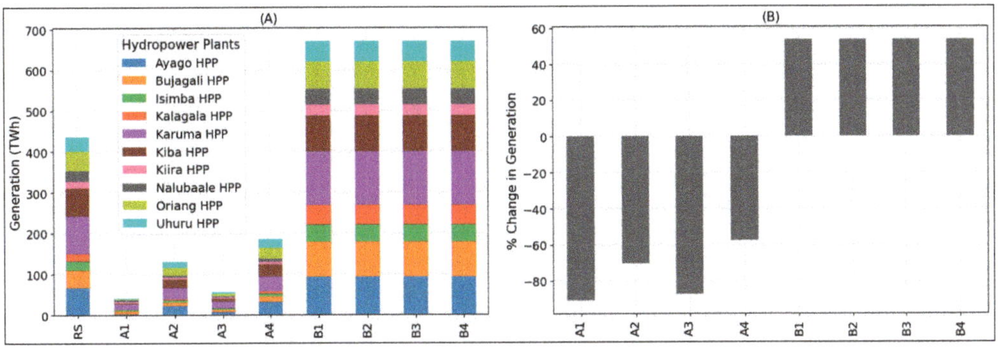


**Figure 18.** Simulated Hydropower Plants Generation for Period 2021–2060: (**A**) under Reference Scenario (RS), Dry (A1, A2, A3, and A4) and Wet Scenarios (B1, B2, B3, and B4), and (**B**) the % Change in Generation of the Dry and Wet Scenarios Compared to the Reference Scenarios.

*3.4. Impact of Adaptation Measures on Hydropower Generation*

Until Section 3.3, SSP5-8.5 ACCESS-CM2 (A4) was regarded as the 'driest' scenario. This was based on it having the lowest CMI value. However, due to intermodel variability from various emission pathways, it can be observed in Figures 17 and 18 that SSP1-2.6 ACCESS-CM2 (A1) was found to have cumulatively the lowest flows and subsequently the lowest projected hydropower generation. It is on this basis that adaptation measures were applied on A1 as duly explained in Section 2.3.2 (E). The general observation, as seen in Figure 19, is that *responsiveness* had a positive significance on generation, no effect from *redundancy*, and mixed results from *robustness*.

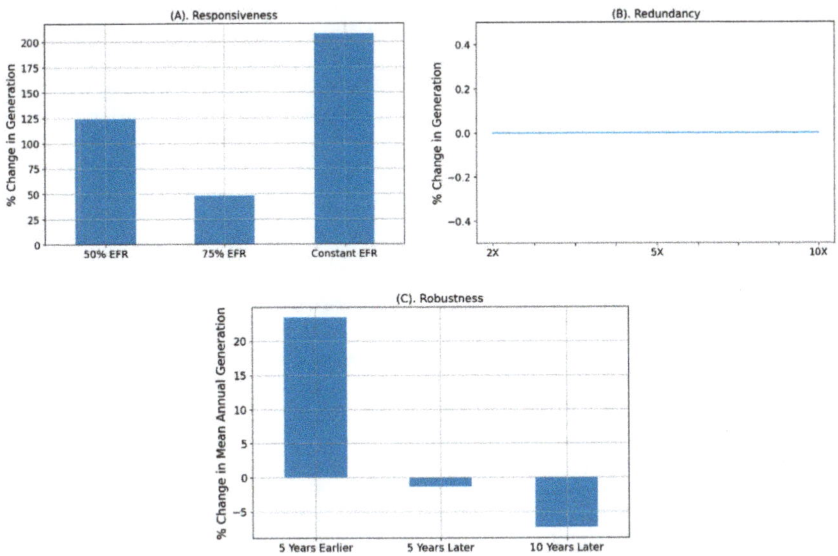


**Figure 19.** Effects of Adaptation Measures on A1 Scenario. (**A**) Is the Responsiveness Measure, which Shows the Change in the Cumulative Generation (2021–2060) when Environment Flow Rates are Varied. (**B**) Demonstrates the Effect of Increased Redundancy in the System by Expanding Reservoir Volumes of Planned Hydropower Plants by 2, 5, and 10 times. (**C**) Shows Results for the Robust Case by Changing the Commercial Operation Dates of Planned Power Plants.

Three *responsive* measures were explored by controlling the EFR at all power plants. The scenario of having a constant EFR of 150 $m^3/s$ as seen in Figure 19A, which is 16% of the mean of historically observed flows, provided the largest increase (207%) in HPP generation. Applying 50% EFR as used in modeling baseline conditions to A1 resulted in a 124% increase and 75% of EFR resulted in a 48% increment of the cumulative generation across the period 2021–2060. Concerning *redundancy*, a change of reservoir volume for planned power plants to 2, 5, and 10 times, demonstrated no significant effect on future generation (Figure 19B). For *robustness*, the study demonstrated that if the commercial operation dates (CODs) of planned power plants were to be moved to 5 years earlier than scheduled, a 23% gain on annual mean generation per power plant would be realized. On the contrary, moving the CODs 5 and 10 years beyond the operational schedule would lead to respective losses of 1.3% and 7% of the normalized average annual plant generation (Figure 19C).

The *responsive* measures increased generation by directing a portion of the water away from ecological purposes to electricity production. In real life, this would be quite an unpopular and perhaps imprudent decision if such measures were to be taken for a long time and if the effects on river biodiversity were to be severe. The consideration of such measures would also depend on the distance between the intake and tailrace. For power plants like Nalubaale, Kiira, Bujagali, and Isimba, the intake and tailrace are only separated by a few meters, in contrast to 8 km for Karuma. In the instance of the former, deviating a portion of water meant for EFR into the generation facilities could do little harm to nature than in the latter's case where not only will the river biodiversity be gravely affected but so would thousands of people who depend on the river for their livelihood. In the context of this study, the 150 $m^3/s$ seems sufficient for the textbook EFR [50] of the R. Nile in Uganda; however, without knowing what is mandated by the regulator or that which is sufficient for river ecology, this EFR serves at best as an indication of the sensitivity of flow on hydropower production during dry conditions.

The study observed that considering cases for larger reservoir volumes for planned power plants did not in any way affect generation during A1 conditions. This is because only two power plants were considered in this scenario (Kalagala and Uhuru) with a combined reservoir storage volume of 48 MCM. If both plants are operational at the same time and all storage is designated to generation, then the reservoir would only support 6 h of hydropower generation under A1 climate conditions, but it would require an average of 47 h to fill up the storage again. These values drastically increase as the volume is expanded such that at 10 times the planned storage, the plants would generate 61 h compared to the 468 h required for a refill. In short, the model does not compute any gains in expanding reservoir volume given that under the A1 scenario, reservoirs never fill up. There is never enough water to be designated for other uses and also for storage. That said, in reality, an expansion of reservoir volume is bound to help increase the generation especially if the flow variability increases. However, these gains could not be quantified in this study since the model was developed on a monthly temporal scale, which encompasses both phenomena of reservoir-aided generation and time required for filling.

If the future is as dry as simulated by the A1 climate scenario, the results seem to suggest that the robust adaptation measure would be to construct Ayago, Kiba, Oriang, Kalagala, and Uhuru five years ahead of their schedule. This means that the scheduled plants would all be constructed by 2034, which falls within the period of relatively high flows, as illustrated in Figure 17A. This could particularly be of interest to developers given that the power plants would stand to generate a substantial amount of their revenues during the early years (2021–2040). However, given that the financing of large hydropower plants is usually locked up in long-term contracts (≈50 years), a guarantee of good revenues of ≈20 years, as predicted by this study, might not be considered a huge incentive.

### 3.5. Conclusions

This study aimed at assessing Uganda's hydropower system's resilience in light of projected climate change and the effectiveness of adaptation measures for extreme aridity. A coupled water-balance and hydropower model was developed in the WEAP tool to model four major long-term scenarios: (i) cycling historical weather to the future, (ii) extreme dry conditions, (iii) effects of emissions pathways, and (iv) effectiveness of adaptation. The study focused on the large power plants constructed and planned along the R. Nile.

It was demonstrated that river flows for the last 70 years have been relatively stable and there is no apparent historical trend between climate change and stream flow. The model projects that if the climate of the past is cycled into the future, the mean flow could increase by 20%, and under the driest conditions, the mean could decrease by 50%. Under the latter's conditions, generation could plummet by 90% compared to the reference scenario (1981–2020). In this case, restricting ecological flow to 150 $m^3/s$ would achieve a 207% improvement in hydropower generation. In addition, developing power plants 5 years ahead of their schedule would improve the normalized plant generation by 23%, while expanding reservoir volumes for planned power plants might yield no significant impact. It was also demonstrated that under arid conditions, stream flows could increase with increased radiative forcing.

To put the results in context, by 2020, 813 MW (64%) of the installed capacity on Uganda's national grid was located on the R. Nile and generated 3453 GWh (78%) [29]. With the planned commissioning of the Karuma hydropower plant in 2023, the production contribution of plants along the Nile will increase to 90%. This implies that the country's high reliance on the R. Nile for power production not only makes it highly susceptible to drought events but also questions the possibility of implementing the adaptive measures modeled in this study. Under the most severe conditions projected in this study, the current available grid capacity would reduce by 72% with a subsequent reduction of 83% in generation. It follows that the per capita consumption would drop from the current 110 kWh to 19 kWh. Restricting EFR to 150 $m^3/s$ would only improve capacity by 10% and generation by 22% and this would result in a per capita consumption of 25 kWh, despite it being just a paltry 1% of the global average according to IEA [72]. It is envisaged that making alterations to EFR will be met with stern opposition from environmentalists and regulators and other geopolitical players (i.e., Egypt, Sudan, and Ethiopia) who have a large stake in the operations of the R. Nile. In essence, the most effective adaptive measure is also the least likely to be implemented. In the long term, effective resilience enhancement strategies would entail a combination of strategies. For example, hydropower plants could be developed with adjustable orifices for controlling EFRs and multiple turbines of varying capacities to be able to operate considerably at high efficiency with varying flows. In this case, low-capacity turbines could run at near-rated flow (and efficiency) during arid conditions.

Several generalizations were made in this study to aid computational expedience or limit the number of uncertainties introduced in the study. Future studies could explore different scenarios that take into consideration generation driven by electrical demand, hydropower production based on turbine efficiency curves, and the performance of the power system with the entire developed and planned generation mix. In addition, given the limitation of selected reanalysis datasets to model runoffs within watersheds with large reservoirs, future studies could explore downscaling and bias-correcting CMIP6 data for the Nile basin and utilizing data from GCMs that close the WBM. Other limitations of this study to be investigated in future works pertain to the inflexibility of the WEAP tool to model the natural flow of rivers emerging from lakes and the development of WEAP water balance models for L. Victoria and L. Kyoga, which do not rely on pseudo groundwater nodes to calibrate lake levels.

**Author Contributions:** Conceptualization, F.M; methodology, F.M.; software, F.M.; validation, F.M.; formal analysis, F.M.; investigation, F.M.; resources, F.M., and R.B.; data curation, F.M.; writing—original draft preparation, F.M.; writing—review and editing, F.M., T.B. and R.B.; visualization, F.M.; supervision, R.B., and T.B. All authors have read and agreed to the published version of the manuscript.

**Funding:** This research received no external funding.

**Institutional Review Board Statement:** Not applicable.

**Informed Consent Statement:** Not applicable.

**Data Availability Statement:** Not applicable.

**Acknowledgments:** Francis Mujjuni acknowledges the Ph.D. scholarship support from the Commonwealth Scholarship Commission in the UK.

**Conflicts of Interest:** The authors declare no conflict of interest.

## Abbreviations

| | |
|---|---|
| CLEWs | Climate Land Environment Water Systems |
| CMI | Climate Moisture Index |
| CMIP5 | Phase 5 Coupled Model Intercomparison Project |
| CMIP6 | Phase 6 Coupled Model Intercomparison Project |
| COD | Commercial Operation Date |
| DEM | Digital Elevation Model |
| ECMWF | European Centre for Medium-Range Weather Forecasts |
| E-flow | Environmental Flow |
| EFR | Environmental Flow Requirement |
| ERA | Electricity Regulatory Authority |
| ERA5 | Fifth Generation ECMWF Atmospheric Reanalysis of the Global Climate |
| FAO | Food and Agricultural Organisation |
| GCMs | General Circulation Models |
| HPP | Hydropower Plant |
| IEA | International Energy Agency |
| kWh | Kilowatt Hour |
| masl | Meters above Sea Level |
| MCM | millions of Cubic Meters |
| MW | Megawatt |
| nRMSE | Normalized Root Means Square Error |
| NSE | Nash–Sutcliffe Efficiency |
| PBIAS | Percentage Bias |
| $R^2$ | Coefficient of Determination |
| RCP | Representative Concentration Pathways |
| UBOS | Uganda Bureau of Statistics |
| USD | United States Dollars |
| WBM | Water Balance Model |
| WEAP | Water Evaluation and Planning |

## References

1. Ministry of Water and Environment. *Uganda National Climate Change Policy*; Ministry of Water and Environment: Kampala, Uganda, 2015.
2. Federal Ministry for Economic Cooperation and Development (BMZ); Deutsche Gesellschaft für Internationale Zusammenarbeit (GIZ). *Climate Risk Profile: Uganda*; GIZ: Bonn, Germany; Potsdam, Germany, 2020.
3. The Government of the Republic of Uganda. *National Irrigation Policy Agricultural Transformation Through Irrigation Development*; Ministry of Water and Environment: Kampala, Uganda, 2017.
4. WBG; GFDRR; ACP-EU. *Disaster Risk Profile Uganda*; WBG: Washington, DC, USA, 2019.
5. Department of Disaster Preparedness and Management. *The National Policy For Disaster Preparedness and Management*; Directorate of Relief, Disaster Preparedness and Refugees Office of the Prime Minister: Kampala, Uganda, 2011.

6. Groen, E.T.; Jacobs, C. *Risk Mapping Uganda*; Sector Disaster Risk Reduction & Emergency Aid; Cordaid: The Hague, The Netherlands, 2012.
7. Kapika, J.; Eberhard, A. *Power-Sector Reform and Regulation in Africa: Lessons from Kenya, Tanzania, Uganda, Zambia, Namibia and Ghana*; HSRC Press: Cape Town, South Africa, 2013; ISBN 9780796924100.
8. Sridharan, V.; Ramos, E.P.; Taliotis, C.; Howells, M.; Basudde, P.; Kinhonhi, I.V. Vulnerability of Uganda's Electricity Sector to Climate Change: An Integrated Systems Analysis. In *Handbook of Climate Change Resilience*; Filho, W.L., Ed.; Springer: Cham, Switzerland, 2019; pp. 1–30. ISBN 978-3-319-71025-9.
9. Sundin, C.; Lindblad, N. Water and Agriculture in Uganda: Supporting a CLEW's Assessment. Bachelor's Thesis, KTH Royal Institute of Technology, Stockholm, Sweden, 2015.
10. Copernicus Emergency Management Service Datasets. Available online: https://cds.climate.copernicus.eu/cdsapp#!/search?type=dataset (accessed on 3 March 2022).
11. Directorate Water Resource Management. *Data Details and Periods*; Directorate Water Resource Management: Kampala, Uganda, 2022.
12. EarthData ASTER Global Digital Elevation Model V003. Available online: https://search.earthdata.nasa.gov/downloads/9382425136 (accessed on 17 February 2022).
13. JICA. *Hydropower Development Master Plan*; Japanese International Cooperation Agency: Kampala, Uganda, 2010.
14. Electricity Regulatory Authority. *The Least Cost Generation Plan*; Electricity Regulatory Authority: Kampala, Uganda, 2015.
15. Electricity Regulatory Authority. Energy Generated to the National Grid. Available online: https://www.era.go.ug/index.php/stats/generation-statistics/energy-generated (accessed on 13 December 2022).
16. European Space Agency. Climate Change Initiative Land Cover Classification Gridded Maps from 1992 to Present Derived from Satellite Observations. Available online: https://cds.climate.copernicus.eu/cdsapp#!/dataset/satellite-land-cover?tab=form (accessed on 13 March 2022).
17. UBOS. *Uganda Bureau of Statistics*; Uganda Bureau of Statistics: Kampala, Uganda, 2021; Volume 1.
18. Uganda Bureau of Statistics. *Uganda Water Accounts Report*; Uganda Bureau of Statistics: Kampala, Uganda, 2021.
19. Stockholm Environment Institute. *WEAP Tutorial*; Stockholm Environment Institute: Stockholm, Sweden, 2016.
20. Technische Universität München. Database for Hydrological Time Series of Inland Waters (DAHITI). Available online: https://dahiti.dgfi.tum.de/en/2/water-level-altimetry/ (accessed on 7 January 2023).
21. Vanderkelen, I.; Van Lipzig, N.P.M.; Thiery, W. Modelling the Water Balance of Lake Victoria (East Africa)-Part 2: Future Projections. *Hydrol. Earth Syst. Sci.* **2018**, *22*, 5527–5549. [CrossRef]
22. Vanderkelen, I.; Van Lipzig, N.P.M.; Thiery, W. Modelling the Water Balance of Lake Victoria (East Africa)-Part 1: Observational Analysis. *Hydrol. Earth Syst. Sci.* **2018**, *22*, 5509–5525. [CrossRef]
23. Brown, E.; Sutcliffe, J.V.; Brown, E.; John, V. The Water Balance of Lake Kyoga, Uganda. *Hydrol. Sci. J.* **2013**, *6667*, 342–353. [CrossRef]
24. Piper, B.S.; Plinston, D.T.; Sutcliffe, J.V. The Water Balance of Lake Victoria. *Hydrol. Sci. J.* **1986**, *31*, 25–37. [CrossRef]
25. Government of Uganda. *Policy Directive with Respect to the Development of Karuma, Isimba and Ayago Hydropower Plants*; Government of Uganda: Kampala, Uganda, 2014.
26. Karubanga, A.; Mugume, R. Design and Construction of the Karuma Hydropower Project in Uganda. *Proc. Inst. Civ. Eng. Civ. Eng.* **2022**, *175*, 12–21. [CrossRef]
27. Isimba Hydro Power Project. *Addendum ESIA Report Addendum Environmental and Social Impacts of Isimba Hydropower Project on the Kalagala Offset Area*; ERMC: Kathmandu, Nepal, 2017.
28. AWE Environmental Engineers. *Environmental Impact Assessment for the Proposed Isimba Hydropower Plant*; Ministry of Energy and Mineral Development: Kampala, Uganda, 2014.
29. Electricity Regulatory Authority. Data & Statistics. Available online: https://www.era.go.ug/index.php/stats (accessed on 1 November 2021).
30. UNMA. Uganda National Meteorological Authority. Available online: https://www.unma.go.ug/ (accessed on 29 December 2022).
31. Gleixner, S.; Demissie, T.; Diro, G.T. Did ERA5 Improve Temperature and Precipitation Reanalysis over East Africa? *Atmosphere* **2020**, *11*, 996. [CrossRef]
32. Khaki, M.; Awange, J. The 2019–2020 Rise in Lake Victoria Monitored from Space: Exploiting the State-of-the-Art GRACE-FO and the Newly Released ERA-5 Reanalysis Products. *Sensors* **2021**, *21*, 4304. [CrossRef] [PubMed]
33. O'Neill, B.C.; Tebaldi, C.; Van Vuuren, D.P.; Eyring, V.; Friedlingstein, P.; Hurtt, G.; Knutti, R.; Kriegler, E.; Lamarque, J.F.; Lowe, J.; et al. The Scenario Model Intercomparison Project (ScenarioMIP) for CMIP6. *Geosci. Model Dev.* **2016**, *9*, 3461–3482. [CrossRef]
34. Riahi, K.; van Vuuren, D.P.; Kriegler, E.; Edmonds, J.; O'Neill, B.C.; Fujimori, S.; Bauer, N.; Calvin, K.; Dellink, R.; Fricko, O.; et al. The Shared Socioeconomic Pathways and Their Energy, Land Use, and Greenhouse Gas Emissions Implications: An Overview. *Glob. Environ. Chang.* **2017**, *42*, 153–168. [CrossRef]
35. Intergovernmental Panel on Climate Change. *Climate Change 2021: The Physical Science Basis—Summary for the Policymakers*; IPCC: Geneva, Switzerland, 2021.

36. Ramos, E.P.; Howells, M.; Sridharan, V.; Engström, R.E.; Taliotis, C.; Mentis, D.; Gardumi, F.; de Strasser, L.; Pappis, I.; Balderrama, G.P.; et al. The Climate, Land, Energy, and Water Systems (CLEWs) Framework: A Retrospective of Activities and Advances to 2019. *Environ. Res. Lett.* **2021**, *16*, 033003. [CrossRef]
37. Sridharan, V. Climate, Land, Energy and Water Strategies (CLEWs): Examples of Integrated Resource Assessments at Different Scales. In Proceedings of the Workshop on Modelling the Water Energy Nexus, Zurich, Switzerland, 13 December 2017.
38. Howells, M.; Hermann, S.; Welsch, M.; Bazilian, M.; Segerström, R.; Alfstad, T.; Gielen, D.; Rogner, H.; Fischer, G.; Van Velthuizen, H.; et al. Integrated Analysis of Climate Change, Land-Use, Energy and Water Strategies. *Nat. Clim. Chang.* **2013**, *3*, 621–626. [CrossRef]
39. Asghar, A.; Iqbal, J.; Amin, A.; Ribbe, L. Integrated Hydrological Modeling for Assessment of Water Demand and Supply under Socio-Economic and IPCC Climate Change Scenarios Using WEAP in Central Indus Basin. *J. Water Supply Res. Technol.—AQUA* **2019**, *68*, 136–148. [CrossRef]
40. Opere, A.O.; Waswa, R.; Mutua, F.M. Assessing the Impacts of Climate Change on Surface Water Resources Using WEAP Model in Narok County, Kenya. *Front. Water* **2022**, *3*, 174. [CrossRef]
41. Abera Abdi, D.; Ayenew, T. Evaluation of the WEAP Model in Simulating Subbasin Hydrology in the Central Rift Valley Basin, Ethiopia. *Ecol Process.* **2021**, *10*, 41. [CrossRef]
42. Sridharan, V.; Shivakumar, A.; Niet, T.; Ramos, E.P.; Howells, M. Land, Energy and Water Resource Management and Its Impact on Ghg Emissions, Electricity Supply and Food Production-Insights from a Ugandan Case Study. *Environ. Res. Commun.* **2020**, *2*, 085003. [CrossRef]
43. Sridharan, V.; Broad, O.; Shivakumar, A.; Howells, M.; Boehlert, B.; Groves, D.G.; Rogner, H.-H.; Taliotis, C.; Neumann, J.E.; Strzepek, K.M.; et al. Resilience of the Eastern African Electricity Sector to Climate Driven Changes in Hydropower Generation. *Nat. Commun.* **2019**, *10*, 302. [CrossRef] [PubMed]
44. Cervigni, R.; Liden, R.; Neumann, J.; Strzepek, K. *Enhancing the Climate Resilience of Africa's Infrastrucutre: The Power and Water Sectors (Technical Appendix)*; World Bank: Washington, DC, USA, 2015; ISBN 978-1-4648-0466-3/978-1-4648-0467-0.
45. Cervigni, R.; Liden, R.; Neumann, J.E.; Strzepek, K.M. *Enhancing the Climate Resilience of Africa's Infrastructure: The Power and Water Sectors*; World Bank: Washington, DC, USA, 2015; Volume 4, ISBN 9781464804663.
46. Sieber, J. *Water Evaluation And Planning (WEAP) System*; Software Version: 2021.0.1.10; Stockholm Environment Institute: Somerville, MA, USA, 2021.
47. Allen, R.G.; Pereira, L.S.; Raes, D.; Smith, M. *FAO Irrigation and Drainage Paper No. 56 Crop Evapotranspiration (Guidelines for Computing Crop Water Requirements)*; FAO: Rome, Italy, 2019; Volume 9.
48. Directorate of Water Resource Management. *Consolidated Hydrological Year Book for Uganda 1978-2014*; DWRM: Kampala, Uganda, 2014.
49. Smakhtin, V.; Revenga, C.; Döll, P. A Pilot Global Assessment of Environmental Water Requirements and Scarcity. *Water Int.* **2009**, *29*, 307–317. [CrossRef]
50. Tennant, D.L. Instream Flow Regimens for Fish, Wildlife, Recreation and Related Environmental Resources. *Fisheries* **1976**, *1*, 6–10. [CrossRef]
51. Tessmann, S.A. *Environmental Use Sector: Reconnaissance Elements of the Western Dakotas Region of South Dakota Study*; Water Resources Institute, South Dakota State University: Brookings, South Dakota, 1979.
52. Pastor, A.V.; Ludwig, F.; Biemans, H.; Hoff, H.; Kabat, P. Accounting for Environmental Flow Requirements in Global Water Assessments. *Hydrol. Earth Syst. Sci.* **2014**, *18*, 5041–5059. [CrossRef]
53. Sridharan, V. Impact of Climate Change on Integrated Resource Systems- Insights from Selected East African Case Studies. Ph.D. Dissertation, KTH Royal Institute of Technology, Stockholm, Sweden, 2020.
54. Chong, K.L.; Huang, Y.F.; Koo, C.H.; Najah Ahmed, A.; El-Shafie, A. Spatiotemporal Variability Analysis of Standardized Precipitation Indexed Droughts Using Wavelet Transform. *J. Hydrol.* **2022**, *605*, 127299. [CrossRef]
55. Thornthwaite, C.W. An Approach toward a Rational Classification of Climate. *Geogr Rev* **1948**, *38*, 55–94. [CrossRef]
56. Willmott, C.J.; Rowe, C.M.; Mintz, Y. Climatology of the Terrestrial Seasonal Water Cycle. *J. Climatol.* **1985**, *5*, 589–606. [CrossRef]
57. Nash, J.E.; Sutcliffe, J.V. River Flow Forecasting through Conceptual Models Part I—A Discussion of Principles. *J. Hydrol.* **1970**, *10*, 282–290. [CrossRef]
58. Eryani, I.G.A.P.; Jayantari, M.W.; Wijaya, I.K.M. Sensitivity Analysis in Parameter Calibration of the WEAP Model for Integrated Water Resources Management in Unda Watershed. *Civ. Eng. Archit.* **2022**, *10*, 455–469. [CrossRef]
59. Esteve, P.; Varela-Ortega, C.; Blanco-Gutiérrez, I.; Downing, T.E. A Hydro-Economic Model for the Assessment of Climate Change Impacts and Adaptation in Irrigated Agriculture. *Ecol. Econ.* **2015**, *120*, 49–58. [CrossRef]
60. Meng, F.; Liu, T.; Wang, H.; Luo, M.; Duan, Y.; Bao, A. An Alternative Approach to Overcome the Limitation of HRUs in Analyzing Hydrological Processes Based on Land Use/Cover Change. *Water* **2018**, *10*, 434. [CrossRef]
61. Hughes, D.A.; Mantel, S.; Farinosi, F. Assessing Development and Climate Variability Impacts on Water Resources in the Zambezi River Basin: Initial Model Calibration, Uncertainty Issues and Performance. *J. Hydrol. Reg. Stud.* **2020**, *32*, 100765. [CrossRef] [PubMed]
62. Dehghanipour, A.H.; Zahabiyoun, B.; Schoups, G.; Babazadeh, H. A WEAP-MODFLOW Surface Water-Groundwater Model for the Irrigated Miyandoab Plain, Urmia Lake Basin, Iran: Multi-Objective Calibration and Quantification of Historical Drought Impacts. *Agric. Water Manag.* **2019**, *223*, 105704. [CrossRef]

63. Fernández-Alberti, S.; Abarca-del-Río, R.; Bornhardt, C.; Ávila, A. Development and Validation of a Model to Evaluate the Water Resources of a Natural Protected Area as a Provider of Ecosystem Services in a Mountain Basin in Southern Chile. *Front. Earth Sci.* **2021**, *8*, 539905. [CrossRef]
64. Sridharan, V.; Ramos, E.P.; Zepeda, E.; Boehlert, B.; Shivakumar, A.; Taliotis, C.; Howells, M. The Impact of Climate Change on Crop Production in Uganda—An Integrated Systems Assessment with Water and Energy Implications. *Water* **2019**, *11*, 1805. [CrossRef]
65. Li, H.; Sheffield, J.; Wood, E.F. Bias Correction of Monthly Precipitation and Temperature Fields from Intergovernmental Panel on Climate Change AR4 Models Using Equidistant Quantile Matching. *J. Geophys. Res. Atmos.* **2010**, *115*, D10101. [CrossRef]
66. Tate, E.; Sutcliffe, J.; Conway, D.; Farquharson, F. Water Balance of Lake Victoria: Update to 2000 and Climate Change Modelling to 2100 Water Balance of Lake Victoria: Update to 2000 and Climate Change Modelling to 2100. *Hydrol. Sci. J.* **2004**, *49*, 563–574. [CrossRef]
67. Sewagudde, S. Lake Victoria's Water Budget and the Potential Effects of Climate Change in the 21st Century. *Afr. J. Trop. Hydrobiol. Fish.* **2010**, *12*, 57367. [CrossRef]
68. Sutcliffe, J.V.; Parks, Y.P. *The Hydrology of the Nile*; Salz, H., Kundzewicz, Z.W., Eds.; International Association of Hydrological Sciences: Wallingford, UK, 1999; Volume 5, ISBN 1-910502-75-9.
69. Zhai, J.; Mondal, S.K.; Fischer, T.; Wang, Y.; Su, B.; Huang, J.; Tao, H.; Wang, G.; Ullah, W.; Uddin, M.J. Future Drought Characteristics through a Multi-Model Ensemble from CMIP6 over South Asia. *Atmos. Res.* **2020**, *246*, 105111. [CrossRef]
70. Falchetta, G.; Gernaat, D.E.H.J.; Hunt, J.; Sterl, S. Hydropower Dependency and Climate Change in Sub-Saharan Africa: A Nexus Framework and Evidence-Based Review. *J. Clean. Prod.* **2019**, *231*, 1399–1417. [CrossRef]
71. Mehta, V.K.; Aslam, O.; Dale, L.; Miller, N.; Purkey, D.R. Scenario-Based Water Resources Planning for Utilities in the Lake Victoria Region. *Phys. Chem. Earth Parts A/B/C* **2013**, *61–62*, 22–31. [CrossRef]
72. International Energy Agency. Electricity Information: Overview—Electricity Consumption. Available online: https://www.iea.org/reports/electricity-information-overview/electricity-consumption (accessed on 17 April 2023).

climate

MDPI

*Article*

# Comparison of Flood Frequency at Different Climatic Scenarios in Forested Coastal Watersheds

**Shreeya Bhattarai, Prem B. Parajuli * and Filip To**

Department of Agricultural and Biological Engineering, Mississippi State University, Mississippi State, Starkville, MS 36762, USA
* Correspondence: pparajuli@abe.msstate.edu

**Abstract:** Climate change-induced extreme precipitation causes coastal flooding. A streamflow simulation in coastal watersheds, Wolf River Watershed (WRW) and Jourdan River Watershed (JRW), was conducted using the Soil and Water Assessment Tool (SWAT) to compare variation in flow at different climates and to analyze the flood frequency. Baseline models were auto-calibrated with SWAT calibration and uncertainty programs (SWAT-CUP). Kling–Gupta efficiency (KGE), defined as the objective function in SWAT-CUP, ranged from 0.8 to 0.7 in WRW and from 0.55 to 0.68 in JRW during the calibration–validation process. Results indicated reliability of the model performances. Monthly averaged baseline flow was 1% greater than historical and 8.9% lower than future climate in WRW. In JRW, monthly averaged baseline flow was 11% greater than historical and 5.7% lower than future climate. Flood frequency analysis showed the highest 1% exceedance probability in annual maximum series (AMS) of baseline model in WRW, whereas AMS of projected model was estimated the highest in JRW. This study aids in preparing for future flood management.

**Keywords:** streamflow; SWAT; climate change; CMIP5; CMHyd; peak flow; annual maximum series; partial duration series; flood frequency analysis; Log Pearson type III

## 1. Introduction

The frequency and severity of weather extremes are rising due to climate change [1]. Climate change causes variations in streamflow [2], triggering more frequent flooding events [3]. In the 21st century, extreme precipitation events are predicted to become more frequent, surpassing the present magnitude [4]. According to the Fifth Assessment Report of the Intergovernmental Panel on Climate Change (IPCC), climate change is projected to increase the risk of storms, extreme precipitation, flooding, and storm surges in coastal regions [5]. Moreover, there is an increasing trend in the population in coastal areas and the global population is projected to be more than 1 billion by the mid-21st century in the low-elevated coastal zone [6]. This poses a high risk of extreme events to the public and property on the coast [7]. Risk assessment of flood-prone areas can be carried out using a hydrological modelling approach [8–12]. Hydrological assessment of coastal plains is essential to study the quantity of streamflow, predict the occurrence of extreme events, analyze its trend in different climatic conditions, and understand the impact of climate change on the discharge of coastal rivers.

Extensive research has been conducted in flood risk assessment, which led to the development of numerous mathematical modelling tools over time [13], for instance, the Soil and Water Assessment Tool (SWAT) [8,11,14], Modélisation de l'Anticipation du Ruissellement et des Inondations pour des évéNements Extrêmes (MARINE) model [8,15], Hydrologic Engineering Center-River Analysis System (HEC-RAS) [9,16,17], Hydrologic Engineering Center-Hydrologic Modeling System (HEC-HMS) [9,18], a two-dimensional watershed rainfall-runoff CASC2D model [10,19], TOPography-based hydrological MODEL (TOPMODEL) [12,20], MIKE FLOOD [12,21], among others. The flood risk analysis by

Ajmal et al. [22] compared the hydrological, hydro-dynamical, and hybrid tools, and concluded that all models can be used in different scenarios depending on the water level area. Hydrologic models are effective tools for watershed management planning and climate change analysis [22].

SWAT is a-semi-distributed hydrologic model [14] capable of modelling future floods [23] on a watershed scale. It can also be used with space-based flood detection components for flood risk analysis in ungagged watersheds [24]. As of December 2022, over 5500 publications were recorded in the SWAT database (https://www.card.iastate.edu/swat_articles/, accessed on 1 December 2022) [25] regarding its development and application. SWAT has been used in numerous global studies at different time steps to investigate the climate change impact on hydrological processes [26–30]. In previous studies, the SWAT model has been applied for hydro-climatological research [31,32] and successfully conducted flood frequency analysis [33,34]. The hydrological modelling tools can account for the complex system of hydrology, quantify the variation in streamflow and predict the flood-causing potential peak flow events. In this study, the SWAT model was applied to coastal watersheds where flood frequency analysis is crucial to assess the risk of weather extreme events in the coastal settlement. Although SWAT is limited to simulating the tidal influence into the coastal watershed [35], watershed delineated further upstream tends to have no influence of the tides; therefore, it is not considered in the context of this study.

The hydrological cycle is affected by climate change [36]. Since the earth's temperature drives evaporation, condensation, and precipitation, climate change is likely to disrupt the hydrological balance. Evidence of a rise in average temperature in recent decades [37] directly relates to an increase in water vapor content in the atmosphere, leading to intense rainfall in the coastal area. The coastal watersheds of Mississippi are vulnerable to extreme weather events. Severe storms, tornadoes, flood incidents, and hurricanes were declared among the 29 major disasters by the Federal Emergency Management Agency between 2005–2020 in the state of Mississippi [38]. High precipitation is recorded over the years in the Mississippi coastal watersheds [39,40] that drain into the Mississippi Sound and eventually onto the Gulf of Mexico. Therefore, the objectives of this study were to: (i) set up baseline SWAT models for two forested coastal watersheds; (ii) set up a SWAT model for historical and projected climate conditions; (iii) compare the variation in streamflow during different climatic conditions with respect to the baseline model; and (iv) analyze the flood frequency of the stream at different climatic extremities.

## 2. Materials and Methods

### 2.1. Study Area

This study was conducted in two forested watersheds: Wolf River Watershed (WRW) and Jourdan River Watershed (JRW) of Mississippi, United States of America. The Wolf River and the Jourdan River run through the coastal plain; thus, these are categorized as coastal streams which fetch fresh water to Saint Louis Bay (SLB). WRW lies between the latitude of 31°5′ N and 30°27′ N and longitude of 89°35′ W and 89°15′ W. JRW lies between the latitude of 30°44′ N and 30°22′ N and longitude of 89°38′ W and 89°22′ W. Both watersheds are close to the Gulf of Mexico and are frequently impacted by extreme weather events. WRW covers an area of 801 $km^2$ and the land use is dominated by forest. The land use of WRW consists of 46.6% forest–evergreen, 24.9% range–brush, 20.8% wetlands–forested, and 7.7% pasture. Similarly, the land use of JRW is also dominated by forest. The total area of JRW is 538 $km^2$ and it consists of 37.9% forest–evergreen, 26.6% range–brush, 18.26% pasture, and 17.3% wetlands–forested. Figure 1 shows the location map of WRW and JRW in the state of Mississippi. Mississippi has a humid subtropical climate. The annual average temperature at SLB is 19.9 °C, with an average precipitation of 1534.2 mm annually occurring on an average of 98 rainy days throughout the year.

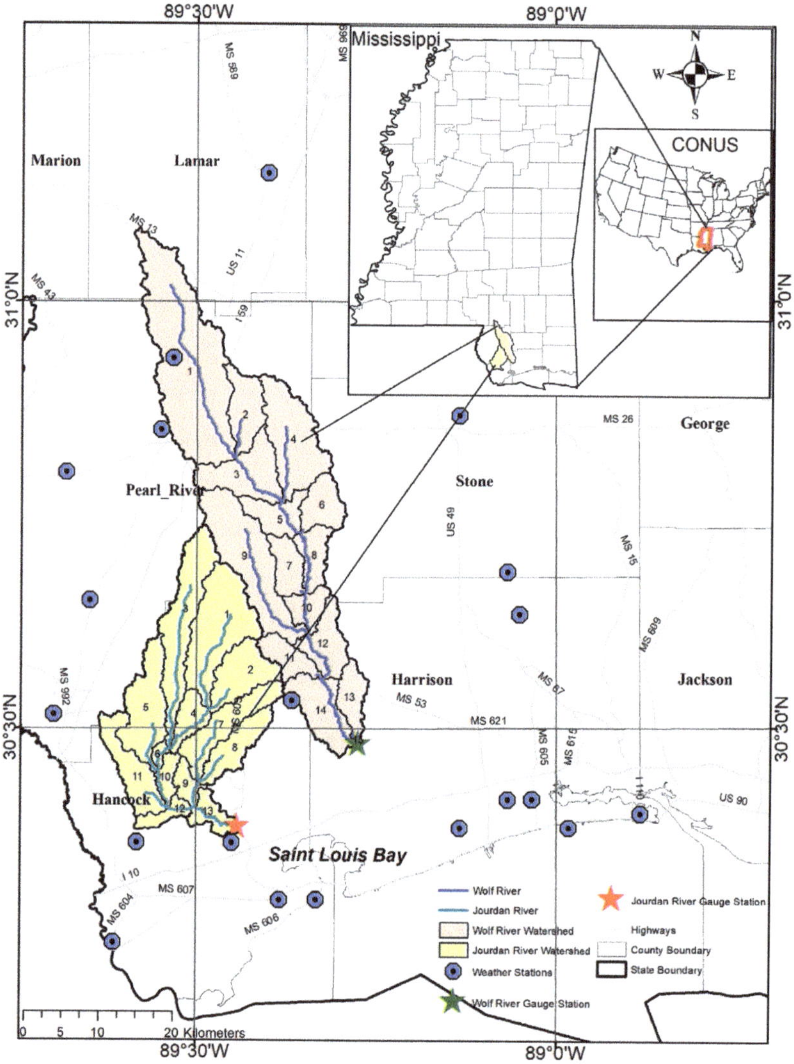


**Figure 1.** Study area map of the Wolf River Watershed and Jourdan River Watershed in Mississippi. (CONUS: continental United States).

### 2.2. Model Setup

Soil and Water Assessment Tool (SWAT) [14], a watershed-scale conceptual process-based model, was applied to the two adjacent watersheds for evaluating the streamflow from the coastal plain. GIS-coupled public domain software ArcSWAT is popular worldwide for hydrological modelling [41]. In the SWAT model, the hydrological analysis is carried out by dividing a watershed into smaller sub-watersheds and further into discrete hydrological response units (HRUs) on the basis of the unique combination of its land use, soil, and slope type [42]. For both watersheds, a baseline model was developed from 1995 to 2010. Additionally, a historical model from 1950 to 1994 and a projected model from 2011 to 2055 were developed to create different climatic conditions for each watershed. Thus, six different SWAT models were developed to study the hydrologic response of past and future climatic conditions ranging 45 years into the past and 45 years into the future from the baseline model. The basic data used to develop the baseline models are listed in Table 1.

**Table 1.** Data for the development of baseline SWAT model.

| S. No. | Data | Source |
|---|---|---|
| 1 | Elevation Data:<br>Digital Elevation Model (DEM) (30 m × 30 m) (2020) | United States Geological Survey (USGS) (http://viewer.nationalmap.gov/viewer/) (accessed on 4 October 2022) |
| 2 | Land-use and Land-cover Data:<br>Cropland Data Layer (CDL) (2010) | United States Department of Agriculture-National Agricultural Statistics Service (USDA-NASS) (http://nassgeodata.gmu.edu/CropScape/) (accessed on 4 October 2022) |
| 3 | Soil Data:<br>USSURGO (2020) | United States Soil Survey Geographic Database (US-SSURGO)<br>SWAT-USSURGO (https://swat.tamu.edu/data/) (accessed on 4 October 2022) |
| 4 | Weather Data:<br>NOAA (1995–2010)<br>Precipitation,<br>Maximum Temperature,<br>Minimum Temperature | National Oceanic and Atmospheric Administration (NOAA)<br>SWAT—Climate Data (https://swat.tamu.edu/data/) (accessed on 4 October 2022) |
| 5 | Discharge Data:<br>-USGS 02481510 (1997–2010)<br>(Wolf River Nr Landon)<br>-USGS 02481660 (2002–2005)<br>(Jourdan River Nr Bay St Louis) | United States Geological Survey (USGS) (https://waterdata.usgs.gov/ms/nwis/) (accessed on 5 October 2022) |

The basic data used to develop the baseline models were the 2020 United States Geological Survey (USGS) Digital Elevation Model (DEM) data (30 m × 30 m resolution) (http://viewer.nationalmap.gov/viewer/, accessed on 4 October 2022) [43], the 2020 United States Soil Survey Geographic Database (US-SSURGO) (https://swat.tamu.edu/data/, accessed on 4 October 2022) [44], the 2010 Cropland Data Layer (CDL) United States Department of Agriculture-National Agricultural Statistics Service (USDA-NASS) (http://nassgeodata.gmu.edu/CropScape/, accessed on 4 October 2022) [45], and the 1995–2010 Weather Data-National Oceanic and Atmospheric Administration (NOAA) Climate Data (https://swat.tamu.edu/data/, accessed on 4 October 2022) [44]. Monthly discharge data were utilized, collected from USGS 02481510 Wolf River near Landon, MS (1997–2010), for WRW flow simulation, whereas daily discharge data were utilized from USGS 02481660 Jourdan River near Bay St Louis, MS (2002–2005), for JRW flow simulation due to the availability of limited data in this gauge station.

DEM, CDL, and USSURGO soil data were fed into the model and the slope classes were defined. A total of 801 km$^2$ WRW was divided into 15 subbasins and 489 HRUs, and 538 km$^2$ JRW was divided into 13 subbasins and 233 HRUs. The meteorological dataset from NOAA, retrieved from the SWAT website, was used as local weather data for precipitation and temperature data (at 21 stations for WRW spanning over five counties, and at 12 stations for JRW spanning over two counties). The baseline model for both WRW and JRW was then set up and run for 16 years (1995–2010) with a two-year warmup period each. The simulation was performed monthly for WRW and daily for JRW. SWAT calibration and uncertainty program (SWAT-CUP) Sufi-2 [46] was used for the uncertainty analysis of hydrological parameters and to calibrate–validate the baseline model of both watersheds. This widely used autocalibration tool was considered in this study because of its simpler calibration procedure, availability of essential hydrological parameters, uncertainty analysis, and sensitivity analysis features [46,47]. Kling–Gupta efficiency (KGE) [48] was used as the objective function in SWAT-CUP for more accurate model calibration. Monthly streamflow calibration was conducted from January 1997 to December 2003 and validation was carried out from January 2004 to December 2010 for WRW at

USGS 02481510. Similarly, daily streamflow calibration was conducted from 10 March 2002 to 31 December 2003 and validation was carried out from 1 January 2004 to 30 September 2004 with all the available data at the USGS 02481660 for JRW. The fitted parameters for the calibrating–validating baseline model of both watersheds are listed in Tables 2 and 3, respectively.

**Table 2.** Parameters used in streamflow calibration for Wolf River Watershed.

| Parameters | Description | Fitted Value |
|---|---|---|
| CN2 | Initial SCS runoff curve number for moisture condition II. | 0.133 |
| ALPHA_BF | Baseflow alpha factor (days). | 0.316 |
| ESCO | Soil evaporation compensation factor. | 0.217 |
| SOL_K | Saturated hydraulic conductivity. | −0.018 |
| SOL_BD | Moist bulk density. | 0.104 |
| SOL_AWC | Available water capacity of the soil layer. | 0.297 |

**Table 3.** Parameters used in streamflow calibration for Jourdan River Watershed.

| Parameters | Description | Fitted Value |
|---|---|---|
| ALPHA_BF | Baseflow alpha factor (days). | 0.917 |
| ESCO | Soil evaporation compensation factor. | 0.910 |
| GW_DELAY | Groundwater delay (days). | 1.126 |
| GWQMN | Threshold depth of water in the shallow aquifer required for return flow to occur (mm). | 874.982 |
| EPCO | Plant uptake compensation factor. | 0.698 |
| CN2 | Initial SCS runoff curve number for moisture condition II. | −0.194 |
| RCHRG_DP | Deep aquifer percolation fraction. | 0.391 |
| REVAPMN | Threshold depth of water in the shallow aquifer for "revap" to occur (mm). | 177.726 |
| SOL_K | Saturated hydraulic conductivity. | −0.139 |
| SOL_BD | Moist bulk density. | 0.223 |
| SOL_AWC | Available water capacity of the soil layer. | 0.038 |
| OV_N | Manning's "n" value for overland flow. | −0.010 |
| CANMX | Maximum canopy storage. | 93.562 |
| SLSUBBSN | Average slope length. | 0.386 |
| LAT_TTIME | Lateral flow travel time. | 3.238 |
| CNCOEF | Plant ET curve number coefficient. | 0.678 |
| CH_N2 | Manning's "n" value for the main channel. | 0.023 |
| CH_K2 | Effective hydraulic conductivity in main channel alluvium. | 65.951 |
| HRU_SLP | Average slope steepness. | 0.361 |

Furthermore, to study the impact of historical and projected climate conditions, two additional models for each watershed were developed for 45 years into the past and 45 years into the future, i.e., 1950–1994 and 2011–2055, respectively, with a warmup period of 1 year for each model. Development of the SWAT model for historical climate events requires historical weather data which were collected from Weather Data-NOAA Climate Data (1950–1994) (https://swat.tamu.edu/data/, accessed on 4 October 2022) [44] for both watersheds. Additionally, for the development of the SWAT model for projected climate events, the website of Coordinated Regional Climate Downscaling Experiment-North America (CORDEX-NA) (https://na-cordex.org/, accessed on 14 October 2022) [49], which archives Coupled Model Intercomparison Project phase 5 (CMIP5), was accessed and Representative Concentration Pathway 4.5 (RCP4.5) data from the EC-EARTH driver, Rossby Centre Regional Atmospheric Model version 4, (RCA4 model) were downloaded to collect daily precipitation, maximum air temperature, and minimum air temperature data. In this study, regional climatic model projected data and RCP4.5 (Representative Concentration Pathway 4.5) emission scenario were used. The regional climatic model generates high-resolution projections of future climate changes at a regional scale and

RCP4.5 is considered the middle-of-the-road scenario, since it lies between lower-emissions scenarios (RCP2.6 and RCP3-PD) and higher-emissions scenarios (RCP6 and RCP8.5). In the context of the United States, the projected mean temperature increases from 2 °C under the RCP 2.6 scenario to 5.5 °C under the RCP 8.5 scenario [49]. The regional climatic model at RCP4.5 does not estimate climate uncertainty but is used to drive climate models to produce projections of future climate change with more detailed information at a regional scale. The downloaded NetCDF formatted data were then downscaled by applying the linear scaling method in the climate model for hydrology (CMhyd) tool available on the SWAT website (https://swat.tamu.edu/software/, accessed on 20 January 2021) [50]. Historical data (1951–2005) and projected data (2006–2100) of 50 km spatial resolution were downloaded from CORDEX-NA and used with NOAA SWAT database's observed data (1950–2010) for bias correction and data extraction in order to acquire data from 2011–2055 for the projected condition. A 90.2% overlap period was achieved with this approach, thus meeting the recommendation of the CMHyd tool [51]. The baseline model with fitted parameters was considered a control model, and the model developed based on historical and projected climate conditions was evaluated using the same set of parameters from the control model to detect the variation of flow in the past and the future. The flowchart of the methodology is shown in Figure 2.

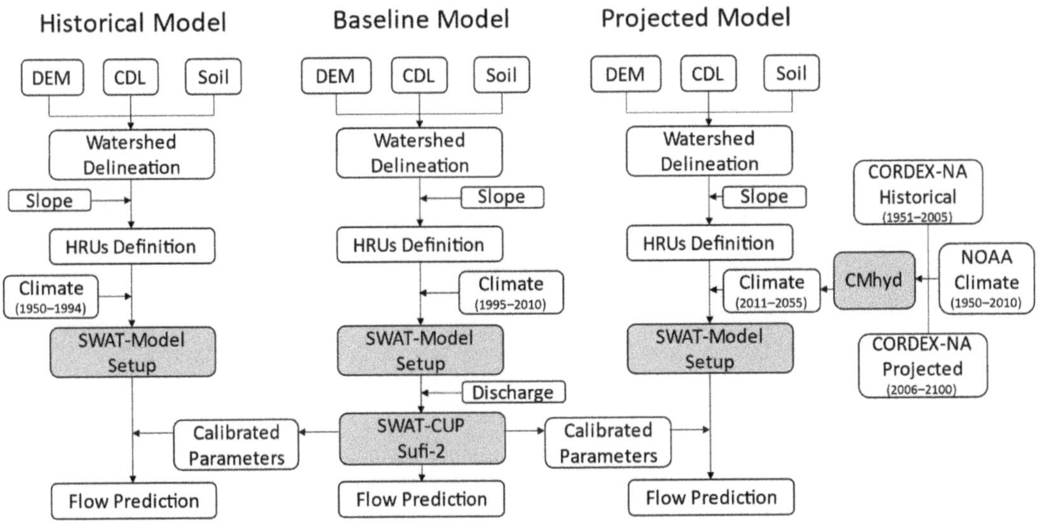


**Figure 2.** Flowchart of the methodology of SWAT model to study impact of climate change on flow. (DEM: Digital Elevation Model, CDL: Cropland Data Layer, HRU: Hydrologic Response Units, Coordinated Regional Climate Downscaling Experiment- North American, CMhyd: Climate Model data for hydrologic modeling, NOAA: National Oceanic and Atmospheric Administration, SWAT-CUP Sufi-2: SWAT Calibration and Uncertainty Programs Sequential Uncertainty Fitting version 2).

### 2.3. *Model Evaluation*

The simulation performance of the baseline models was evaluated using coefficient of determination ($R^2$) [52] (Equation (1)), Nash-Sutcliffe Efficiency (NSE) [53] (Equation (2)), Percentage Bias (PBAIS) [54] (Equation (3)), and KGE [48] (Equation (4)).

$R^2$ is the measure of model fit that shows the linear relation between observed and simulated data which ranges from 0 to 1. $R^2$ result approaching 1 indicates good model performance. NSE defines how well the observed and simulated data fits. NSE ranges from $-\infty$ to 1, 1 indicating a perfect fit. The average tendency of the simulated data to be greater or smaller than the observed data is measured by PBIAS. PBAIS negative value indicates overestimation and positive indicates underestimation. KGE is based on the decomposition of the NSE and mean square error into correlation, bias error, variation coefficient error. KGE ranges between $-\infty$ to 1, where 1 is the optimal value.

$$R^2 = \left( \frac{\sum_{i=1}^{N}(O_i - \overline{O})\ (S_i - \overline{S})}{\sqrt{(\sum_{i=1}^{N}(O_i - \overline{O})^2)} \cdot \sqrt{(\sum_{i=1}^{N}(S_i - \overline{S})^2)}} \right)^2 \tag{1}$$

$$\text{NSE} = 1 - \frac{\sum_{i=1}^{N}(O_i - S_i)^2}{\sum_{i=1}^{N}(O_i - \overline{O}_i)^2} \tag{2}$$

$$\text{PBAIS} = \frac{\sum_{i=1}^{N}(O_i - S_i)}{\sum_{i=1}^{N} O_i} \times 100 \tag{3}$$

where, $O_i$ = observed value, $\bar{O}$ = mean of observed values, $S_i$ = simulated value, $\overline{S}$ = mean of simulated values, and $N$ = total number of observations [50–52].

And,

$$\text{KGE} = 1 - \sqrt{(r-1)^2 + (\alpha - 1)^2 + (\beta - 1)^2} \tag{4}$$

where; $r$ = Linear correlation coefficient between simulated and observed data; $\alpha = \frac{\sigma s}{\sigma o}$, a measure of relative variability of the predicted and observed runoff; $\sigma s$ and $\sigma o$ = Standard deviation of predicted and observed runoff; and $\beta$ = Bias factor $\beta = \frac{\mu s}{\mu o}$ ; where $\mu$ is the mean of observed and simulated flows [48].

Moriasi et al. [55] hydrologic model calibration criteria for $R^2$, NSE, and PBIAS were referred to determine the efficiency of the developed model, and Brighenti et al. [56] was referred for the evaluation criteria of KGE. After satisfactory calibration and validation of baseline model, the flow in the past (1951–1994 with one year warmup period) and in the future (2012–2055 with one year warmup period) were compared with respect to the baseline flow. Historical hydrological extreme events and future hydrological extreme events were also analyzed. Calibrated-validated period and onwards peak flow events of baseline model were considered for flood frequency analysis in this study.

### 2.4. Flood Frequency Analysis

The aim of analyzing the flood frequency was to use probability distribution to relate the magnitude of extreme flow events to the frequency of occurrence [57]. Log Pearson type III (LP3) distribution is recommended by USGS to fit extreme events for flood frequency analysis [58]. Generally, Annual Maximum Series (AMS) is considered as the flow extreme events for LP3 distribution. In this study, two different approaches were considered to identify extreme events. The first approach was to sort out AMS, and the second approach was to generate the Partial Duration Series (PDS) considering the 99th percentile ranked flow event during the respective period of different climatic conditions. These two data series of extreme events were fitted in LP3 distribution to generate flood frequency graphs. Detailed descriptions of the preparation of flood frequency curves using LP3 distribution can be found in Bulletin 17B guidelines (https://water.usgs.gov/osw/bulletin17b/dl_flow.pdf, accessed on 9 November 2022) [58].

## 3. Results

### 3.1. Streamflow Calibration and Validation

Figure 3 shows the trend of observed and simulated data for WRW monthly streamflow simulation after the model calibration and validation with SWAT-CUP. The $R^2$ ranged from 0.82 to 0.75 indicating good results, NSE ranged from 0.81 to 0.73 indicating very good to good results, PBAIS ranged from (−4.9) to (−3) indicating very good results and KGE ranged from 0.80 to 0.70 indicating good to intermediate results during model calibration and validation.

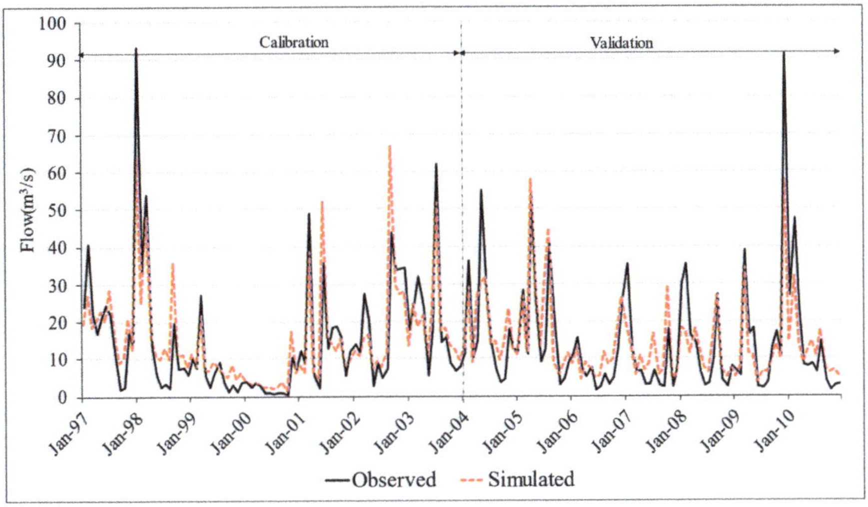


**Figure 3.** Monthly streamflow calibration-validation of Wolf River Watershed.

Similarly, Figure 4 shows the trend of observed and simulated data for JRW daily streamflow simulation after the model calibration and validation with SWAT-CUP. The $R^2$ ranged from 0.59 to 0.72 indicating not satisfactory to satisfactory results, NSE ranged from 0.42 to 0.71 indicating not satisfactory to good results, PBAIS ranged from 36.7 to 4.7 indicating not satisfactory to very good results whereas KGE ranged from 0.55 to 0.68 indicating intermediate results during model calibration-validation. Less than satisfactory results for $R^2$, NSE, and PBAIS were observed. However, since KGE was defined as an objective function in calibration, the performance of the model was intermediate. Underestimation of the result might have occurred due to the lack of continuous observed streamflow data or due to the low-gradient topography of the JRW. The results of both model performances are summarized in Table 4.

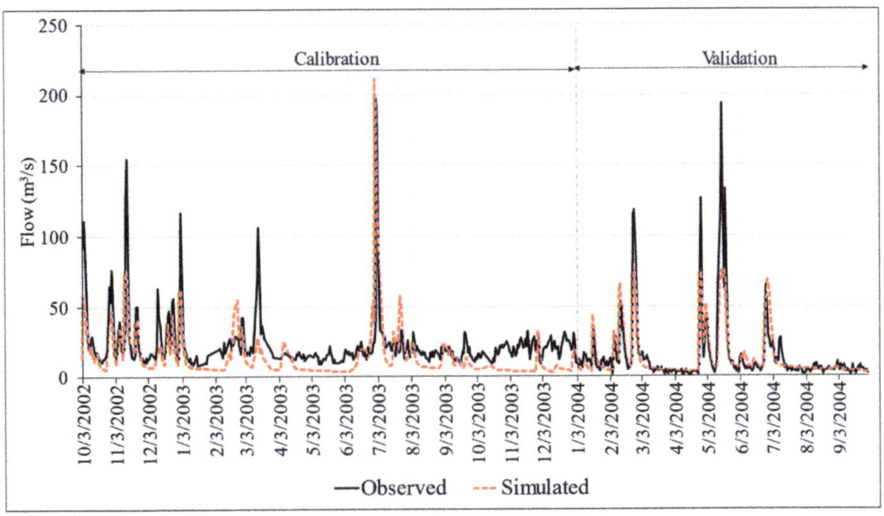


**Figure 4.** Daily streamflow calibration-validation of Jourdan River Watershed.

**Table 4.** Model performance of baseline SWAT model to simulate streamflow at Wolf River Watershed and Jourdan River Watershed.

| Watershed | Model Development | $R^2$ | NSE | PBIAS | KGE |
|---|---|---|---|---|---|
| WRW | Calibration (January 1997–December 2003) | 0.82 | 0.81 | −4.9 | 0.80 |
| | Validation (January 2004–December 2010) | 0.75 | 0.73 | −3.0 | 0.70 |
| JRW | Calibration (10 March 2002–31 December 2003) | 0.59 | 0.42 | 36.7 | 0.55 |
| | Validation (1 January 2004–30 September 2004) | 0.72 | 0.71 | 4.7 | 0.68 |

### 3.2. Flow Comparison of Different Climate Conditions

The baseline average maximum temperature ($Tmax_{avg}$) was 24.60 °C, the average minimum temperature ($Tmin_{avg}$) was 13.67 °C and average precipitation ($PPT_{avg}$) was 3.77 mm for WRW. The acquired dataset shows that the historical $Tmax_{avg}$ was 0.71 °C warmer and projected $Tmax_{avg}$ will be 1.60 °C warmer as compared to the baseline condition. Whereas the historical $Tmin_{avg}$ was 0.01 °C cooler and projected $Tmin_{avg}$ will be 1 °C warmer than the baseline. Additionally, the historical $PPT_{avg}$ was higher by 0.59 mm and projected $PPT_{avg}$ will increase by 0.78 mm than the baseline condition. Although $PPT_{avg}$ was higher in the historical condition, $Flow_{avg}$ was observed to be 0.14 mm less than baseline. Whereas, with increased $PPT_{avg}$ in projected condition, the $Flow_{avg}$ was observed to increase by 1.23 mm in comparison to the baseline $Flow_{avg}$ of 13.89 mm.

Similarly, the baseline $Tmax_{avg}$ was 24.54 °C, the $Tmin_{avg}$ was 14.71 °C and $PPT_{avg}$ was 3.90 mm for JRW. The historical $Tmax_{avg}$ was 0.58 °C warmer and projected $Tmax_{avg}$ will be 1.55 °C warmer compared to baseline. Whereas historical $Tmin_{avg}$ was 0.46 °C cooler and projected $Tmin_{avg}$ will be 0.65 °C warmer than the baseline. Moreover, the historical PPTavg was 0.51 mm higher and projected $PPT_{avg}$ will be 0.77 mm higher than the baseline. Due to the increase in PPTavg in both climatic conditions, the increase in flow was observed in both climates in JRW. The predicted historical $Flow_{avg}$ was observed to be 1.25 mm high and projected $Flow_{avg}$ was observed to increase by 0.65 mm as compared to the baseline $Flow_{avg}$ of 11.47 mm.

The trend of streamflow obtained by plotting the monthly averaged flow of SWAT flow output from the historical and projected models, presented in Figures 5 and 6, was analyzed with respect to the baseline model. In the context of WRW, there was an increment during spring and a reduction during fall in both climatic conditions, whereas during summer, the projected streamflow appeared larger than the historical and, in most cases, exceeded the baseline, too. Historical flow exceeded the baseline flow in 6 months and was lower than that in 6 months (Dec–May), whereas projected flow exceeded the baseline flow in 7 months (Jan–May and July–Aug) and was lower than that in 5 months (Jun and Sep–Dec). Overall, with respect to the baseline monthly average of streamflow, a reduction in the historical flow and an increment in the projected flow were estimated in WRW.

In the context of JRW, there too was an increment during spring and a reduction during fall in both climatic conditions, whereas in summer, the projected streamflow appeared higher in July–August and historical appeared higher in May. However, in the rest of the cases, both conditions' streamflow was lower than that of the baseline. Historical flow exceeded the baseline flow in 6 months and was lower in 6 months (Jan–May and Aug), whereas projected flow exceeded the baseline flow in 7 months (Jan–Mar and May–Aug) and was lower in 5 months (April and Sep–Dec). Overall, with respect to the baseline monthly averaged stream flow, increments in both historical and projected flow were estimated in JRW.

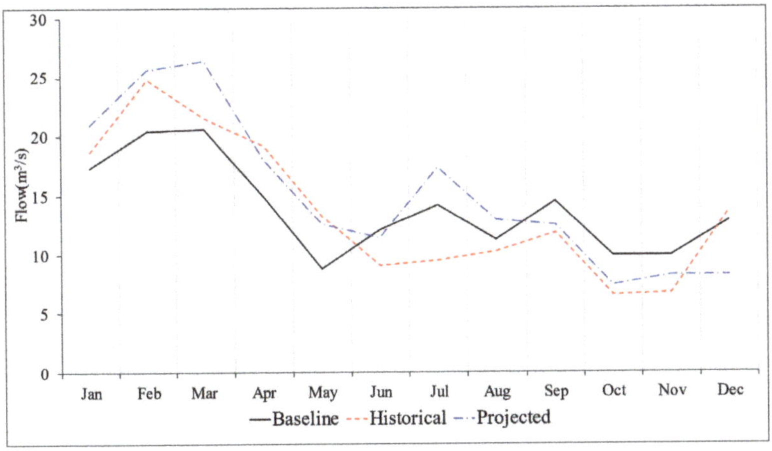


**Figure 5.** Monthly averaged streamflow flow comparison of Wolf River Watershed. (Historical: 1951–1994, Baseline: 1997–2010, Projected: 2012–2055).

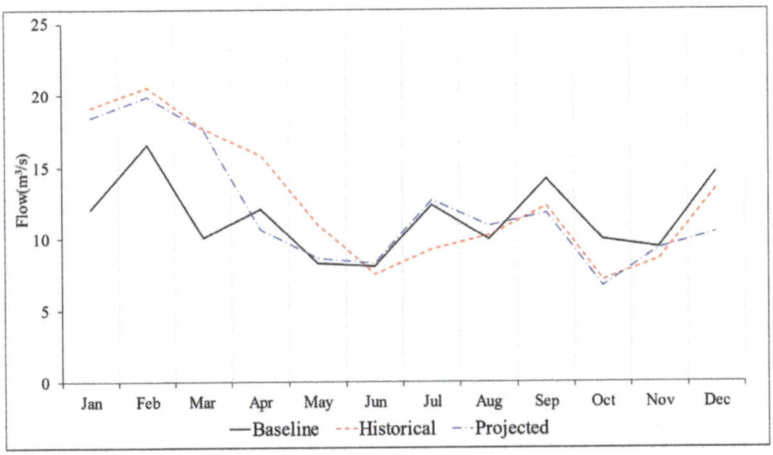


**Figure 6.** Monthly averaged streamflow flow comparison of Jourdan River Watershed. (Historical: 1951–1994, Baseline: 1997–2010, Projected: 2012–2055).

### *3.3. Flow Extreme Events*

Annual maximum peak flow series: The maximum flow events of each year from all climatic conditions were sorted to produce AMS, as shown in Figure 7. In total, 14 peaks from baseline (1997–2010) and 44 peaks each from historical (1951–1994) and projected (2012–2055) conditions were captured to obtain AMS for WRW. Similarly, for JRW, 9 peaks from baseline (2002–2010) and 44 peaks each from historical (1951–1994) and projected (2012–2055) conditions were captured to obtain AMS.

Partial duration peak flow series: Partial duration peak flow series for every condition in this study was obtained by defining the 99th percentile ranking; it is plotted in Figure 8. For WRW, 52 peak flow events in 14 years of the baseline condition, 161 peak flow events in 44 years of the historical climatic condition, and 161 peak flow events in 44 years of the projected climatic condition were defined as extreme values for the streamflow. Similarly, 33 peak flow events in 8 years of the baseline condition, 161 peak flow events in 44 years of the historical climatic condition, and 161 peak flow events in 44 years of the projected climatic condition were identified in JRW.

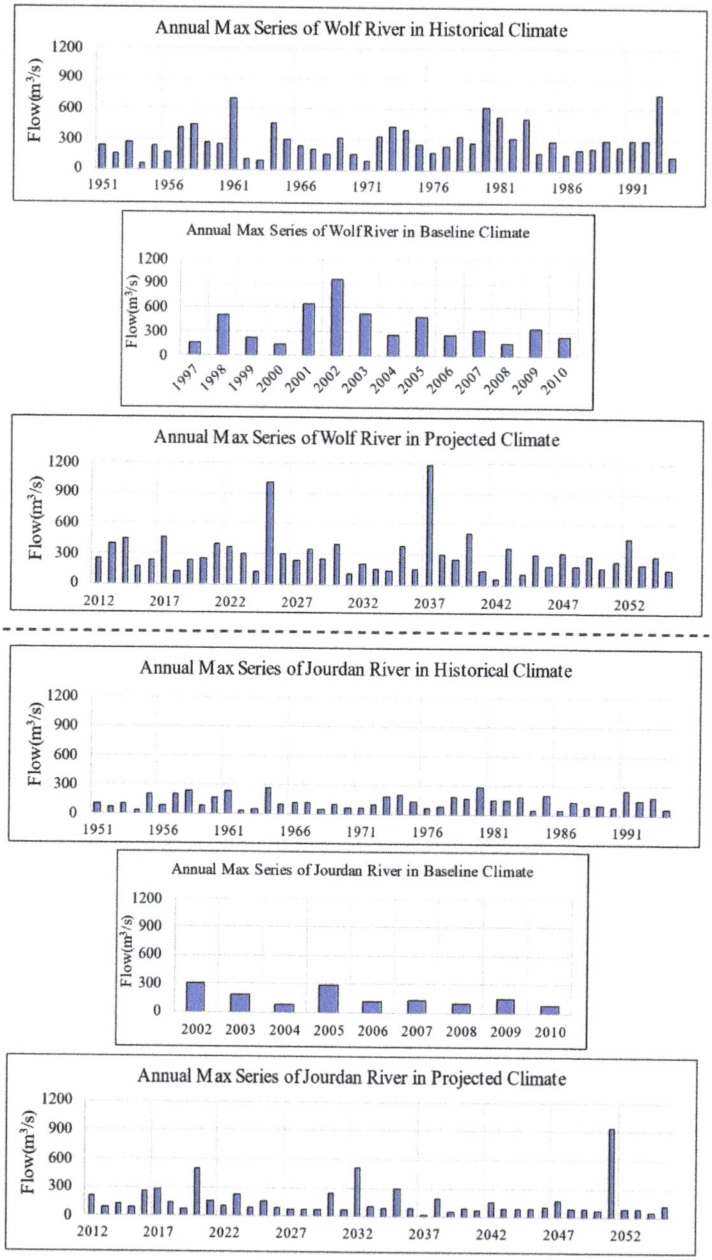


**Figure 7.** Annual maximum series graph for different climatic conditions.

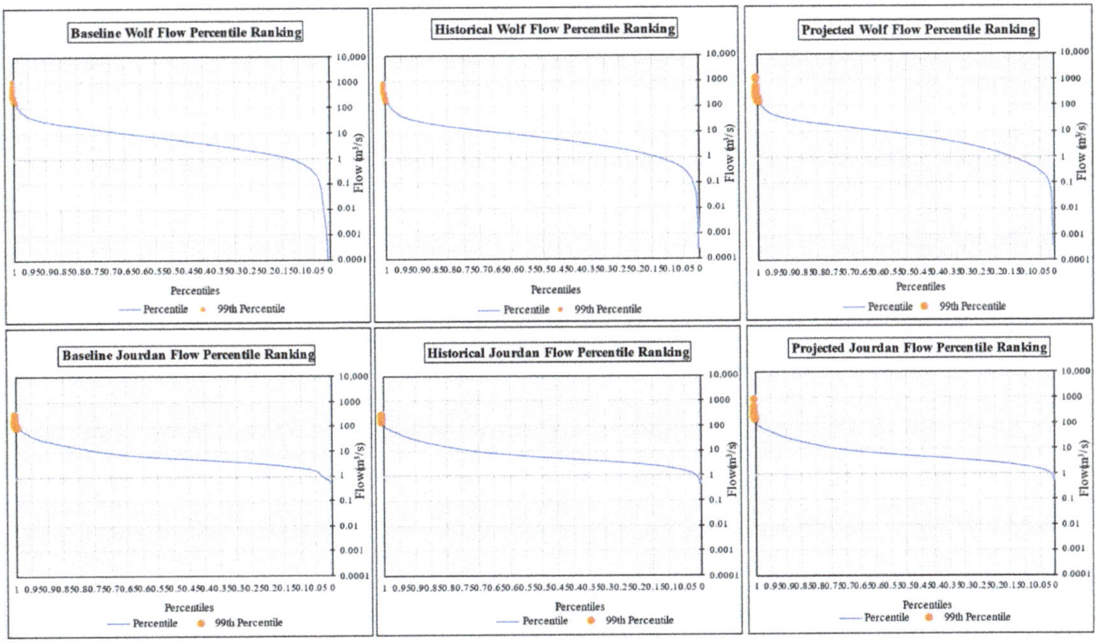


**Figure 8.** 99th percentile ranking of streamflow at different climatic conditions.

The gamma distribution plot using the shape–scale parameterization for 99th percentile peak flow PDS is shown in Figure 9. The plot shows the relation between peak flows and its probability of occurrence. The probability of occurrence was highest during the historical condition in both WRW and JRW, whereas the probability of occurrence was the lowest for the baseline PDS in WRW and projected PDS for JRW. Since the highest amongst the lowest flow magnitude in all climatic condition PDS of WRW was 146.7 $m^3/s$ in historical PDS, the graph was plotted from 150 $m^3/s$ through 1200 $m^3/s$ to include 1185 $m^3/s$ flow of highest right-skewed projected PDS in Figure 9a. Similarly for JRW, the highest PDS amongst the lowest flow magnitude in all climatic conditions was 92.17 $m^3/s$. Therefore, Figure 9b was plotted from 100 $m^3/s$ to 1000 $m^3/s$ in order to include 934.4 $m^3/s$ highest right-skewed projected PDS. From this distribution, the largest magnitude of the flow extreme series was predicted during the projected condition for both watersheds.

### *3.4. Flood Frequency Analysis*

The flood frequency analysis was carried out with Log Pearson type III (LP3) distribution graph using AMS and PDS as shown in Figures 10 and 11 for WRW and JRW, respectively. A spreadsheet model was developed to compute the exceedance probability and to plot flood frequency curve using LP3 distribution. Pearson type III deviation was calculated using standard normal deviation and a skewness coefficient of the logarithm of peak flow [58]. The 95% confidence interval capturing the uncertainty in the estimate of flood probabilities of the extreme flow events is also shown in Figures 10 and 11. Lower and upper bound curves are generated using a probability coefficient with a 95% confidence interval based on the peak flow records. Plotting the extreme flow values against the exceedance probability using standardized methods such as LP3 distribution can be used to extrapolate flooding events beyond the experienced values [58].

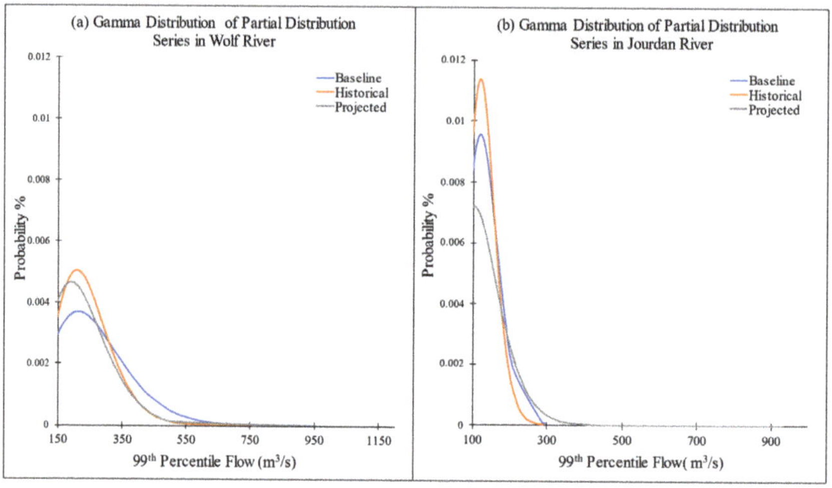


**Figure 9.** (**a**,**b**) Gamma distribution plot of partial distribution series.

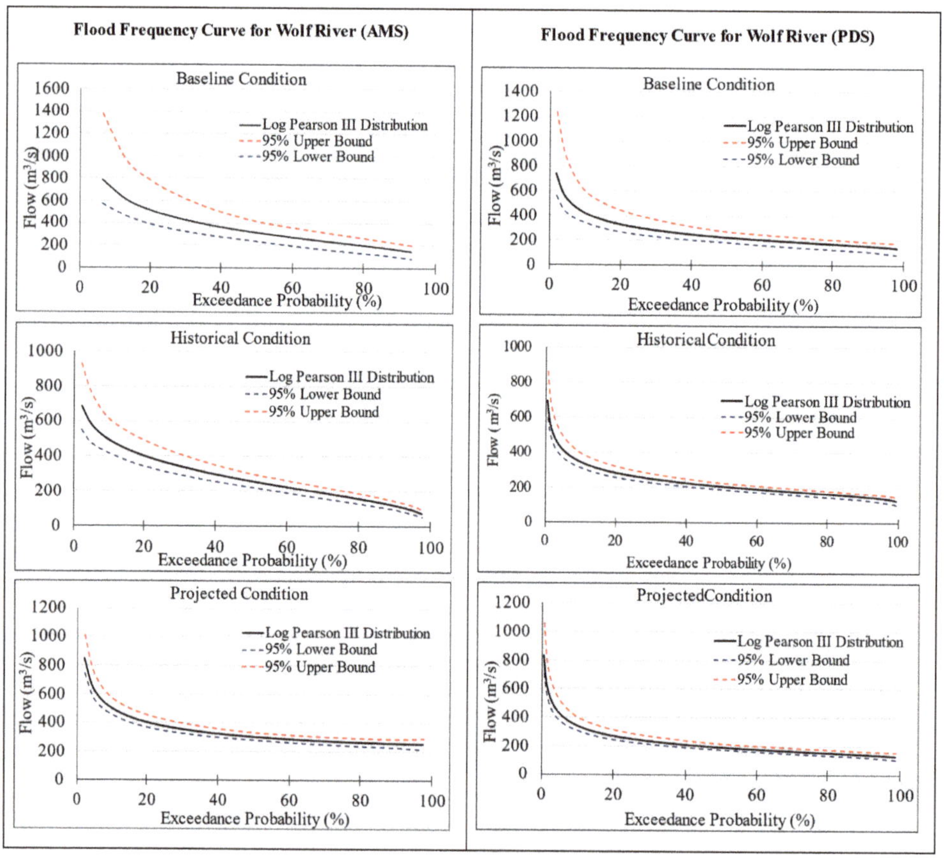


**Figure 10.** Wolf River flood frequency curves for the annual maximum series and partial duration series.

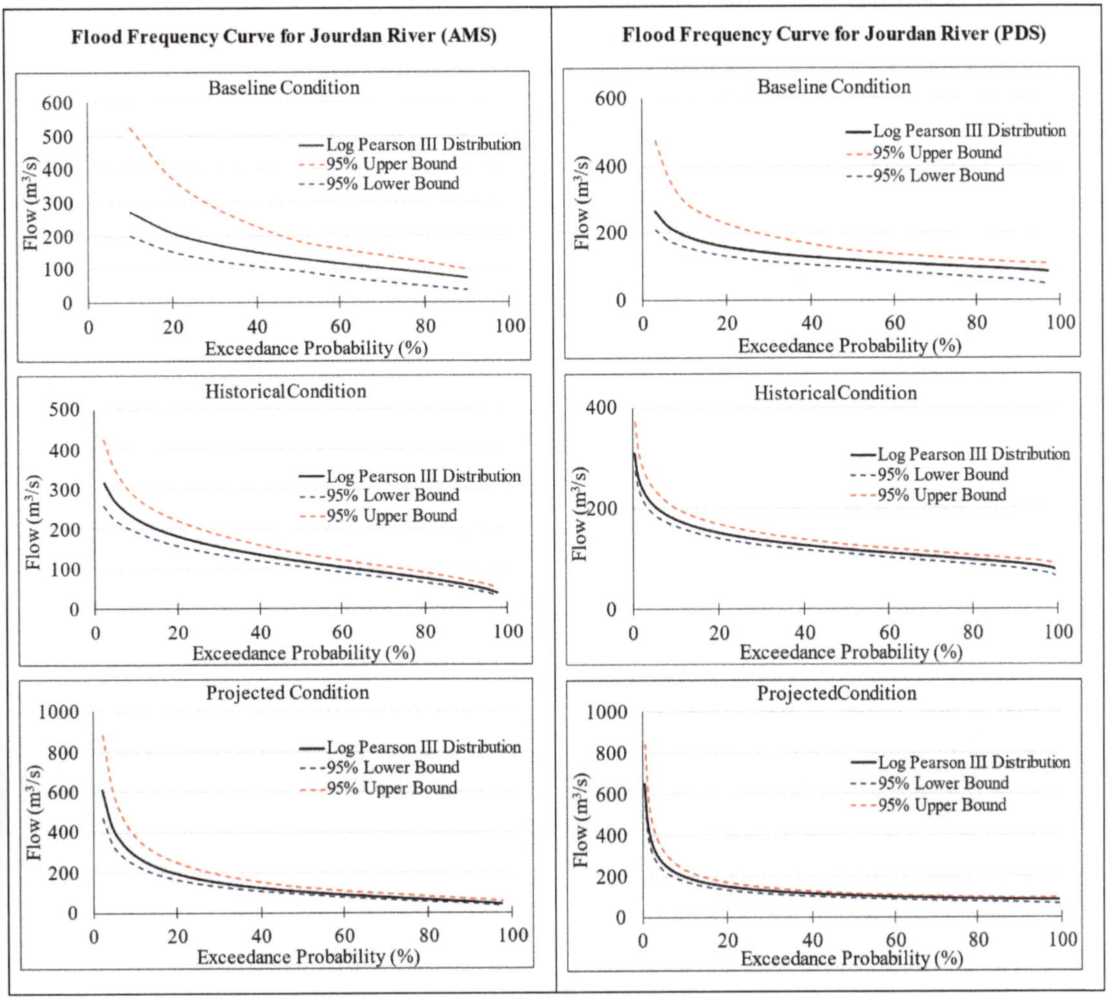


**Figure 11.** Jourdan River flood frequency curves for the annual maximum series and partial duration series.

In Figures 12 and 13, both AMS and PDS are plotted together to compare the flood frequency curve at different climatic conditions using different peak flow series. Comparing Figures 12 and 13, the peak flow estimated by the AMS series is higher than the one estimated by the PDS series in both watersheds. While comparing the 1% exceedance probability, i.e., 100-year return flood, it increased by 27% from historical to baseline condition with AMS but decreased by 22% from baseline to projected condition in WRW. Similarly, with PDS, the 1% exceedance probability increased by 17% from historical to baseline, and decreased again by 15% from baseline to projected condition.

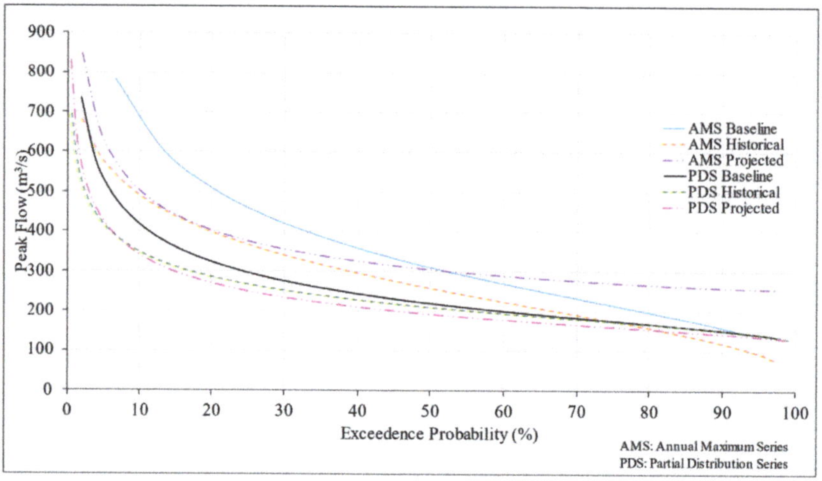


**Figure 12.** Wolf River Flood frequency curve at different climatic conditions with different peak flow series.

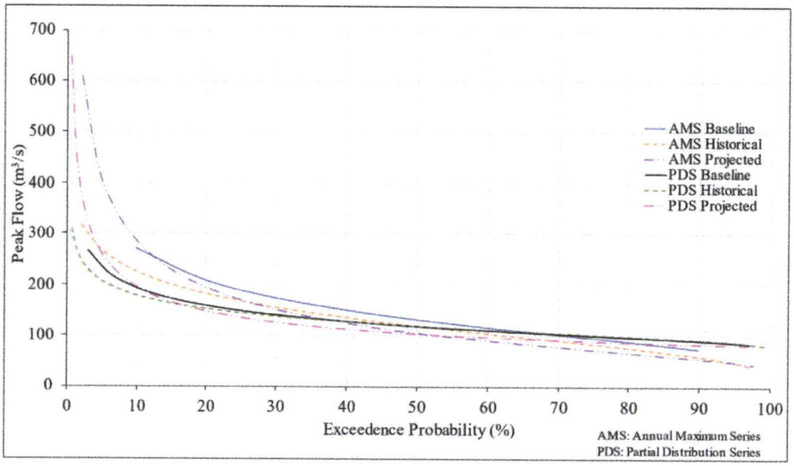


**Figure 13.** Jourdan River flood frequency curve at different climatic conditions with different peak flow series.

In JRW, the 1% exceedance probability increased by 8.1% from historical to baseline and further increased by 17% from baseline to projected with AMS. Similarly, with PDS, the 1% exceedance probability increased from 4.4% from historical to baseline and 10% from baseline to projected condition.

## 4. Discussion

This study aims to investigate the impact of climate change on the streamflow from the coastal watersheds of Mississippi draining into the Saint Louis Bay by comparing historical and projected climate events. The findings on the model development, and the calibration–validation process suggest that the SWAT model is capable of simulating the hydrological phenomenon in the forested coastal watershed as described by Upadhyay et al. [35]. Moreover, the results of statistical measures were considerably better in the watershed with continuous discharge data. Both models were able to generate satisfactory KGE values, indicating the reliability of the model performance [56]. The comparison of flow between different climatic conditions depicted an increment of monthly averaged flow in comparison to the baseline and historical climate. This might have been a result of the increased precipitation in the coastal watershed of Mississippi as mentioned in Sankar et al.'s [39] and Parra et al.'s [40] works. Since IPCC [5] highlights the risk of extreme precipitation and flooding in coastal regions due to climate change, riverine flooding risk was also analyzed in this study. Flood frequency analysis using LP3 distribution revealed that the 1% exceedance probability was highest in the AMS of the baseline condition in WRW, whereas for JRW, the AMS of the projected condition was highest. The increment in exceedance probability in the future condition in JRW aligns with the recent study carried out on the coastal plain of northeastern United States [59]. Contrary to JRW, the 1% exceedance probability is the highest in the baseline condition of WRW. Although both watersheds are adjacent to each other, the flood frequency at the outlet of these watersheds varies significantly given the same future climate condition. In addition, during the flood frequency analysis of JRW, we used the AMS baseline condition with only nine years for peak flow data. Therefore, in the estimation of 1% exceedance probability, there are higher uncertainties.

Since flood frequency analysis can predict the likelihood of flood occurrence, flood warning systems and emergency response plans for vulnerable areas can be developed beforehand. Infrastructure design and construction could consider the magnitude and frequency of predicted flooding events in order to minimize the unforeseeable damage to life and property.

## 5. Conclusions

This study compared the variation in flow in different climates and analyzed the flood frequencies in the two forest-dominated coastal watersheds, WRW and JRW. Baseline models for two adjacent coastal watersheds were developed using the SWAT model and auto-calibrated with the SWATCUP-Sufi2 tool applying KGE as an objective function. The fitted parameters used to develop an acceptable baseline model were further used to set up different climatic models to study the impact of climate change on the streamflow of both watersheds in different periods. The historical climatic model was developed using observed weather data from NOAA, and for the development of the projected climatic model, the CMIP5 CORDEX-NA data were downloaded and downscaled using the CMhyd tool to acquire projected weather data. The streamflow during different climatic conditions in the past and the future was compared with the baseline streamflow and the variation in the magnitude was analyzed. Log Pearson type III distribution was used to fit the AMS and PDS to analyze the flood frequency of the stream at different climatic conditions extremities. This study quantifies the hydrologic response of Wolf River Watershed and Jourdan River Watershed due to historical and future projected climatic conditions, which aids in the management of future flooding.

**Author Contributions:** S.B.: conceptualization, data curation, formal analysis, methodology, software, visualization, writing—original draft. P.B.P.: conceptualization, funding acquisition, methodology, resources, supervision, validation, writing—review and editing. F.T.: writing—review and editing. All authors have read and agreed to the published version of the manuscript.

**Funding:** This project was paid for in part with federal funding from the U.S. Department of the Treasury, the Mississippi Department of Environmental Quality, and the Mississippi Based RESTORE Act Center of Excellence under the Resources and Ecosystems Sustainability, Tourist Opportunities, and Revived Economies of the Gulf Coast States Act of 2012 (RESTORE Act). The statements, findings, conclusions, and recommendations are those of the authors and do not necessarily reflect the views of the Department of the Treasury, the Mississippi Department of Environmental Quality, or the Mississippi Based RESTORE Act Center of Excellence.

**Institutional Review Board Statement:** Not applicable.

**Informed Consent Statement:** Not applicable.

**Data Availability Statement:** Not applicable.

**Acknowledgments:** We would like to acknowledge the partial support of the Bagley College of Engineering, Mississippi Agricultural and Forestry Experiment Services (MAFES), and College of Agriculture and Life Sciences at Mississippi State University, United States Geological Survey (USGS), and all our collaborators in this study.

**Conflicts of Interest:** The authors declare no conflict of interest.

## References

1. Pörtner, H.O. (Ed.) *Climate Change 2022: Impacts, Adaptation and Vulnerability Working Group II Contribution to the Sixth Assessment Report of the Intergovernmental Panel on Climate Change*; Cambridge University Press: Cambridge, UK, 2022. [CrossRef]
2. Lee, C.H.; Yeh, H.F. Impact of Climate Change and Human Activities on Streamflow Variations Based on the Budyko Framework. *Water* **2019**, *11*, 2001. [CrossRef]
3. Chiang, L.C.; Liao, C.J.; Lu, C.M.; Wang, Y.C. Applicability of Modified SWAT Model (SWAT-Twn) on Simulation of Watershed Sediment Yields under Different Land Use/Cover Scenarios in Taiwan. *Environ. Monit. Assess.* **2021**, *193*, 1–23. [CrossRef] [PubMed]
4. Giorgi, F.; Raffaele, F.; Coppola, E. The Response of Precipitation Characteristics to Global Warming from Climate Projections. *Earth Syst. Dyn.* **2019**, *10*, 73–89. [CrossRef]
5. Pachauri, R.K.; Allen, M.R.; Barros, V.R.; Broome, J.; Cramer, W.; Christ, R.; Church, J.A.; Clarke, L.; Dahe, Q.; Dasgupta, P.; et al. *Climate Change 2014: Synthesis Report*; Contribution of Working Groups I, II and III to the Fifth Assessment Report of the Intergovernmental Panel on Climate Change; IPCC: Geneva, Switzerland, 2014; p. 151.
6. Merkens, J.L.; Reimann, L.; Hinkel, J.; Vafeidis, A.T. Gridded Population Projections for the Coastal Zone under the Shared Socioeconomic Pathways. *Glob. Planet. Change* **2016**, *145*, 57–66. [CrossRef]
7. Kron, W. Coasts: The High-Risk Areas of the World. *Nat. Hazards* **2013**, *66*, 1363–1382. [CrossRef]
8. Boithias, L.; Sauvage, S.; Lenica, A.; Roux, H.; Abbaspour, K.C.; Larnier, K.; Dartus, D.; Sánchez-Pérez, J.M. Simulating Flash Floods at Hourly Time-Step Using the SWAT Model. *Water* **2017**, *9*, 929. [CrossRef]
9. Zhang, K.; Shalehy, M.H.; Ezaz, G.T.; Chakraborty, A.; Mohib, K.M.; Liu, L. An Integrated Flood Risk Assessment Approach Based on Coupled Hydrological-Hydraulic Modeling and Bottom-up Hazard Vulnerability Analysis. *Environ. Model. Softw.* **2022**, *148*, 105279. [CrossRef]
10. Chao, L.; Zhang, K.; Li, Z.; Wang, J.; Yao, C.; Li, Q. Applicability Assessment of the CASCade Two Dimensional SEDiment (CASC2D-SED) Distributed Hydrological Model for Flood Forecasting across Four Typical Medium and Small Watersheds in China. *J. Flood Risk Manag.* **2019**, *12*, e12518. [CrossRef]
11. Jodar-Abellan, A.; Valdes-Abellan, J.; Pla, C.; Gomariz-Castillo, F. Impact of Land Use Changes on Flash Flood Prediction Using a Sub-Daily SWAT Model in Five Mediterranean Ungauged Watersheds (SE Spain). *Sci. Total Environ.* **2019**, *657*, 1578–1591. [CrossRef]
12. Li, W.; Lin, K.; Zhao, T.; Lan, T.; Chen, X.; Du, H.; Chen, H. Risk Assessment and Sensitivity Analysis of Flash Floods in Ungauged Basins Using Coupled Hydrologic and Hydrodynamic Models. *J. Hydrol.* **2019**, *572*, 108–120. [CrossRef]
13. Nkwunonwo, U.C.; Whitworth, M.; Baily, B. A Review of the Current Status of Flood Modelling for Urban Flood Risk Management in the Developing Countries. *Sci. Afr.* **2020**, *7*, e00269. [CrossRef]
14. Arnold, J.G.; Srinivasan, R.; Muttiah, R.S.; Williams, J.R. Large Area Hydrologic Modeling and Assessment Part I: Model Development 1. *J. Am. Water Resour. Assoc.* **1998**, *34*, 73–89. [CrossRef]
15. Roux, H.; Labat, D.; Garambois, P.A.; Maubourguet, M.M.; Chorda, J.; Dartus, D. A Physically-Based Parsimonious Hydrological Model for Flash Floods in Mediterranean Catchments. *Nat. Hazards Earth Syst. Sci.* **2011**, *11*, 2567–2582. [CrossRef]
16. Brunner, G.W. HEC-RAS River Analysis System. Hydraulic Reference Manual. Version 1.0. Hydrologic Engineering Center Davis CA. 1995. Available online: https://www.hec.usace.army.mil/confluence/rasdocs/ras1dtechref/latest (accessed on 2 December 2022).
17. Kastridis, A.; Stathis, D. Evaluation of Hydrological and Hydraulic Models Applied in Typical Mediterranean Ungauged Watersheds Using Post-Flash-Flood Measurements. *Hydrology* **2020**, *7*, 12. [CrossRef]
18. Feldman, A.D. *Hydrologic Modeling System HEC-HMS Technical Reference Manual*; US Army Corps of Engineers, Hydrologic Engineering Center: Davis, CA, USA, 2000.
19. Julien, P.Y.; Saghafian, B. CASC2D User's Manual: A Two-Dimensional Watershed Rainfall-Runoff Model; Diss. Colorado State University. *Libraries* **1991**, 1–66.
20. Beven, K.J.; Kirkby, M.J. A Physically Based, Variable Contributing Area Model of Basin Hydrology. *Hydrol. Sci. J.* **1979**, *24*, 43–69. [CrossRef]
21. MIKE FLOOD 1D-2D and 1D-3D Modelling User Manual. *DHI Water Environ.* **2021**, 1–154.
22. Ajmal, A.; Irfan, S.; Sabahat, N.; Saleem, S.B. Analysis of Flood Risk Management in the Context of Mathematical Models. In Proceedings of the 2019 International Conference on Innovative Computing (ICIC)2019, Lahore, Pakistan, 1–2 November 2019; pp. 1–6. [CrossRef]

23. Iqbal, M.S.; Dahri, Z.H.; Querner, E.P.; Khan, A.; Hofstra, N. Impact of Climate Change on Flood Frequency and Intensity in the Kabul River Basin. *Geosciences* **2018**, *8*, 114. [CrossRef]
24. Khan, S.I.; Hong, Y.; Wang, J.; Yilmaz, K.K.; Gourley, J.J.; Adler, R.F.; Brakenridge, G.R.; Policelli, F.; Habib, S.; Irwin, D. Satellite Remote Sensing and Hydrologic Modeling for Flood Inundation Mapping in Lake Victoria Basin: Implications for Hydrologic Prediction in Ungauged Basins. *IEEE Trans. Geosci. Remote Sens.* **2010**, *49*, 85–95. [CrossRef]
25. SWAT Literature Database for Peer-Reviewed Journal Articles. Center for Agricultural and Rural Development, Ames, IA, USA. 2022. Available online: https://www.card.iastate.edu/swat_articles/ (accessed on 1 December 2022).
26. Parajuli, P.B.; Risal, A. Evaluation of Climate Change on Streamflow, Sediment, and Nutrient Load at Watershed Scale. *Climate* **2021**, *9*, 165. [CrossRef]
27. Bhatta, B.; Shrestha, S.; Shrestha, P.K.; Talchabhadel, R. Evaluation and Application of a SWAT Model to Assess the Climate Change Impact on the Hydrology of the Himalayan River Basin. *Catena* **2019**, *181*, 104082. [CrossRef]
28. Kiesel, J.; Gericke, A.; Rathjens, H.; Wetzig, A.; Kakouei, K.; Jähnig, S.C.; Fohrer, N. Climate Change Impacts on Ecologically Relevant Hydrological Indicators in Three Catchments in Three European Ecoregions. *Ecol. Eng.* **2019**, *127*, 404–416. [CrossRef]
29. Nguyen, H.H.; Recknagel, F.; Meyer, W.; Frizenschaf, J.; Shrestha, M.K. Modelling the Impacts of Altered Management Practices, Land Use and Climate Changes on the Water Quality of the Millbrook Catchment-Reservoir System in South Australia. *J. Environ. Manag.* **2017**, *202*, 1–11. [CrossRef] [PubMed]
30. Rabezanahary Tanteliniaina, M.F.; Rahaman, M.H.; Zhai, J. Assessment of the Future Impact of Climate Change on the Hydrology of the Mangoky River, Madagascar Using ANN and SWAT. *Water* **2021**, *13*, 1239. [CrossRef]
31. Singh, V.; Sharma, A.; Goyal, M.K. Projection of Hydro-Climatological Changes over Eastern Himalayan Catchment by the Evaluation of RegCM4 RCM and CMIP5 GCM Models. *Hydrol. Res.* **2019**, *50*, 117–137. [CrossRef]
32. Menna, B.Y. Simulation of Hydro Climatological Impacts Caused by Climate Change: The Case of Hare Watershed, Southern Rift Valley of Ethiopia. *Hydrol Curr. Res.* **2017**, *8*, 2. [CrossRef]
33. Devkota, L.P.; Gyawali, D.R. Impacts of Climate Change on Hydrological Regime and Water Resources Management of the Koshi River Basin, Nepal. *J. Hydrol. Reg. Stud.* **2015**, *4*, 502–515. [CrossRef]
34. Maghsood, F.F.; Moradi, H.; Massah Bavani, A.R.; Panahi, M.; Berndtsson, R.; Hashemi, H. Climate Change Impact on Flood Frequency and Source Area in Northern Iran under CMIP5 Scenarios. *Water* **2019**, *11*, 273. [CrossRef]
35. Upadhyay, P.; Linhoss, A.; Kelble, C.; Ashby, S.; Murphy, N.; Parajuli, P.B. Applications of the SWAT Model for Coastal Watersheds: Review and Recommendations. *J. ASABE* **2022**, *65*, 453–469. [CrossRef]
36. Kundzewicz, Z.W. Climate Change Impacts on the Hydrological Cycle. *Ecohydrol. Hydrobiol.* **2008**, *8*, 195–203. [CrossRef]
37. Wuebbles, D.; Fahey, D.; Takle, E.; Hibbard, K.; Arnold, J.; DeAngelo, B.; Doherty, S.; Easterling, D.; Edmonds, J.; Edmonds, T.; et al. *Climate Science Special Report: Fourth National Climate Assessment (NCA4), Volume I*; U.S. Global Change Research Program: Washington, DC, USA, 2017.
38. Kunkel, K.E.; Frankson, R.; Runkle, J.; Champion, S.M.; Stevens, L.E.; Easterling, D.R.; Stewart, B.C.; McCarrick, A.; Lemery, C.R. State Climate Summaries for the United States 2022. NOAA Technical Report NESDIS 150. NOAA NESDIS. 2022. Available online: https://www.nesdis.noaa.gov/about/documents-reports/technical-reports (accessed on 4 October 2022).
39. Sankar, M.S.; Dash, P.; Lu, Y.; Hu, X.; Mercer, A.E.; Wickramarathna, S.; Beshah, W.T.; Sanders, L.; Arslan, Z.; Dyer, J.; et al. Seasonal Changes of Trace Metal-Nutrient-Dissolved Organic Matter Conveyance along with Coastal Acidification over the Largest Oyster Reef in Western Mississippi Sound, Northern Gulf of Mexico. *North. Gulf Mex.* **2021**. [CrossRef]
40. Parra, S.M.; Sanial, V.; Boyette, A.D.; Cambazoglu, M.K.; Soto, I.M.; Greer, A.T.; Chiaverano, L.M.; Hoover, A.; Dinniman, M.S. Bonnet Carré Spillway Freshwater Transport and Corresponding Biochemical Properties in the Mississippi Bight. *Cont. Shelf Res.* **2020**, *199*, 104114. [CrossRef]
41. Dile, Y.T.; Daggupati, P.; George, C.; Srinivasan, R.; Arnold, J. Introducing a New Open Source GIS User Interface for the SWAT Model. *Environ. Model. Softw.* **2016**, *85*, 129–138. [CrossRef]
42. Neitsch, S.L.; Arnold, J.G.; Kiniry, J.R.; Srinivasan, R.; Williams, J.R. *Soil and Water Assessment Tool User's Manual, Version 2000*; Texas Water Resources Institute: College Station, TX, USA, 2000; pp. 1–27.
43. USGS The National Map-Advanced Viewer. Available online: https://apps.nationalmap.gov/viewer/ (accessed on 4 October 2022).
44. Global Data | SWAT | Soil & Water Assessment Tool. Available online: https://swat.tamu.edu/data/ (accessed on 4 October 2022).
45. CropScape-NASS CDL Program. Available online: https://nassgeodata.gmu.edu/CropScape/ (accessed on 4 October 2022).
46. Abbaspour, K.C. *Swat-Cup 2012. SWAT Calibration and Uncertainty Program—A User Manual*; Swiss Federal Institute of Aquatic Science and Technology: Dübendorf, Switzerland, 2013.
47. Sao, D.; Kato, T.; Tu, L.H.; Thouk, P.; Fitriyah, A.; Oeurng, C. Evaluation of Different Objective Functions Used in the SUFI-2 Calibration Process of SWAT-CUP on Water Balance Analysis: A Case Study of the Pursat River Basin, Cambodia. *Water* **2020**, *12*, 2901. [CrossRef]
48. Gupta, H.V.; Kling, H.; Yilmaz, K.K.; Martinez, G.F. Decomposition of the Mean Squared Error and NSE Performance Criteria: Implications for Improving Hydrological Modelling. *J. Hydrol.* **2009**, *377*, 80–91. [CrossRef]

49. Mearns, L.; McGinnis, S.; Korytina, D.; Arritt, R.; Biner, S.; Bukovsky, M.; Chang, H.; Christensen, O.; Herzmann, D.; Jiao, Y.; et al. 2017: The NA-CORDEX Dataset, Version 1.0. NCAR Climate Data Gateway, Boulder CO. 2017. Available online: https://na-cordex.org/ (accessed on 14 October 2022). [CrossRef]
50. Software | SWAT | Soil & Water Assessment Tool. Available online: https://swat.tamu.edu/software/ (accessed on 20 January 2021).
51. Rathjens, H.; Bieger, K.; Srinivasan, R.; Chaubey, I.; Arnold, J.G. CMhyd User Manual Documentation for Preparing Simulated Climate Change Data for Hydrologic Impact Studies. 2016. Available online: http://swat.tamu.edu/software/cmhyd (accessed on 15 October 2022).
52. Krause, P.; Boyle, D.P.; Bäse, F. Comparison of Different Efficiency Criteria for Hydrological Model Assessment. *Adv. Geosci.* **2005**, *5*, 89–97. [CrossRef]
53. Nash, J.E.; Sutcliffe, J.V. River Flow Forecasting through Conceptual Models Part I-A Discussion of Principles. *J. Hydrol.* **1970**, *10*, 282–290. [CrossRef]
54. Gupta, H.V.; Sorooshian, S.; Yapo, P.O. Status of Automatic Calibration for Hydrologic Models: Comparison with Multilevel Expert Calibration. *J. Hydrol. Eng.* **1999**, *4*, 135–143. [CrossRef]
55. Moriasi, D.N.; Gitau, M.W.; Pai, N.; Daggupati, P.; Gitau, M.W.; Member, A.; Moriasi, D.N. Hydrologic and Water Quality Models: Performance Measures and Evaluation Criteria. *Trans. ASABE* **2015**, *58*, 1763–1785. [CrossRef]
56. Brighenti, T.M.; Bonumá, N.B.; Grison, F.; de Almeida Mota, A.; Kobiyama, M.; Chaffe, P.L.B. Two Calibration Methods for Modeling Streamflow and Suspended Sediment with the SWAT Model. *Ecol. Eng.* **2019**, *127*, 103–113. [CrossRef]
57. Chow, V.T.; Maidment, D.R.; Mays, L.W. *Applied Hydrology McGraw-Hill International Editions*; McGraw-Hill Science/Engineering/Math: New York, NY, USA, 1988.
58. Subcommittee, H. *USGS Interagency Advisory Committee on Water Data*; Office of Water Data Coordination, US Geologic Survey: Reston, VA, USA, 1986. Available online: https://water.usgs.gov/osw/bulletin17b/dl_flow.pdf (accessed on 9 November 2022).
59. Mayo, T.L.; Lin, N. Climate Change Impacts to the Coastal Flood Hazard in the Northeastern United States. *Weather Clim. Extrem.* **2022**, *36*, 100453. [CrossRef]

climate

MDPI

Article

# Improving Future Estimation of Cheliff-Mactaa-Tafna Streamflow via an Ensemble of Bias Correction Approaches

Mohammed Renima [1], Ayoub Zeroual [2,*], Yasmine Hamitouche [2], Ali Assani [3], Sara Zeroual [4], Ahmed Amin Soltani [5], Cedrick Mulowayi Mubulayi [2,6], Sabrina Taibi [7], Senna Bouabdelli [2], Sara Kabli [2], Allal Ghammit [2], Idris Bara [2], Abdennour Kastali [2] and Ramdane Alkama [8]

1 Laboratory of Chemistry Vegetable-Water-Energy, Agronomy Department, University of Hassiba Benbouali, Chlef 02000, Algeria
2 Water Engineering and Environment Laboratory, National Higher School for Hydraulics (ENSH-Blida), Blida 09000, Algeria
3 Department of Environmental Sciences, University of Quebec at Trois-Rivières, 3351 Boulevard des Forges, Trois-Rivières, QC G9A 5H7, Canada
4 LEGHYD Laboratory, Faculty of Civil Engineering, University of Sciences and Technology Houari Boumediene (USTHB), BP 32 El Alia Bab Ezzouar, Algiers 16111, Algeria
5 VESDD Laboratory, Hydraulic Department, University of M'sila, M'sila 28000, Algeria
6 Department of Natural Resources Management, Faculty of Agronomic Sciences, University of Kinshasa, Kinshasa P.O. Box 1031, Democratic Republic of the Congo
7 LPPRE, Département des Sciences de l'Eau et l'Environnement, Université de Blida 1, Blida 09000, Algeria
8 European Commission, JRC, Directorate D-Sustainable Resources, Bio-Economy Unit, TP124 Via E. Fermi, 2749, 21027 Ispra, Italy
* Correspondence: zeroualayoub34@yahoo.fr

**Abstract:** The role of climate change in future streamflow is still very uncertain, especially over semi-arid regions. However, part of this uncertainty can be offset by correcting systematic climate models' bias. This paper tries to assess how the choice of a bias correction method may impact future streamflow of the Cheliff-Mactaa-Tafna (CMT) rivers. First, three correction methods (quantile mapping (QM), quantile delta mapping (QDM), and scaled distribution mapping (SDM)) were applied to an ensemble of future precipitation and temperature coming from CORDEX-Africa, which uses two Representative Concentration Pathways: RCP4.5 and RCP8.5. Then, the Zygos model was used to convert the corrected time series into streamflow. Interestingly, the findings showed an agreement between the three methods that revealed a decline in future streamflow up to [−42 to −62%] in autumn, [+31% to −11%] in winter, [−23% to −39%] in spring, and [−23% to −41%] in summer. The rate of decrease was largest when using QM-corrected model outputs, followed by the raw model, the SDM-corrected model, and finally, the QDM-corrected model outputs. As expected, the RCP presents the largest decline especially by the end of the 21st Century.

**Keywords:** future streamflow; semi-arid river; bias correction methods; RCP4.5 and RCP8.5

## 1. Introduction

Global climate change and drought events are likely to have a significant impact on water resources worldwide [1]. Access to drinking water for the greatest amount of the population, as well as securing this often over-exploited and poorly managed resource, due to the impacts of global warming [2,3], have an impact on hydrological cycles at multiple scales [4,5]. Thus, watershed hydrological behavior modelling is essential for anyone concerned with natural hazards (floods, drought, groundwater salinization), as well as the prospect of global warming in this century and beyond [6–8].

Global climate model (GCM) outputs are widely used to study not only the global climate system response to both natural and anthropogenic radiative forcing, but also to assess the impact of climate change on environmental systems (e.g., hydrological systems).

However, global models with resolutions ranging from 100 to 200 km are unable to capture sufficient detail at the global regional scale. Therefore, bias adjustment techniques should be used to adjust the biased outputs of global and regional models because the use of regional climate models (RCMs) or statistical downscaling of global models can provide reliable climate information at the regional and local scales [9]. For instance, the precipitation and temperatures simulated by these climatic models in hydrological assessments can have significant systematic biases when compared to observed data [10,11]. Faulty conceptualization, spatial averaging across grid cells, and discretization can all be blamed for systematic errors in climate model outputs [12]. As a result, almost all studies assessing the impact of climate change require bias adjustment as a post-processing step.

Several bias adjustment algorithms for removing systematic errors from model outputs have recently been described in the literature. These algorithms establish a statistical relationship between a modelled and an observed climate variable on a regional scale or specific site over a calibration period (historical). Such relationships are then applied on a grid cell per point basis to correct some aspects of the biases in future climate simulations. The algorithms are designed to adjust one of the climate variables' statistical properties, such as the mean, variance, quantiles, number of rain days, and so on.

In comparison to alternative algorithms, several recent studies [12–14] emphasize the quantile mapping (QM) fitting algorithm as the most appropriate algorithm [15,16]. The QM algorithm, on the other hand, can artificially alter raw climate change signals, distorting future model simulation trends. According to [17], this alternative to signals of climate evolution is mainly due to the assignment of the same cumulative distribution function (CDF) used for observations of future simulations even though this distribution may change in future projections. Furthermore, the QM algorithm assumes that the modelled and observed distributions' biases remain constant over time [18,19]. Grenier [19] mentioned that this algorithm can assign to a particular variable an impossible value. These flaws in the QM algorithm necessitate techniques that effectively preserve changes in the simulated quantiles, address the problem of simulating impossible situations, and deal with the difficulty of removing temporal pattern bias (e.g., interannual variability). One adaptation of the standard QM method that preserves the raw signals of climate change is the detrended QM method. It is a non-parametric method using empirical quantiles [13,20,21].

Previous research has shown that while this method preserves the trend on a monthly scale, it alters the signals of raw climate change on a daily scale [22,23]. Quantile delta mapping (QDM) is a recently introduced adjustment to the QM algorithm developed by [24]. The QDM algorithm is based on the delta method and the detrended QM method [24]. This method, according to [25], preserves the trend for all quantiles while modifying the extreme indices. Switanek et al. [18] recently developed a new method that is structurally similar to QDM and considers the frequency of many days, climatic indices, and the probability of specific events. The three algorithms have been evaluated in several studies for correcting model-simulated bias of temperature and precipitation data.

Although several bias corrections have implications when corrected variables are used to assess impacts at the local scale, researchers have always focused their studies on comparing the corrected variables from simulations to the observed variables. In this context, numerous studies in the literature have attempted to improve the potential of RCM simulations in the analysis of the impact of climate change on water resources by developing or comparing bias correction algorithms that are applicable to the meteorological parameters required for the hydrological model [12,26–29]. Mpelasoka & Chiew. [26] demonstrated that the effect of three bias correction algorithms, namely the delta change method, daily scaling, and daily translation (DT), on computing the mean annual runoff is very small. Van Roosmalen et al. [27] discovered similar results when comparing four bias correction algorithms. Nguyen et al. [30] highlighted the relevance of the multivariate frequency bias correction approach compared to traditional correction approaches for hydrological modelling.

Based on the above insights, the main objective of this paper is to study how these three bias correction algorithms modulate the climate change signal of precipitation and temperature over six mountainous watersheds of northwestern Algeria and the resulting impact on their runoff using a lumped conceptual Zygos hydrological model.

Although a number of studies have been conducted in Algeria to investigate the harmful effects of climate change on the availability of water resources and the identification of regional and local drought episodes [31,32], only a small number of studies have been conducted to investigate the future availability of water resources, particularly in northwestern Algeria regions, which already are experiencing water scarcity [31]. Existing studies have shown a drastic decrease in rainfall of about 30% in western Algeria, which caused serious hydrological crises and significantly affected the plains of this region known for its fertility [31–33]. Water stress has been experienced over the past few decades. In the Tafna and Macta basins in the country's extreme west, [33] discovered a significant dry trend in base flows between 1972 and 1992 on annual and seasonal scales of 50% to 71%. This reduction was marked on the basins located in plains as on the basins in relief (Beni-Bahdel, Pierre de chat, Chouly and Khemis), although they are the least disturbed by human activities given their strong gradient of altitude [34]. The western region of northern Algeria has been identified as a region prone to increasing temperatures and aridity in the future as a result of decreased precipitation and increased temperature.

## 2. Study Area

The research was based on three basins in Algeria's far northwest, which are located in the southern Mediterranean area, namely the Cheliff, Macta, and Tafna basins (Figure 1), in which the Cheliff basin occupies an area of 44,630 km$^2$ and is located between the geographic coordinates 34° to 36°30′ north latitude and 0° to 3°30′ east longitude, exhibiting the shape of an axe-blade running north–south. It has an arid to semi-arid Mediterranean climate in the south, with warm Saharan influences in the north and east, and a mild climate in the north and east. Precipitation is very regular in time and space, with two extreme zones: one is wet with an annual average of 524 mm to 658 mm, and the other is less rainy with an annual average of 350 mm. The Cheliff watershed is located in the semi-arid to moderately temperate climate zone, with an average inter-annual precipitation of 571 mm and average monthly temperatures ranging from 10° in January to 28° in July and August, with an average yearly temperature of 18 °C.

The Cheliff furrow is compartmentalized into three basins (higher, middle, and lower Cheliff) separated by tow thresholds such as bedrock, the threshold of Ain Defla and, therefore, the threshold of Oum D'rou further west. Many permeable geological formations contain groundwater; the oldest are assigned to the Jurassic period, and therefore, the most up-to-date correspond to the quaternary alluvium. Within the northern part of the study area, the two Tellian chains are poor resources, and it is difficult to take advantage of them directly; the permeable levels (limestone and sandstone) are generally less developed and encased in powerful formations that have a very low permeability.

### *2.1. The Basin of Macta*

The basin of Macta occupies a complete area of 14,410 km$^2$; its geographical position is between −1.25° west and 0.60° east in longitude and between 34° and 36° north in latitude. It is limited within the northwest by the mountain ranges of Tessala, within the south by the highlands of Maalif, within the west by the plateaus of Telagh, and within the east by the Saida Mountains [35]. The Macta basin is bordered in the north by the Mediterranean Sea, to the south by the mountains of Saida (1201 m) and the Daya Mountains (1356 m), and to the southwest by the mountains of Tlemcen, including the mountains of Beni Chougrane and, therefore, the plain of Mohamadia [35]. The typical annual rainfall ranges from 206 mm in the southern part to 380 mm in the Saïda Mountains. The mean monthly temperature ranges from 11° in January to 26 °C in July and August. The Macta watershed is drained by two major rivers, namely: El-Hammam Wadi in the east and Wadi Mekkera (called Wadi Mebtouh downstream) in the west [35].

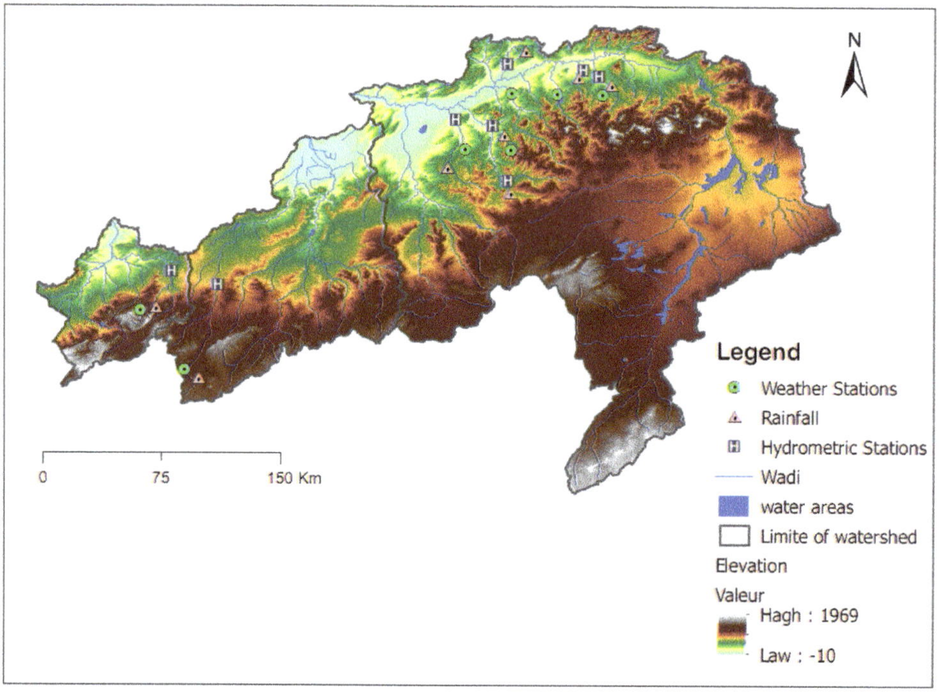


**Figure 1.** Cheliff-Mactaa-Tafna basin climate and streamflow stations.

### 2.2. The Basin of Tafna

The Tafna's catchment basin is located in Algeria's extreme northwest. One of the most important wadis in the west crosses it. Tafna, with over 6900 km$^2$, flows from west to east, from Morocco to the Mediterranean (near Beni Saf), and the length of the greatest river bed is 759 km (Figure 1). The basin is dominated to the south by a WSE–ENE-oriented mountainous bar (800–1400 m), while the plain areas of Maghnia, Hannaya, and Sidi Abdelli dominate to the north. This orographic structure, which is dominated in the north by the small-scale Traras mountains (1081 m a.s.l.), creates an effective precipitation barrier, explaining the aridity of the Maghnia plain [36]. The Tafna basin's climate is equivalent to that of the Northern Africa Mediterranean region, which is warm and humid, with the two hottest months being July and August, with an average temperature of 26 °C [33]. The Tafna River's hydrographic network consists mostly of two arteries: Wadi Tafna in the west and Wadi Isser in the east, and it originates in the Tlemcen Mountains.

## 3. Dataset Used

Various climate data were used for climate change impact assessment studies using a hydrological model at the monthly scale, including streamflow, precipitation, temperature, and evaporation.

### 3.1. Hydrological Model Data

The Zygos model requires the monthly time series of precipitation, potential evapotranspiration, runoff, as well as groundwater extraction (if these data are available) to simulate the watershed response. The Algerian National Hydraulic Resources Agency provided climate data for seven stations located within and near the Cheliff-Mactaa-Tafna

(CMT) basins from 1975 to 2012 (Table 1 and Figure 1). Nevertheless, the gaps in the overall precipitation and temperature time series do not exceed 8%. To detect outliers and fill gaps in the data series, three steps were used: visual inspection, comparison with the nearest station in the same area, and regression relationships between neighboring stations.

**Table 1.** Rainfall stations from the National Agency ANRH.

| Stations Code | Name Stations | Watershed | Latitude | Longitude | Measurement Period |
|---|---|---|---|---|---|
| PV011901 | El Touaibia | Cheliff | 1°94′ | 36°12′ | 1990–2012 |
| PV012004 | Tikezal | Cheliff | 1°75′ | 36°19′ | 1989–2012 |
| PV012201 | Larabaa Ouled Fares | Cheliff | 1°24′ | 36°24′ | 1971–2012 |
| PV012507 | Oued Lili | Cheliff | 1°26′ | 35°52′ | 1975–2005 |
| PV012703 | Kenanda Ferme | Cheliff | 0°82′ | 35°65′ | 1978–2005 |
| PV110102 | Ras Elma | Mactaa | −0°83′ | 34°46′ | 1980–2010 |
| PV160601 | Chouly | Tafna | −1°13′ | 34°86′ | 1975–2012 |

### 3.2. *Streamflow/River Discharge Data*

The Algerian National Hydraulic Resources Agency manages several gauging stations in the CMT basin, which are in small tributaries of the CMT River and cover small watersheds. All but eight of these stations (Table 2 and Figure 1) are outside the scope of this research analysis due to significant data gaps, and that the majority of stations have not been operational for an extended period.

**Table 2.** Characteristics of hydrometric stations.

| Station Code | Name Stations | Watershed | Wadi | Latitude | Longitude | Surface (km$^2$) | Measurement Period |
|---|---|---|---|---|---|---|---|
| Qm011905 | Bir Ouled Tahar | Cheliff | Zeddine | 36°19′ | 1°85′ | 450 | 1990–2008 |
| Qm012004 | Tikezal | Cheliff | Tikezal | 36°19′ | 1°75′ | 130 | 1990–2012 |
| Qm012201 | Larabaa Ouled Fares | Cheliff | Ouahrane | 36°22′ | 1°21′ | 262 | 1983–2011 |
| Qm012501 | Oued Lilli | Cheliff | Tiguiguest | 35°59′ | 1°24′ | 1612 | 1975–2006 |
| Qm012601 | Ammi Moussa | Cheliff | Rhiou | 35°86′ | 1°12′ | 1937 | 1975–2006 |
| Qm012701 | Djidiouia | Cheliff | Djidiouia | 35°92′ | 0°88′ | 836 | 1975–2006 |
| Qm110101 | Haciabia | Mactaa | Mekerra | 34°69′ | −0°75′ | 941 | 1980–2001 |
| Qm160601 | Chouly | Tafna | Chouly | 34°86′ | −1°13′ | 167 | 1975–2006 |

### 3.3. *Climate Scenario Data and Bias Correction Method*

To assess the impact of climate change on the Cheliff-Mactaa-Tafna (CMT) basins' hydrology, the monthly precipitations and monthly temperature data simulated from the Rossby Centre Regional Climate Model (RCA4) driven by the MPI-ESM-LR General circulation model from the Coupled Model Intercomparison Project—Phase 5 (IPCC5) [37], available within the CORDEX project, were extracted for each climate station during the 1971–2100 period. The entire dataset simulated spans 1971–2100, consisting of a historical period (1971–2005) and two projection periods (2025–2050 and 2075–2100). The projection period was forced by two Representative Concentration Pathway (RCP) scenarios, RCP4.5 and RCP8.5. The data extracted from the RCA4-MPI-ESM-LR climate model (monthly precipitations, monthly temperature) deviate from the data observed at the climate stations. Therefore, bias adjustment is a required post-processing step in almost all studies assessing the impact of climate change. We adjusted the data bias generated from the RCA4-MPI-ESM-LR regional climate model using three methods: quantile mapping, scaled distribution mapping, and quantile delta mapping before predicting the possible change in the future hydrology of the Cheliff-Mactaa-Tafna (CMT) basins. Afterwards, the same procedure that was used in the historical period for validation was applied for the two future periods for discharge simulation. The three bias correction algorithms are summarized in the Supplementary Material S3 to avoid filling up the text with formulas.

### 3.4. Hydrological Modeling Using Zygos

The Zygos model is a conceptual rainfall–runoff modeling tool that employs a series of reservoirs (Figure 2) to represent the soil and subsoil schematically. The total flow is composed by four principal components [38] including the direct discharge, QDt, caused by the presence of impervious formations, through which the proportion of rainfall is transformed directly into runoff; the surface discharge, QQt, which results from an immediate reaction due to soil saturation; the subsurface discharge, QIt, which is a slow response caused by the lateral (horizontal) movement of water, infiltrating into the soil; and the base discharge, Qbt, being the lower soil layers' (aquifer) response, by means of employing springs. Simulating subwatershed flow is a similar approach to Thornthwaite's model [38] (National Technical University of Athens (NTUA) Research Team, 2010). It is a lumped conceptual water balance model that runs on a monthly time step in most cases. On the one hand, the model's input data are monthly time series of precipitation Pt, potential evapotranspiration Ept, runoff Qt, and the extraction rate from groundwater PUMPt; on the other hand, the model's outputs are runoff at the watershed outlet Qt (surface water and groundwater), actual evapotranspiration Et, and the watershed outlets. A detailed description of this model is presented in Charizopoulos and Psilovikos. [39], among others.

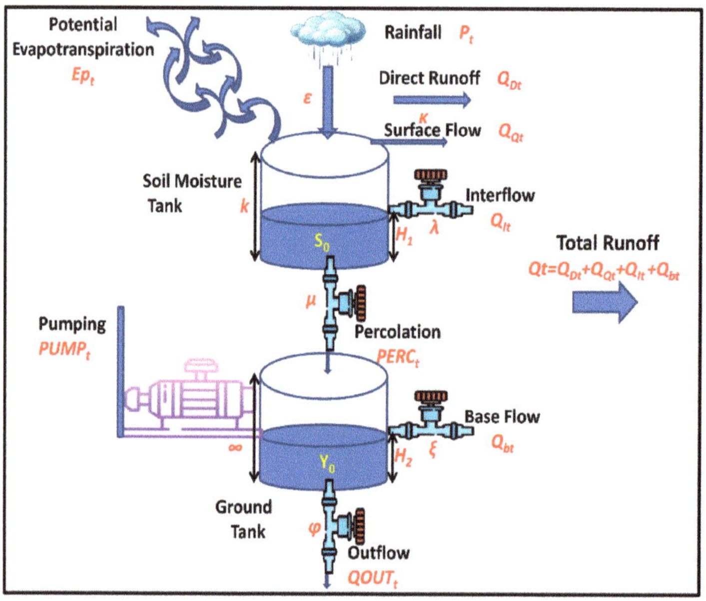


**Figure 2.** Schematic representation of the hydrologic processes of the conceptual hydrological model Zygos (Adapted from Kozanis et al. [38]).

### 3.5. Parameter's Description

The Zygos model includes eleven parameters (Table 3) that define the flow distribution or reservoir characteristics (initial level and capacity, $H_1$ or $H_2$ values) [39]. This may result in the possibility of over-parameterization of the same hydrosystem using Zygos, as well as the ability of compensation between the different models' parameters used to model the output data. Several parameters can take values corresponding to the limit values allowed in calibration [39]. The model's state variables are soil moisture and groundwater storage, which require information about the initial conditions S0, K, and Yo, respectively. To avoid weighing down the text with formulas, the descriptions of the model parameters are summarized in Table 3 [39]. The detailed mathematical description of the model operation is presented in Charizopoulos and Psilovikos [39] and Charizopoulos et al. [40].

**Table 3.** Parameters of the manual calibration of the rainfall–runoff model Zygos (Reprinted from Kozanis et al. 2010 [38] and Charizopoulos and Psilovikos [39]).

| Parameters | Description |
|---|---|
| $\varepsilon$ | Rainfall proportion available for the achievement of direct evapotranspiration. |
| $\kappa$ | The rainfall excess proportion, appearing as direct runoff, caused by the occurrence of impermeable formations. Through them, the rainfall proportion is transformed directly into runoff. Essentially, it is the percentage of impermeable surface and expresses the percentage that runs off directly without percolating the soil. |
| k | The capacity of the soil moisture tank, which expresses the maximum storage capacity of the ground (mm). |
| So | Initial reserve of the soil moisture. |
| $\lambda$ | Discharge rate of the soil moisture tank, for the creation of subsurface flow. |
| $H_1$ | Reserve threshold of the soil moisture tank, for the creation of subsurface flow. |
| $\mu$ | Discharge rate of the soil moisture tank, for the creation of infiltration. |
| $\xi$ | Discharge rate of the groundwater tank, for the creation of base flow. |
| $H_2$ | Reserve threshold of the groundwater tank, for the creation of base flow. |
| $\varphi$ | Discharge rate of the groundwater tank, for the creation of subsurface outflow. |
| Yo | Initial reserve of the groundwater tank. |

### 3.6. Performance Criteria

Each reservoir represents an essential physical process carried out during the water flow within the watershed. The Nash coefficient [41] was selected as the criterion for evaluating the error between simulated and observed runoff. It is a transformed and normalized measure of the RMSE overall (normalized to the variance of the observed hydrograph) [41]. The Nash–Sutcliffe coefficient (NSE) is considered a typical indicator of a good fit for hydrologic models, given by Equation (1).

$$\mathrm{NSE} = 1 - \frac{\sum_{i=1}^{N}(\mathrm{Qo} - \mathrm{Q_E})^2}{\sum_{i=1}^{N}(\mathrm{Qo} - \mathrm{Qoave})^2} \tag{1}$$

where: $Q_o$ = the observed runoff, $Q_E$ = the runoff estimated by the model, $Q_{oave}$ = the mean value of observed runoffs, N = the total number of observations. The coefficient value ranges from $-\infty$ to 1. If $R < 0$, the fit of the model is considered poor, whereas when the value is near 1, the simulated time series provides a better fit compared to the mean observed value $Q_{oave}$.

## 4. Results

The Zygos model must be tested before being used for research or operational reasons. This procedure is known as model calibration/validation, and it was used for each basin in the study area. The Zygos model was automatically calibrated and validated using the reference periods' monthly precipitation, evaporation, and discharge data; the results are given in Table 4.

Validation was performed for the stations of Bir Ouled Tahar, Ammi Moussa, and Chouly from 2003 to 2008, 1998 to 2004, and 1997 to 2004, respectively. For the two phases "calibration/validation", the Pearson correlation coefficient ranged between 0.66 and 0.95.

The average size of these basins explains the model's good fit to the observed data. In terms of the general dynamics of river flows at the stations of Larabaa Ouled Fares, Oued Lilli, and Djediouia, we found that this dynamic is well reproduced because the validation yielded Pearson correlation coefficients ranging from 0.45 to 0.68, reflecting the model's average quality in reproducing the hydrological reality that characterizes these basins. Furthermore, the short water path results in a quick concentration time. This reduces losses due to seepage, evaporation, and absorption.

**Table 4.** Simulation parameters.

| | Stations | Ammi Moussa | Chouly | Djediouia | Haciaba | L. Ouled Fares | Oued Lilli | Tikezal | Bir Ouled Tahar |
|---|---|---|---|---|---|---|---|---|---|
| Calibration | Period | 1980–1997 | 1979–1996 | 1979–1996 | 1980–1995 | 1983–2000 | 1979–1996 | 1990–2004 | 1990–2002 |
| | NSE (calibration) | 0.56 | 0.98 | −0.84 | 0.16 | −5.81 | 0.55 | 0.62 | 0.60 |
| | RMSE | 40.67 | 3.61 | 9.29 | 5.53 | 60.92 | 4.67 | 4.29 | 24.91 |
| Zygos modelparameters | κ | 0.247 | 0.04 | 0.154 | 0.013 | 0.694 | 0.131 | 0.01 | 0.023 |
| | μ | 0.023 | 0.99 | 0.22 | 0.839 | 0.4 | 0.188 | 0.886 | 0.017 |
| | ε | 0.547 | 0.99 | 0.99 | 0.099 | 0.01 | 0.399 | 0.813 | 0.189 |
| | $H_1$ | 39.42 | 133.99 | 13.52 | 101.23 | 3.80 | 40.09 | 6.00 | 0.74 |
| | $H_2$ | 68.99 | 96.88 | 263.29 | 158.71 | 5.00 | 115.60 | 72.23 | 60.40 |
| | λ | 0.104 | 0.889 | 0.145 | 0.378 | 0.99 | 0.318 | 0.99 | 0.029 |
| | ξ | 0.341 | 0.225 | 0.77 | 0.659 | 0.699 | 0.89 | 0.99 | 0.63 |
| | φ | 0.01 | 0.03 | 0.35 | 0.23 | 0.01 | 0.03 | 0.02 | 0.15 |
| | k | 120.28 | 156.9 | 111.85 | 182.73 | 170.14 | 100.01 | 195.61 | 111.28 |
| | So | 14.62 | 17.68 | 11.11 | 16.6 | 20.26 | 10.37 | 9.61 | 8.39 |
| | Yo | 5.09 | 5.00 | 226.84 | 271.23 | 295.31 | 20.03 | 290.00 | 116.20 |
| | Objective function | 0.594 | 0.767 | 0.41 | 0.071 | 0.0736 | 0.696 | 0.018 | 0.171 |
| Validation | Period | 1998–2004 | 1997–2004 | 1997–2004 | 1996–2001 | 2001–2007 | 1997–2004 | 2005–2011 | 2003–2008 |
| | NSE (validation) | 1.00 | 0.96 | −1.48 | 0.85 | −0.45 | 0.02 | 0.37 | 0.29 |
| | RMSE | 0.84 | 2.56 | 10.65 | 0.82 | 27.19 | 3.42 | 23.34 | 42.36 |

### 4.1. Parameter of Simulation

The results of rainfall–runoff modeling can be more dependent on the quality of the input data than on the model [42]. The optimal value of the Nash coefficient (Equation (1)) or the objective function that expresses the differences between simulated and observed values was varied between −2.74 and 0.77, and it was reached with the combination of the parameters presented in Table 4. This value is high, indicating that the simulated runoff adapted well to the measured runoff. Because of variations in altitude and average yearly temperatures, the proportion of rain available for direct evapotranspiration, indicated by the parameter ε, varied (from 0.01 to 0.99) from one basin to another. Due to the location of the area study in the most extensive karst system in northern Algeria and which presents the widest natural groundwater reservoir in the west (S0: 8.39 to 20.26 mm), the karst aquifers of this region are considered as the largest natural reservoirs of rainfall in north Algeria. The direct runoff, maximum storage capacity, subsurface flow (groundwater), and discharge rate from the soil moisture reservoir for the development of infiltration, as expressed by the parameters κ, k, λ, $H_1$, and μ, suggest that runoff is superior to infiltration. This is related to the occurrence of semi-permeable formations in the area. In addition, the parameter ξ varies from 0.225 to 0.99 and $H_2$ from 5 to 263.29 mm, which are crucial in generating the base flow. The outflow coefficient φ (0.01 to 0.35) is related to the karstic mass of Chlef-Relizane, Saida, and Telemcen, which flows towards the plains of Cheliff, Ghriss-Mascara, and Maghnia.

### 4.2. Changes in Evapotranspiration

The projected temperature increase in the Cheliff, Mactaa, and Tafna catchments would result in an increase in ETP from 33% to 41% and from 29% to 38% by 2050 for the RCP4.5 and RCP8.5 scenarios, respectively, and from 47% to 57% and 77% to 94% by 2100 for the two scenarios RCP4.5 and RCP8.5 (Table S1 and Figure 3). The largest rise is reported by the stations of Ammi Moussa, Oued Lilli, and Haciaba, with 41% by 2050 and 94% by 2100. In general, the average rate of rise for the 2050 and 2100 timeframes ranges from 34% to 37% and from 53% to 85%, respectively.

### 4.3. Projected Precipitation

Future precipitation data for Bir Ouled Tahar, Tikezal, Larabaa Ouled Fares, Ammi Moussa, Oued Lilli, Kenanda Farme, Ras Elma, and Chouly in the CMT basins were corrected using three bias correction techniques: the QM, SDM, and QDM algorithms. In our study, the RCP4.5 and RCP8.5 climate scenarios were evaluated for two future periods, 2050 and 2100, compared to the reference period (1975–2012). After adjusting the biases of the raw precipitation sample, the study found that average annual precipitation in the CMT basins would generally decline in the future (Table 5 and Figure S1).

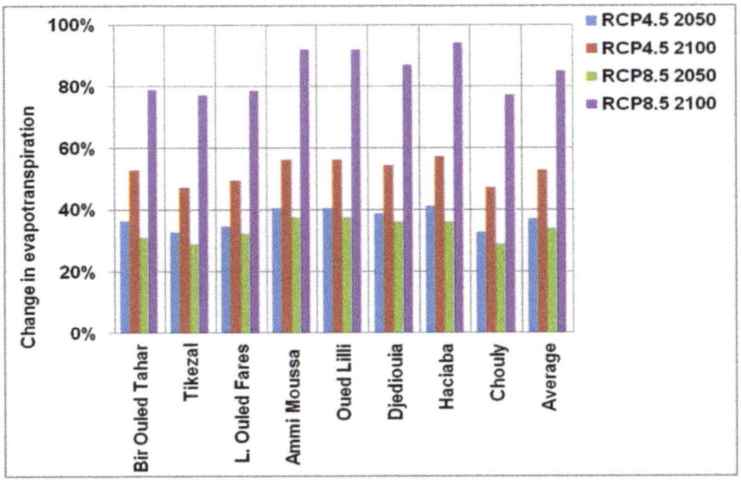


**Figure 3.** Changes in evapotranspiration under two scenarios (RCP4.5 and RCP8.5).

**Table 5.** Annual precipitation mean changes in the CMT basins from the baseline period (1975–2012).

| STATIONS | Quantile Mapping (QM) | | | | Scaled Distribution Mapping (SDM) | | | |
|---|---|---|---|---|---|---|---|---|
| | RCP4.5 2050 | RCP4.5 2100 | RCP8.5 2050 | RCP8.5 2100 | RCP4.5 2050 | RCP4.5 2100 | RCP8.5 2050 | RCP8.5 2100 |
| Bir Ouled Tahar | −27% | −30% | −31% | −51% | −22% | −25% | −16% | −38% |
| Tikezal | −57% | −55% | −44% | −53% | −42% | −38% | −25% | −37% |
| Larabaa Ouled Fares | −16% | −15% | −15% | −40% | −17% | −14% | −14% | −30% |
| Ammi Moussa | −20% | −30% | −25% | −48% | −5% | −17% | −16% | −27% |
| Oued Lilli | −16% | −27% | −27% | −49% | −3% | −6% | −15% | −27% |
| Djediouia | −22% | −28% | −27% | −37% | −17% | −30% | −12% | −20% |
| Haciaba | −42% | −41% | −46% | −54% | −23% | −30% | −33% | −41% |
| Chouly | −18% | −30% | −29% | −47% | 0% | −6% | −11% | −24% |
| **STATIONS** | **Quantile Delta Mapping (QDM)** | | | | **Model (RAW)** | | | |
| | **RCP4.5 2050** | **RCP4.5 2100** | **RCP8.5 2050** | **RCP8.5 2100** | **RCP4.5 2050** | **RCP4.5 2100** | **RCP8.5 2050** | **RCP8.5 2100** |
| Bir Ouled Tahar | −9% | −12% | −15% | −38% | −24% | −26% | −24% | −46% |
| Tikezal | −43% | −39% | −27% | −38% | −50% | −48% | −35% | −47% |
| Larabaa Ouled Fares | −11% | −11% | −11% | −38% | −21% | −28% | −27% | −47% |
| Ammi Moussa | −12% | −21% | −16% | −37% | −9% | −19% | −20% | −40% |
| Oued Lilli | −4% | −16% | −16% | −37% | −9% | −19% | −20% | −40% |
| Djediouia | −16% | −9% | −22% | −32% | −13% | −23% | −25% | −40% |
| Haciaba | 51% | 67% | 64% | 66% | −15% | −16% | −20% | −27% |
| Chouly | 71% | 56% | 58% | 35% | −21% | −23% | −21% | −41% |

For the QM-RCP4.5 climate scenario, the change in mean annual precipitation ranges from −15 to −57%, while for the QM-RCP8.5 high-level climate scenario, the change is from −15 to −54%. For the RCP4.5 climate scenario, decreases range from −3% to 42%, whereas for the RCP8.5 climate scenario, decreases range from −11% to −41%. For the

RCP4.5 climate scenario, the correction via the QDM approach revealed decreases ranging from −4% to −43% and increases up to +71%. The RCP8.5 climatic scenario showed similar results, with decreases ranging from −11% to −38% and increases of up to +66%. Finally, over all future periods, the decreases in raw climate model outputs range from −9% to −50% for the RCP4.5 climate scenario and from −20% to −47% for the RCP8.5 climate scenario (Table 5 and Figure S1).

### 4.4. Streamflow Projected

The Zygos hydrological model calibrated and validated at the monthly scale for the reference period (1975–2012) was used to simulate future flows in the CMT basin for two future periods under two scenarios. The outputs of the Zygos model were compared to those of the reference period for the two projected periods 2050 (2025–2050) and 2100 (2075–2100) under the two climatic scenarios. The two climate scenarios resulted in moderate reductions in the average annual deficit for all future periods, due to the anticipated decrease in precipitation. However, the RCP4.5 and RCP8.5 climate scenarios for the year 2100 showed a decrease in the deficit for all methods due to a projected increase in temperature, which causes an increase in evaporation rather than a decrease in precipitation, except for the QDM method, which showed an increase in the deficit (Table 6; Figure S2). We detected a trend of decreasing average monthly flows in the RCP4.5 scenario and a mixed pattern in the RCP8.5 scenario for the two periods analyzed and all stations studied. The change rate in annual mean flows spans from −91% to −3% for the QM-RCP4.5 climate scenario and from −92% to −6% for the QM-RCP8.5 climate scenario.

**Table 6.** Annual streamflow mean changes in the CMT basins from the baseline period (1975–2012).

| STATIONS | Quantile Mapping (QM) | | | | Scaled Distribution Mapping (SDM) | | | |
|---|---|---|---|---|---|---|---|---|
| | RCP4.5 2050 | RCP4.5 2100 | RCP8.5 2050 | RCP8.5 2100 | RCP4.5 2050 | RCP4.5 2100 | RCP8.5 2050 | RCP8.5 2100 |
| Bir Ouled Tahar | −10% | −3% | −27% | −51% | −18% | −19% | −6% | −44% |
| Tikezal | −90% | −91% | −36% | −32% | −38% | −52% | 27% | 42% |
| Larabaa Ouled Fares | −7% | −4% | −6% | −28% | −10% | −7% | −3% | −21% |
| Ammi Moussa | −22% | −23% | −9% | −50% | −16% | −15% | −6% | −31% |
| Oued Lilli | −27% | −23% | −24% | −48% | −9% | −18% | −8% | −29% |
| Djediouia | −18% | −19% | −7% | −11% | −18% | −24% | −2% | −7% |
| Haciaba | −75% | −75% | −77% | −80% | −69% | −71% | −73% | −75% |
| Chouly | −61% | −58% | −60% | −92% | −55% | −35% | −58% | −93% |

| STATIONS | Quantile Delta Mapping (QDM) | | | | Model (RAW) | | | |
|---|---|---|---|---|---|---|---|---|
| | RCP4.5 2050 | RCP4.5 2100 | RCP8.5 2050 | RCP8.5 2100 | RCP4.5 2050 | RCP4.5 2100 | RCP8.5 2050 | RCP8.5 2100 |
| Bir Ouled Tahar | 21% | 24% | 26% | −35% | −1% | −1% | −5% | −53% |
| Tikezal | −36% | −77% | 46% | 22% | −88% | −91% | −22% | −41% |
| Larabaa Ouled Fares | −1% | 0% | −2% | −25% | −13% | −18% | −44% | −37% |
| Ammi Moussa | −14% | −17% | 5% | −39% | −5% | −5% | −6% | −47% |
| Oued Lilli | −20% | −19% | −11% | −36% | −23% | −19% | −31% | −50% |
| Djediouia | −4% | −72% | 5% | 6% | −10% | −22% | −27% | −58% |
| Haciaba | −43% | −37% | −38% | −38% | −66% | −66% | −68% | −70% |
| Chouly | 118% | 63% | −70% | −55% | −70% | −45% | −52% | −92% |

The decreases for the SDM technique range from −71% to −7% for the RCP4.5 climate scenario and from −93% to −2% for the RCP8.5 climate scenario. For this last scenario, an increase in flow of around +42% is expected in the Tikezal station. In addition, for the RCP4.5 climatic scenario, the QDM approach revealed drops in annual mean flows ranging from −77% to −1% and rises to +118%. The same results were reported for the RCP8.5 climatic scenario, with losses ranging from −2 to −70% and gains reaching +46%. Finally, for the raw climate model output, flow decreases range from −1% to −91% for the RCP4.5 climate scenario and from −92% to −5% for the RCP8.5 climate scenarios for all future periods (Table 6 and Figure S2).

### *4.5. Projected Season Precipitation*

To evaluate the impact of climate change seasonal runoff, we established the seasons of autumn from September to November, winter from December to February, spring from March to May, and summer from June to August in this study. Under both scenarios (RCP4.5 and RCP8.5), the outputs from the three corrected techniques QM, SDM, and QDM, as well as the uncorrected version of the models (raw) showed that precipitation in the study region will continue to decline during the period 2075–2100. The decrease in precipitation will be more important in the RCP8.5 scenario than under the RCP4.5 scenario (Figures 4–7 and Tables S2–S5). According to the RCP 4.5 scenario (Figures 4–7 and Tables S2–S5), the MPI model predicted a decrease in precipitation between −17 and −58% in autumn, −3 and −56% in winter, and −25 and −74% in spring by the end of the 21st Century, compared to the QM method.

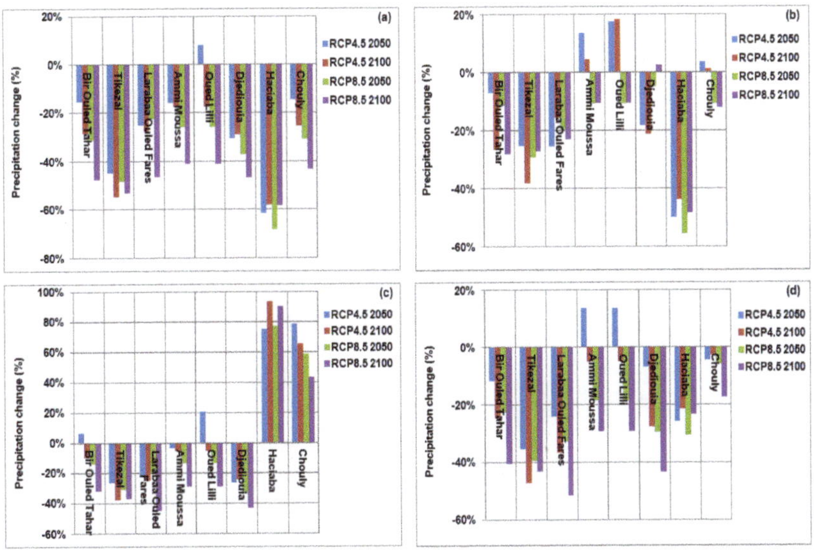


**Figure 4.** Changes in autumn precipitation relative to the historical period under two scenarios (RCP4.5 and RCP8.5): (**a**) quantile mapping, (**b**) scaled distribution mapping, (**c**) quantile delta mapping, and (**d**) model (RAW).

The SDM approach showed a decrease in precipitation ranging from 1% to −44% in autumn, from −18% to −43% in winter, and from −14% to −58% in spring. The QDM approach, on the other hand, showed a smaller rate of change (increase and decrease) of precipitation than the other methods, with this rate varying between −38% and 93% in autumn, −43% and 17% in winter, and −41% and 89% in spring. The pessimistic scenario RCP 8.5 reduced the precipitation more than the pessimistic scenario RCP4.5. The MPI model likewise showed larger precipitation reductions for the period 2075–2100, ranging from −41 to −59% in autumn, from −17 to −70% in winter, and from −54 to −94% in spring for the QM technique. The SDM approach (Figures 4–7 and Tables S2–S5) revealed a rate of change in precipitation ranging from 3% to −48% in autumn, from −24% to −48% in winter, and from −37% to −80% in spring. The QDM approach, on the other hand, showed the smallest decrease in precipitation when compared to the other two methods. This rate of decrease/increase fluctuated between −45% and 91% in autumn, −33% and

98% in winter, and −61%t and 17% in spring. It should also be noticed that the simulated precipitation after bias correction is much higher than the raw data for the prediction periods 2025–2050 and 2075–2100. This is due to the model data being corrected in relation to the observed precipitation, which was significantly underestimated by the MPI model during the historical period.

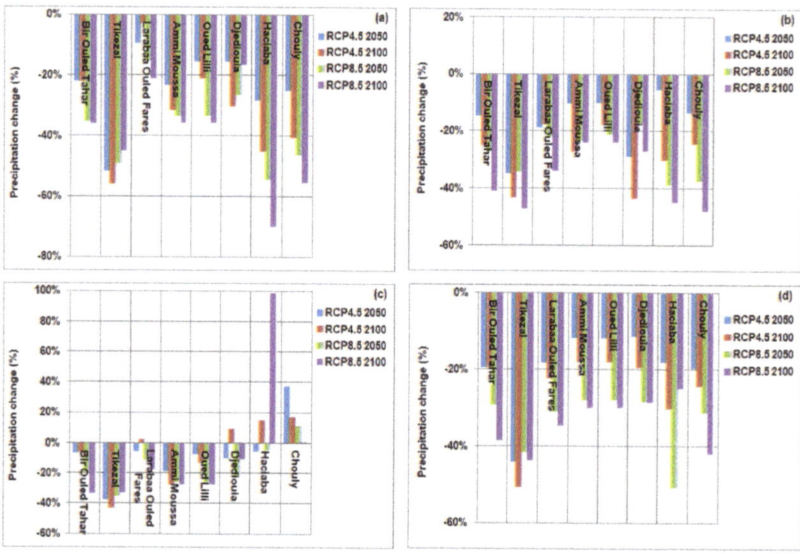


**Figure 5.** Changes in winter precipitation relative to the historical period under two scenarios (RCP4.5 and RCP8.5): **(a)** quantile mapping, **(b)** scaled distribution mapping, **(c)** quantile delta mapping, and **(d)** model (RAW).

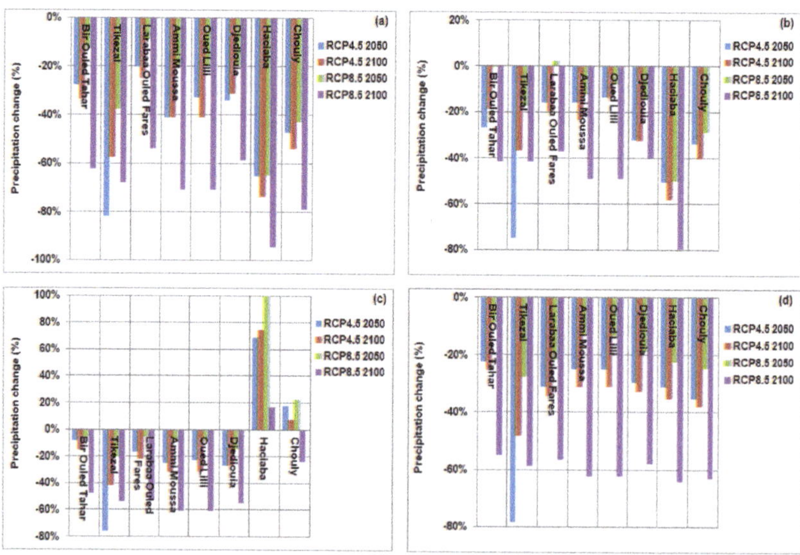


**Figure 6.** Changes in spring precipitation relative to the historical period under two scenarios (RCP4.5 and RCP8.5): **(a)** quantile mapping, **(b)** scaled distribution mapping, **(c)** quantile delta mapping, and **(d)** RAW.

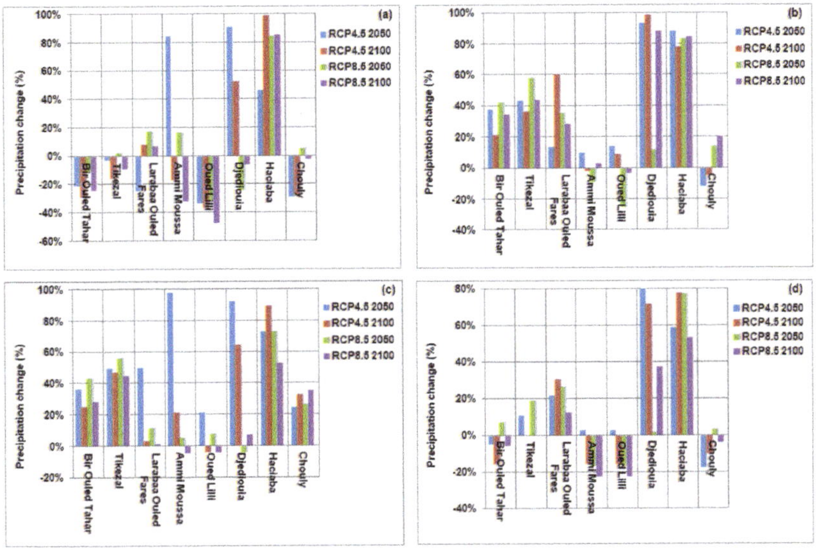


**Figure 7.** Changes in summer precipitation relative to the historical period under two scenarios (RCP4.5 and RCP8.5): (**a**) quantile mapping, (**b**) scaled distribution mapping, (**c**) quantile delta mapping, and (**d**) RAW.

### *4.6. Seasonal Streamflow Projections*

Tables S6–S9 show the percentage change rates of seasonal flow in the CMT basin simulated by the Zygos hydrological models based on the outputs corrected by the three methods QM, SDM, and QDM and the raw model data MPI-ESM-LR model under the RCP4.5 and RCP8.5 scenarios for the two future periods 2025–2050 and 2075–2100. The results of the simulations of the three methods QM, SDM, and QDM, as well as the raw model output showed that the seasonal flow has a tendency to decrease during the period 2075–2100 under the two scenarios and during all seasons except summer under the RCP4.5 and RCP8.5 scenarios.

Under the RCP4.5 climate scenario, the rate of change of autumn flows (Table S6 and Figure 8) varies from +20% to −91% in the short term (2050s) and from −13% to −92% in the long term (2100s). The rate of change under the RCP8.5 scenario, on the other hand, ranges from −9% to −91% by 2050 and from −35% to −92% over the 2100 horizon. Under the RCP4.5 scenario, the adjustment using the same algorithm (QM) revealed a rate of change in winter (Figure 9)ranging from −89% to 60% and from −90% to +29% by 2100. With the RCP8.5 climate scenario, this rate of change is somewhat positive, ranging from −72% to +71% and −84% to 79% by 2100. The rate of change for future spring flows (Table S8 and Figure 10) varies between −94% and 63 percent for the two climatic scenarios. Using the QM-algorithm-corrected models, an increase in summer flows (Table S9 and Figure 11) was found at three stations. This increase varied between +12% and +168% for the two time periods (2050 and 2100) and two scenarios. Summer flow is expected to decrease in the other stations (reaching −100 percent).

Under the RCP4.5 climate scenario, the rate of change of the predicted autumn flows varies between −89% and +12% in the near term (2050s) and +4% to −91% in the long term (2100s) when utilizing precipitation data corrected by the SDM approach (Table S6 and Figure 8). The rate of change in the RCP8.5 scenario, on the other hand, ranges from −82% to −5% and from −3% to −89% by 2100. The correction using the same algorithm (SDM) (Table S7 and Figure 9) reveals a rate of change in winter ranging from −47% to +46% and from −63% to +30% by 2100 under the RCP4.5 scenario. With the RCP8.5 climate scenario, this rate of change is somewhat positive, ranging from −42% to +105%

and −88% to +48% by 2100. Using the SDM-algorithm-corrected models, an increase in future summer flows was observed at two stations (Ammi Moussa and Oued Lilli station) (Table S9 and Figure 11). The rate of change for future spring flows (Table S8 and Figure 10) varies between −98 percent and +91 percent for the two climate scenarios. This rise ranges between +12% and +181% for the two time periods (2050 and 2100) and two scenarios. Summer flow is expected to fall at the other stations (varying between −6 and −99%).

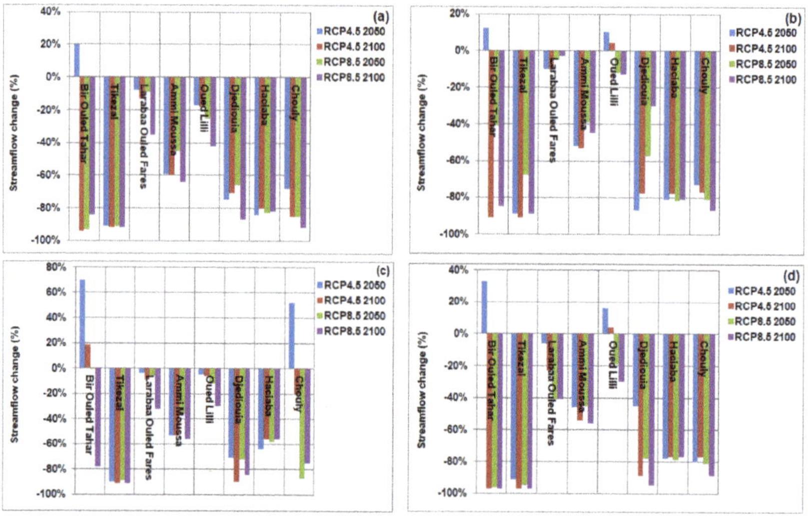


**Figure 8.** Changes in autumn streamflow relative to the historical period under two scenarios (RCP4.5 and RCP8.5): (**a**) quantile mapping, (**b**) scaled distribution mapping, (**c**) quantile delta mapping, and (**d**) model RAW.

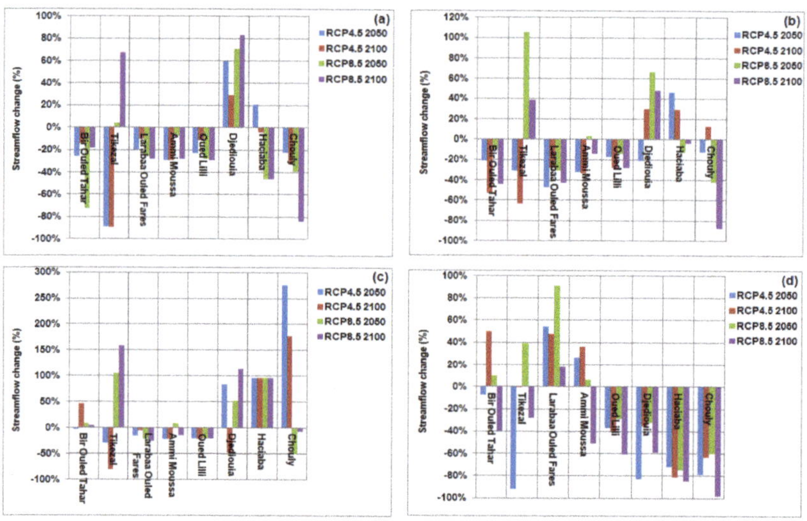


**Figure 9.** Changes in winter streamflow relative to the historical period under two scenarios (RCP4.5 and RCP8.5): (**a**) quantile mapping, (**b**) scaled distribution mapping, (**c**) quantile delta mapping, and (**d**) model RAW.

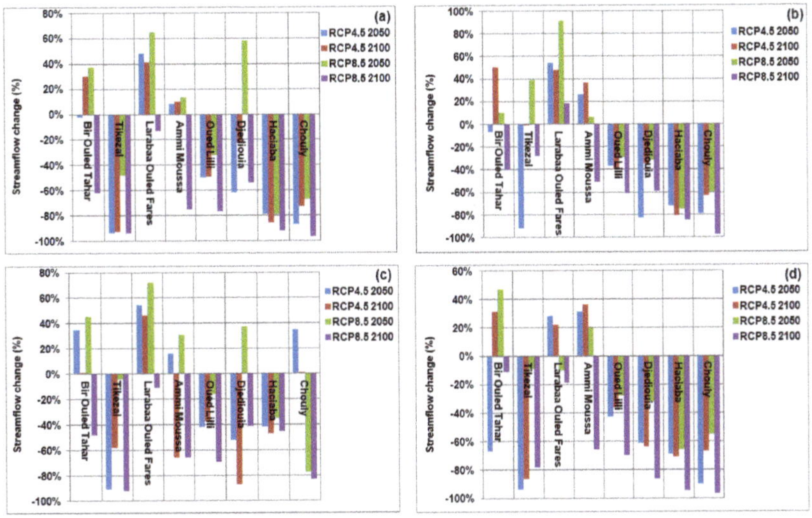


**Figure 10.** Changes in spring streamflow relative to the historical period under two scenarios (RCP4.5 and RCP8.5): (**a**) quantile mapping, (**b**) scaled distribution mapping, (**c**) quantile delta mapping, and (**d**) model RAW.

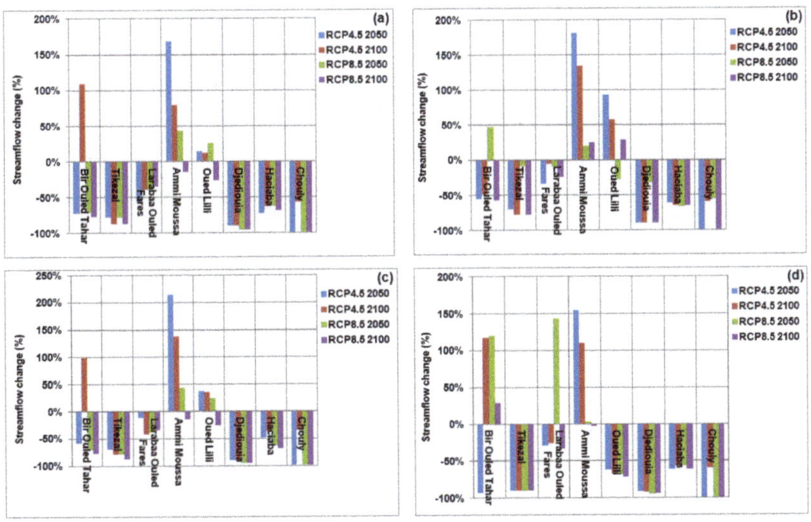


**Figure 11.** Changes in summer streamflow relative to the historical period under two scenarios (RCP4.5 and RCP8.5): (**a**) quantile mapping, (**b**) scaled distribution mapping, (**c**) quantile delta mapping, and (**d**) model RAW.

Under the RCP4.5 climate scenario, the rate of change of the predicted autumn flows ranges from −90% to +70% in the near term (2050s) and from −91% to +18% in the long term (2100s) when utilizing precipitation data corrected using the QDM approach (Table S6 and Figure 8). The rate of change under the RCP8.5 scenario, on the other hand, ranges from −1% to −89% and from −30% to −91% by 2100. Correction using the same algorithm (QDM) (Table S7 and Figure 9) revealed a rate change in winter ranging from −29% to +275% and from −80% to +177% by 2100 under the RCP4.5 scenario. With the RCP8.5 climate scenario, this rate of change is somewhat positive, ranging from −51% to

+105% and −25% to +158% by 2100. For the two climate scenarios, the rate of variation of future spring streamflow (Table S8 and Figure 10) varies from −92% to +72%. Using the QDM-algorithm-corrected models, an increase in summer flows (Table S9 and Figure 11) was found at two stations (Ammi Moussa and Oued Lilli station). This rise varies between +17% and +214% for the two periods (2050 and 2100) and the two scenarios. Summer flow is expected to decrease at the other stations (varying between −99 to −1%).

Finally, using raw precipitation data (Table S6 and Figure 8), the rate of change of expected autumn flows ranges between −91 and +33% in the near term (2050s) and −97 and +4% in the long term (2100s) under the RCP4.5 climate scenario. The change rate in the RCP8.5 scenario, on the other hand, fluctuates from −96% to −19% and from −97% to −30% over the 2100 horizon. Using raw data (Table S7 and Figure 9), the change rate in winters under the RCP4.5 scenario varies from −86 to +58% and −90 to +63% by 2100. With the RCP8.5 climate scenario, this goes from −74% to +53% by 2050 and −85% to +37% by 2100; this rate of change is slightly positive. For the two climate scenarios, the rate of variation of future spring streamflow (Table S9 and Figure 10) ranges from −94% to +47%. Using the QDM-algorithm-corrected models, an increase in summer flows (Table S9 and Figure 11) was found at three stations. This increase varies between +3% and +143% for the two time periods (2050 and 2100) and the two scenarios. Summer flow is expected to decrease at the other locations (varying between −3 and −99 percent).

In comparison with the corrective methods, a decrease in future flows in autumn, winter, and spring is projected in all of the watersheds analyzed; however, the rates of decrease are not synchronous. The rate of decrease is greatest when utilizing model outputs corrected by the QM approach, then raw model data, model outputs adjusted by the SDM method, and finally, model outputs corrected by the QDM method. In the summer, the average rate for all stations indicates an increase in future flows, albeit this rise is greater when model outputs corrected by the QDM approach are used than when the SDM and QM methods are used. The raw model data yielded the lowest rate of increase, followed by that corrected by the QM approach.

## 5. Discussion

The Zygos model was used in this study to investigate the impact of climate change on the CMT watersheds' water resources. The calibration/validation of the Zygos components on the datasets observed for the CMT basins reference period (1975–2012) allowed us to predict future flow evolution. After calibrating and validating the model by adjusting the model parameters, precipitation, and evapotranspiration, time series data from the Rossby Center Regional Climate Model (RCA4) RCA4-MPI-ESM-LR were substituted into the model and the flows were estimated. Before predicting probable changes in the Cheliff-Mactaa-Tafna (CMT) basins' hydrology, we adjusted the bias of the simulated precipitation data from the regional climate model RCA4-MPI-ESM-LR using three algorithms: quantile mapping, scaled distribution mapping, and quantile delta mapping.

### 5.1. Change in Precipitation

The results of this study showed that all bias correction methods improved the raw regional climate model (RCM) data to some extent. Nonetheless, the quality of RCM-adjusted precipitation is heavily dependent on the correction algorithm chosen, both for current and future climate conditions. The quantile delta algorithm, for example, is a stable and robust method that generates future time series with dynamics comparable to current conditions because it is based on observations [43]. In the case of the average, the QDM approach is regarded as the best method of correction. According to our analysis, the amount of precipitation varies from one station to another. Under the two scenarios, they fluctuate between −14 and +35% for all stations. This variation is between −32 and +34% for the SDM method, −45 and 05% for the QM method, and −37 and 14% for raw data.

In comparison to the corrective methods (Figure 12), a drop in future precipitation in autumn, winter, and spring is projected in all of the watersheds analyzed. At the same time, the rates of decrease are not synchronous among the correction methods. The rate of decline is greatest when utilizing model outputs corrected by the QM method, then raw model data, model outputs adjusted by the SDM method, and finally, model outputs corrected by the QDM method. In the summer, the average rate for all stations indicates an increase in future flows, albeit this rise is greater when model outputs corrected by the QDM approach are used than when the SDM and QM methods are used. The raw model data yielded the lowest rate of increase, followed by that corrected by the QM approach. Furthermore, we would like to point out that none of the correction methods used to account for the physical reasons of precipitation biases were used (e.g., temporal errors in the main circulation systems or errors in the settings of the clouds' processes and precipitation).

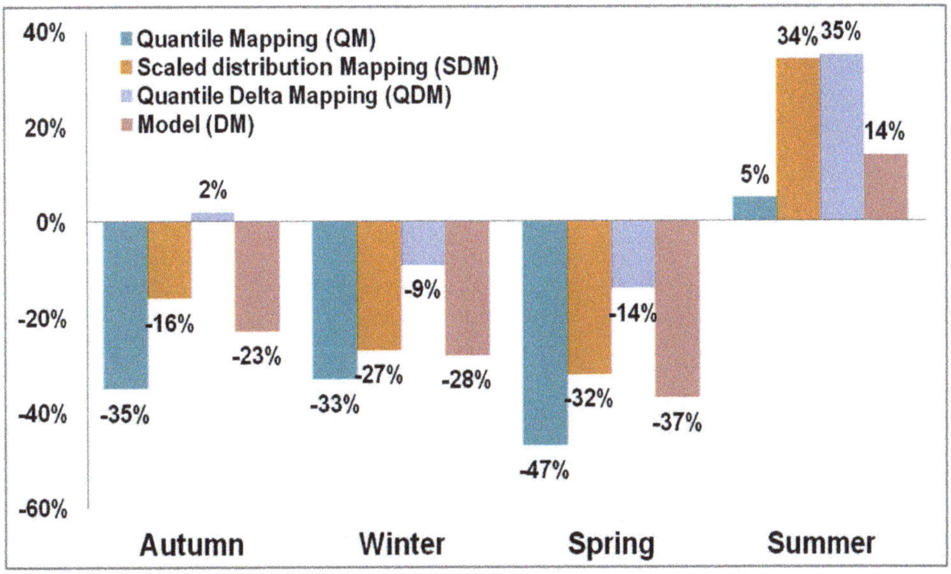


**Figure 12.** Seasonal precipitation mean variation over the CMT basins under the RCP4.5 and RCP8.5 climate scenarios.

## *5.2. Change in Streamflow*

Figure 13 depicts the results of a comparison of the flows across two time periods, 2050 and 2100, using three correction methods: QM, SDM, and QDM, under two scenarios, RCP4.5 and RCP8.5. As indicated in Figure 13, all correction methods and all scenarios forecast a decrease in future seasonal flow, except for the rainy season (winter), which predicts an increase for the QDM correction method. The rate of decline is highest when utilizing model outputs corrected by the QM method, followed by model outputs corrected by the SDM method, raw model data, and finally, model outputs corrected by the QDM method.

Bias correction methods have a direct impact on the resulting hydrological simulations since they change the quality of the fit RCM data. Our findings show that it is possible to adjust climate model simulations in such a way that the features of the resulting mean for monthly flows are greatly reduced. Furthermore, by using higher-performing correction algorithms, variability ranges can be greatly lowered. Variability in the flow simulations, however, is caused not only by the various bias correction approaches, but also, to a significant degree, by the parameters that determine which winter flood peaks will be dominant and occur earlier due to excessive precipitation. Variations in the autumn flow simulations, on the other hand, are predicted to diminish dramatically, owing in part to the RCMs predicting less precipitation and more heat. Over the period from 1950 to 2016, the trend toward dryness between 0.8 and 0.9 °C since the 1980s in the coastal districts of Algeria and since the 1990s on the high plateaus of Algeria was also observed [44]. Climate models predict that this warming will often exceed 1 °C on an annual scale, particularly in summer, between 1945 and 2100 under the RCP8.5 scenario [45]. The findings concerning precipitation decreases, seasonal, and yearly flows reflect previous findings in the literature for Algerian basins. According to the RCP 4.5 scenario, the prediction of the future evolution of rainfall in western Algeria shows a decline from −12% to −38% by the end of the 21st Century [46]. According to the two future scenarios, [47] discovered that drought episodes in the northwest areas are anticipated to be more severe and of longer lengths than in the past, particularly during the hot season (between May and September) between 2021 and 2071. Taïbi et al [48] discovered that the availability of surface water collected at the Ain Dalia dam in northeastern Algeria is likely to decline by 5% to 13% by 2050 and by 21% to 44% by 2100. Hadour et al [49] found similar results in some basins in the northwestern Algeria region, and Zeroual et al [50] found similar results in the Algerian-Hodna-Soummam basins.

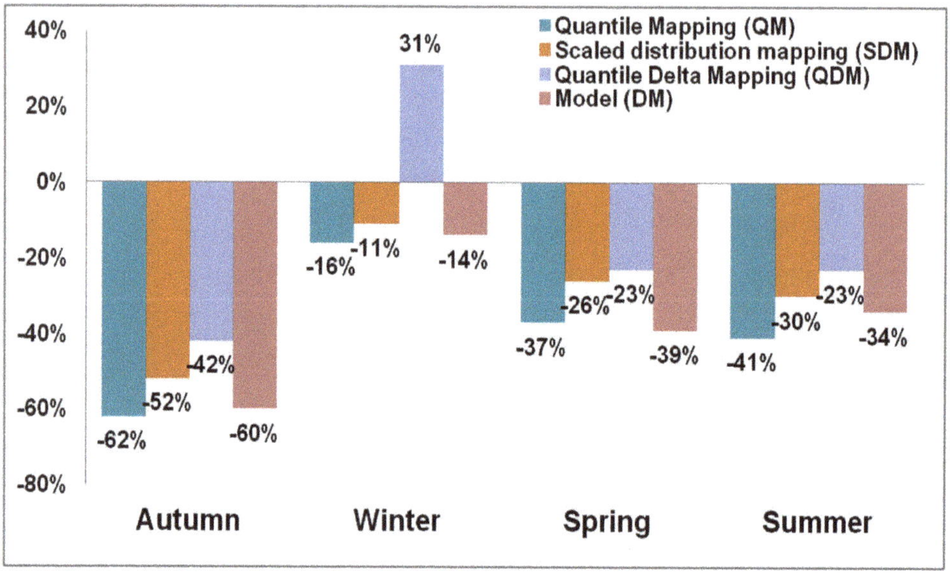


**Figure 13.** Seasonal streamflow mean variation over the CMT basins under the RCP4.5 and RCP8.5 climate scenarios.

## 6. Conclusions

The impact of climate change on the hydrology of the Cheliff-Mactaa-Tafna (CMT) basins was investigated using a CORDEX-Africa adjusted bias set on two climate projection scenarios, RCP 4.5 and RCP 8.5, of the concentration pathway representative of precipitation and temperature. Zygos simulates streamflow and the distribution mapping bias correction methods of quantile mapping (QM), quantile delta mapping (QDM), and scaled distribution mapping (SDM) to improve precipitation, temperature, and streamflow simulations utilizing a lumped approach. The bias correction algorithm chosen is critical in assessing the effects of climate change. The performance of the three bias correction algorithms (QM, QDM, and SDM) modulating the climate change signal of precipitation over six mountainous watersheds in northeastern Algeria was compared in this study. The capability of these corrective methods was assessed by modeling flows using the Zygos conceptual hydrological model. This led to various conclusions. Improvement was obtained for future data using all bias correction approaches, with varying rates of change. The RCM's raw outputs are highly skewed, making them unsuitable for direct use in analyzing the consequences of climate change. The RCM simulations portrayal is largely dependent on region and season. Although bias correction methods have the potential to improve the performance of precipitation and temperature reproduction, their final findings are heavily influenced by the bias correction approaches. In our study area, all correction methods and both scenarios predicted a decrease in future seasonal flow, except for the rainy season (winter), which predicts an increase for the QDM method. The rate of decline is greater when utilizing model outputs corrected by the QM method, followed by model outputs corrected by the SDM method, raw model data, and finally, the QDM method. Future climate-corrected precipitation projections revealed significant variations over time. Most years will see high rainfall; however, certain years will experience low average precipitation when compared to the observed data.

Predicted rainfall and flow will also influence crop choices, cropping patterns, crop rotations, crop management frameworks, planting times, cropping area extent, agricultural yields, and so on. Changes in land use and cover, changes in groundwater, river, and surface water levels, flooding, and soil erosion will be highly posed in urban areas. Furthermore, the study area is well known for its heritage and tourism. Thus, the study supplies environmentalists, urban planners, and water resource managers with clear information on future rainfall and flows. Rainwater harvesting, aquifer recharge, reforestation, and channeling excess water to the river through proper channels are all viable options for dealing with future excess rainfall.

**Supplementary Materials:** The following supporting information can be downloaded at: https://www.mdpi.com/article/10.3390/cli10080123/s1, Figure S1: Changes in annual precipitation under tow scenarios (RCP4.5 and RCP8.5), (a) Quantile Mapping, (b) Scaled Distribution Mapping, (c) Quantile Delta Mapping and (d) Model (RAW); Figure S2: Changes in annual precipitation under tow scenarios (RCP4.5 and RCP8.5), (a) Quantile Mapping, (b) Scaled Distribution Mapping, (c) Quantile Delta Mapping and (d) Model (RAW); Table S1: Evapotranspiration changes on future time horizons: 2050 and 2100; Table S2: Change in mean seasonal (autumn) precipitation projection in CMT basins from the baseline period (1975–2012); Table S3: Change in mean seasonal (Winter) precipitation projection in CMT basins from the baseline period (1975–2012); Table S4: Change in mean seasonal (Spring) precipitation projection in CMT basins from the baseline period (1975–2012); Table S5: Change in mean seasonal (Summer) precipitation projection in CMT basins from the baseline period (1975–2012); Table S6: Change in mean seasonal (Autumn) projected streamflow in CMT basins from the baseline period (1975–2012); Table S7: Change in mean seasonal (Winter) projected streamflow in CMT basins from the baseline period (1975–2012); Table S8: Change in mean seasonal (Spring) projected streamflow in CMT basins from the baseline period (1975–2012); Table S9: Change in mean seasonal (Summer) projected streamflow in CMT basins from the baseline period (1975–2012); Supplementary Material S3: Improving Future Estimation of Cheliff-Mactaa-Tafna Stream-flow via an Ensemble of Bias Correction Approaches [18,24,51–53].

**Author Contributions:** Conceptualization, M.R., A.Z., A.A., S.T. and R.A.; methodology, M.R., A.Z., A.A., S.T. and R.A.; formal analysis, M.R., A.Z., Y.H., A.A., S.Z. and S.T.; investigation, M.R., Y.H., S.Z., A.A.S., C.M.M., S.B., S.K., A.G., I.B. and A.K.; funding acquisition, A.Z.; software, M.R., Y.H., S.Z., A.A.S., C.M.M., S.B., S.K., A.G., I.B. and A.K.; writing—original draft, M.R. and Y.H.; writing—review & editing, A.Z., Y.H., A.A., S.Z., A.A.S., C.M.M., S.T., S.B., S.K., A.G., I.B. and A.K. All authors have read and agreed to the published version of the manuscript.

**Funding:** This research received no external funding.

**Institutional Review Board Statement:** Not applicable.

**Informed Consent Statement:** Not applicable.

**Data Availability Statement:** All data are available from the corresponding author upon reasonable request.

**Acknowledgments:** This paper is dedicated to the memory of our dear co-author, Cedrick Mulowayi Mubulayi, who passed away while this paper was being peer reviewed. The authors wish to thank the National Agency of Water Resources for providing material and data required in this study, as well as to the anonymous reviewers and editors for their valuable comments and suggestions, which greatly improved the paper.

**Conflicts of Interest:** The authors declare no conflict of interest.

## References

1. Change, Intergovernmental Panel On Climate. *Climate Change*; IPCC: Geneva, Switzerland, 2014.
2. Mukheibir, P. Water access, water scarcity, and climate change. *Environ. Manag.* **2010**, *45*, 1027–1039.
3. Schewe, J.; Heinke, J.; Gerten, D.; Haddeland, I.; Arnell, N.W.; Clark, D.B.; Dankers, R.; Eisner, S.; Fekete, B.M.; Colón-González, F.J.; et al. Multimodel assessment of water scarcity under climate change. *Proc. Natl. Acad. Sci. USA* **2014**, *111*, 3245–3250.
4. Arnell, N.W.; Gosling, S.N. The impacts of climate change on river flow regimes at the global scale. *J. Hydrol.* **2013**, *486*, 351–364.
5. Hattermann, F.F.; Krysanova, V.; Gosling, S.N.; Dankers, R.; Daggupati, P.; Donnelly, C.; Flörke, M.; Huang, S.; Motovilov, Y.; Buda, S.; et al. Cross-scale intercomparison of climate change impacts simulated by regional and global hydrological models in eleven large river basins. *Clim. Chang.* **2017**, *141*, 561–576.
6. Theodossiou, N. Assessing the Impacts of Climate Change on the Sustainability of Groundwater Aquifers. Application in Moudania Aquifer in N. Greece. *Environ. Process.* **2016**, *3*, 1045–1061.
7. Hamed, Y.; Hadji, R.; Redhaounia, B.; Zighmi, K.; Bâali, F.; El Gayar, A. Climate impact on surface and groundwater in North Africa: A global synthesis of findings and recommendations. *Euro-Mediterr. J. Environ. Integr.* **2018**, *3*, 25.
8. Mokadem, N.; Redhaounia, B.; Besser, H.; Ayadi, Y.; Khelifi, F.; Hamad, A.; Hamed, Y.; Bouri, S. Impact of climate change on groundwater and the extinction of ancient "Foggara" and springs systems in arid lands in North Africa: A case study in Gafsa basin (Central of Tunisia). *Euro-Mediterr. J. Environ. Integr.* **2018**, *3*, 28.
9. Meehl, G.A.; Covey, C.; Delworth, T.; Latif, M.; McAvaney, B.; Mitchell, J.F.B.; Stouffer, R.J.; Taylor, K.E. The WCRP CMIP3 Multimodel Dataset: A New Era in Climate Change Research. *Bull. Am. Meteorol. Soc.* **2007**, *88*, 1383–1394.
10. Mearns, L.O.; Arritt, R.; Biner, S.; Bukovsky, M.S.; McGinnis, S.; Sain, S.; Caya, D.; Correia, J.; Flory, D.; Gutowski, W.; et al. The North American regional climate change assessment program: Overview of phase I results. *Bull. Am. Meteorol. Soc.* **2012**, *93*, 1337–1362.
11. Sillmann, J.; Kharin, V.V.; Zwiers, F.W.; Zhang, X.; Bronaugh, D. Climate extremes indices in the CMIP5 multimodel ensemble: Part 2. Future climate projections. *J. Geophys. Res. Atmos.* **2013**, *118*, 2473–2493. [CrossRef]
12. Teutschbein, C.; Seibert, J. Bias correction of regional climate model simulations for hydrological climate-change impact studies: Review and evaluation of different methods. *J. Hydrol.* **2012**, *456*, 12–29.
13. Themeßl, M.J.; Gobiet, A.; Heinrich, G. Empirical-statistical downscaling and error correction of regional climate models and its impact on the climate change signal. *Clim. Chang.* **2012**, *112*, 449–468.
14. Räisänen, J.; Räty, O. Projections of daily mean temperature variability in the future: Cross-validation tests with ENSEMBLES regional climate simulations. *Clim. Dyn.* **2013**, *41*, 1553–1568.
15. Teutschbein, C.; Seibert, J. Is bias correction of regional climate model [RCM] simulations possible for non-stationary conditions? *Hydrol. Earth Syst. Sci.* **2013**, *17*, 5061–5077.
16. Fang, G.H.; Yang, J.; Chen, Y.N.; Zammit, C. Comparing bias correction methods in downscaling meteorological variables for a hydrologic impact study in an arid area in China. *Hydrol. Earth Syst. Sci.* **2015**, *19*, 2547–2559.
17. Ahmed, K.F.; Wang, G.; Silander, J.; Wilson, A.M.; Allen, J.M.; Horton, R.; Anyah, R. Statistical downscaling and bias correction of climate model outputs for climate change impact assessment in the U.S. northeast. *Glob. Planet. Chang.* **2013**, *100*, 320–332.
18. Switanek, M.B.; Troch, P.A.; Castro, C.L.; Leuprecht, A.; Chang, H.-I.; Mukherjee, R.; Demaria, E.M.C. Scaled distribution mapping: A bias correction method that preserves raw climate model projected changes. *Hydrol. Earth Syst. Sci.* **2017**, *21*, 2649–2666.
19. Grenier, P. Two Types of Physical Inconsistency to Avoid with Univariate Quantile Mapping: A Case Study over North America Concerning Relative Humidity and Its Parent Variables. *J. Appl. Meteorol. Clim.* **2018**, *57*, 347–364.
20. Willems, P.; Vrac, M. Statistical precipitation downscaling for small-scale hydrological impact investigations of climate change. *J. Hydrol.* **2011**, *402*, 193–205.
21. Gudmundsson, L.; Bremnes, J.B.; Haugen, J.E.; Engen-Skaugen, T. Downscaling RCM precipitation to the station scale using statistical transformations–a comparison of methods. *Hydrol. Earth Syst. Sci.* **2012**, *16*, 3383–3390.
22. Hempel, S.; Frieler, K.; Warszawski, L.; Schewe, J.; Piontek, F. A trend-preserving bias correction–the ISI-MIP approach. *Earth Syst. Dyn.* **2013**, *4*, 219–236.
23. Bürger, G.; Murdock, T.Q.; Werner, A.T.; Sobie, S.R.; Cannon, A.J. Downscaling Extremes—An Intercomparison of Multiple Statistical Methods for Present Climate. *J. Clim.* **2012**, *25*, 4366–4388.
24. Cannon, A.J.; Sobie, S.R.; Murdock, T.Q. Bias correction of GCM precipitation by quantile mapping: How well do methods preserve changes in quantiles and extremes? *J. Clim.* **2015**, *28*, 6938–6959.
25. Casanueva, A.; Herrera, S.; Iturbide, M.; Lange, S.; Jury, M.; Dosio, A.; Maraun, D.; Gutiérrez, J.M. Testing bias adjustment methods for regional climate change applications under observational uncertainty and resolution mismatch. *Atmos. Sci. Lett.* **2020**, *21*, e978.

26. Mpelasoka, F.S.; Chiew, F.H.S. Influence of Rainfall Scenario Construction Methods on Runoff Projections. *J. Hydrometeorol.* **2009**, *10*, 1168–1183.
27. Van Roosmalen, L.; Sonnenborg, T.O.; Jensen, K.H.; Christensen, J.H. Comparison of Hydrological Simulations of Climate Change Using Perturbation of Observations and Distribution-Based Scaling. *Vadose Zone J.* **2011**, *10*, 136–150.
28. Muerth, M.J.; Gauvin St-Denis, B.; Ricard, S.; Velázquez, J.A.; Schmid, J.; Minville, M.; Caya, D.; Chaumont, D.; Ludwig, R.; Turcotte, R. On the need for bias correction in regional climate scenarios to assess climate change impacts on river runoff. *Hydrol. Earth Syst. Sci.* **2013**, *17*, 1189–1204.
29. Teng, J.; Potter, N.J.; Chiew, F.H.S.; Zhang, L.; Wang, B.; Vaze, J.; Evans, J.P. How does bias correction of regional climate model precipitation affect modelled runoff? *Hydrol. Earth Syst. Sci.* **2015**, *19*, 711–728.
30. Nguyen, H.; Mehrotra, R.; Sharma, A. Assessment of Climate Change Impacts on Reservoir Storage Reliability, Resilience, and Vulnerability Using a Multivariate Frequency Bias Correction Approach. *Water Resour. Res.* **2020**, *56*, e2019WR026022.
31. Achour, K.; Meddi, M.; Zeroual, A.; Bouabdelli, S.; Maccioni, P.; Moramarco, T. Spatio-temporal analysis and forecasting of drought in the plains of northwestern Algeria using the standardized precipitation index. *J. Earth Syst. Sci.* **2020**, *129*, 42.
32. Zerouali, B.; Mesbah, M.; Chettih, M.; Djemai, M. Contribution of cross time-frequency analysis in assessment of possible relationships between large-scale climatic fluctuations and rainfall of northern central Algeria. *Arab. J. Geosci.* **2018**, *11*, 392.
33. Bouabdelli, S.; Meddi, M.; Zeroual, A.; Alkama, R. Hydrological drought risk recurrence under climate change in the karst area of Northwestern Algeria. *J. Water Clim. Chang.* **2020**, *11*, 164–188.
34. Zeroual, A.; Assani, A.A.; Meddi, M.; Alkama, R. Assessment of climate change in Algeria from 1951 to 2098 using the Köppen–Geiger climate classification scheme. *Clim. Dyn.* **2019**, *52*, 227–243.
35. Benzater, B.; Elouissi, A.; Benaricha, B.; Habi, M. Spatio-temporal trends in daily maximum rainfall in northwestern Algeria (Macta watershed case, Algeria). *Arab. J. Geosci.* **2019**, *12*, 370.
36. Bouchelkia, H.; Belarbi, F.; Rémini, B. Estimated flows of suspended solids by the statistical analysis of outfall drainage basin of Tafna (Algeria). *Soil Water Res.* **2013**, *8*, 63–70.
37. Jones, C.; Giorgi, F.; Asrar, G. The Coordinated Regional Downscaling Experiment: CORDEX–an international downscaling link to CMIP5. *CLIVAR Exch.* **2011**, *16*, 34–40.
38. Kozanis, S.; Christoforides, A.; Efstratiadis, A. Scientific Documentation of Hydrognomon software [version 4]. Development of Database and Software Application in a Web Platform for the "National Database and Meterological Information". ITIA Research Team, National Technical University of Athens. Available online: https://www.itia.ntua.gr/en/docinfo/928/1/Documents/HydrognomonV4TheoryGR-V1,2 (accessed on 6 July 2022).
39. Charizopoulos, N.; Psilovikos, A. Hydrologic processes simulation using the conceptual model Zygos: The example of Xynias drained Lake catchment (central Greece). *Environ. Earth Sci.* **2016**, *75*, 777. [CrossRef]
40. Charizopoulos, N.; Psilovikos, A.; Zagana, E. A lumped conceptual approach for modeling hydrological processes: The case of Scopia catchment area, Central Greece. *Environ. Earth Sci.* **2017**, *76*, 632. [CrossRef]
41. Madsen, H. Automatic calibration of a conceptual rainfall–runoff model using multiple objectives. *J. Hydrol.* **2000**, *235*, 276–288.
42. Khazaei, M.R.; Zahabiyoun, B.; Saghafian, B.; Ahmadi, S. Development of an Automatic Calibration Tool Using Genetic Algorithm for the ARNO Conceptual Rainfall-Runoff Model. *Arab. J. Sci. Eng.* **2013**, *39*, 2535–2549.
43. Anandhi, A.; Frei, A.; Pierson, D.C.; Schneiderman, E.M.; Zion, M.S.; Lounsbury, D.; Matonse, A.H. Examination of change factor methodologies for climate change impact assessment. *Water Resour. Res.* **2011**, *47*. [CrossRef]
44. Taïbi, S.; Zeroual, A.; Meddi, M. Effect of autocorrelation on temporal trends in air-temperature in Northern Algeria and links with teleconnections patterns. *Theor. Appl. Climatol.* **2022**, *147*, 959–984. [CrossRef]
45. Zeroual, A.; Assani, A.A.; Meddi, H.; Bouabdelli, S.; Zeroual, S.; Alkama, R. Assessment of Projected Precipitations and Temperatures Change Signals over Algeria Based on Regional Climate Model: RCA4 Simulations. In *Water Resources in Algeria—Part I. The Handbook of Environmental Chemistry*; Negm, A.M., Bouderbala, A., Chenchouni, H., Barceló, D., Eds.; Springer: Cham, Switzerland, 2020; Volume 97, pp. 135–159. [CrossRef]
46. Taïbi, S.; Anza, F.Z.H.; Zeroual, S. Etude de l'impact des changements climatiques sur la disponibilité des ressources en eau basée sur les simulations du modèle climatique régional RCA4: Cas du bassin de Ain DALIA (Algérie). *Alger. J. Environ. Sci. Technol.* **2021**, *7*, 1860–1869.
47. Bouabdelli, S.; Zeroual, A.; Meddi, M.; Assani, A. Impact of temperature on agricultural drought occurrence under the effects of climate change. *Theor. Appl. Climatol.* **2022**, *148*, 191–209.
48. Taïbi, S.; Zeroual, A.; Melhani, N. Evaluation de deux méthodes de correction de biais des sorties de modèles climatiques régionaux Cordex-Africa pour la prévision des pluies: Cas du bassin côtier oranais. *Proc. Int. Assoc. Hydrol. Sci.* **2021**, *384*, 213–218. [CrossRef]
49. Hadour, A.; Mahé, G.; Meddi, M. Watershed based hydrological evolution under climate change effect: An example from North Western Algeria. *J. Hydrol. Reg. Stud.* **2020**, *28*, 100671.
50. Zeroual, A.; Meddi, M.; Bensaad, S. The impact of climate change on river flow in arid and semi-arid rivers in Algeria. *IAHS-AISH Proc. Rep.* **2013**, *359*, 1–6.
51. Piani, C.; Haerter, J.O.; Coppola, E. Statistical bias correction for daily precipitation in regional climate models over Europe. *Theor. Appl. Climatol.* **2010**, *99*, 187–192.

52. Ringard, J.; Seyler, F.; Linguet, L. A Quantile Mapping Bias Correction Method Based on Hydroclimatic Classification of the Guiana Shield. *Sensors* **2017**, *17*, 1413. [CrossRef]
53. Li, C.; Sinha, E.; Horton, D.E.; Diffenbaugh, N.S.; Michalak, A.M. Joint bias correction of temperature and precipitation in climate model simulations. *J. Geophys. Res. Atmos.* **2014**, *119*, 13–153.

climate

MDPI

Article

# Extreme Precipitation Events During the Wet Season of the South America Monsoon: A Historical Analysis over Three Major Brazilian Watersheds

**Aline Araújo de Freitas *, Vanessa Silveira Barreto Carvalho and Michelle Simões Reboita**

Institute of Natural Resources, Federal University of Itajubá (UNIFEI), Itajubá 37500-903, Brazil; vanessa.silveira@unifei.edu.br (V.S.B.C.); reboita@unifei.edu.br (M.S.R.)

* Correspondence: alinefreitas@unifei.edu.br

**Abstract:** Most of South America, particularly the region between the southern Amazon and southeastern Brazil, as well as a large part of the La Plata Basin, has its climate regulated by the South American Monsoon System. Extreme weather and climate events in these areas have significant socioeconomic impacts. The Madeira, São Francisco, and Paraná river basins, three major watersheds in Brazil, are especially vulnerable to wet and drought periods due to their importance as freshwater ecosystems and sources of water for consumption, energy generation, and agriculture. The scarcity of surface meteorological stations in these basins makes meteorological studies challenging, often using reanalysis and satellite data. This study aims to identify extreme weather (wet) and climate (wet and drought) events during the extended wet season (October to March) from 1980 to 2022 and evaluate the performance of two gridded datasets (CPC and ERA5) to determine which best captures the observed patterns in the Madeira, São Francisco, and Paraná river basins. Wet weather events were identified using the 95th percentile, and wet and drought periods were identified using the Standardized Precipitation Index (SPI) on a 6-month scale. In general, CPC data showed slightly superior performance compared to ERA5 in reproducing statistical measures. For extreme day precipitation, both datasets captured the time series pattern, but CPC better reproduced extreme values and trends. The results also indicate a decrease in wet periods and an increase in drought events. Both datasets performed well, showing they can be used in the absence of station data.

**Keywords:** extreme rainfall; drought events; Brazilian basins; CPC; ERA5

## 1. Introduction

Most of South America, between the southern Amazon and southeastern Brazil and a large part of the La Plata Basin, has its climate regulated by a combination of systems and processes that are named the South America Monsoon System (SAMS) [1–6]. The lifecycle of the SAMS is also influenced by several factors such as the complex topography and land uses of the region and the variability in different time scales over the Pacific and Atlantic Oceans, among others [4].

The occurrence of extreme rainfall events leading to floods and rainfall deficits causing droughts has been extensively reported in the SAMS region [7–12], where watersheds are extremely vulnerable to its impacts. Since the present study focuses on extreme events, it is helpful to define what extreme weather and climate events are. Extreme weather events are rare events in a specific location and time of year and extreme climate events are occurrences where a climate variable value exceeds a threshold near the upper or lower limits of observed values and persists for an extended period, such as prolonged droughts or heavy rainfall throughout the season [13].

In Brazil, major watersheds such as the Madeira, São Francisco, and Paraná river basins are extremely vulnerable to extreme weather (wet) and climate (wet and drought) events because of their relevance as freshwater ecosystems and as water sources for human consumption, energy generation, and agriculture [11,12,14]. Extreme precipitation events, on a weather scale, can lead to flash floods and river flooding [15], while on a climate scale, seasonal meteorological droughts can cause hydrological droughts, which can impact the water supply, energy production, the transportation sector, and water quality [16]. The north of Brazil has been suffering from a lack of rainfall, especially in the Amazon, which has faced drought episodes in recent years, notably in 2005, 2010, 2015–2016, and most recently in 2023/2024 [17,18].

It is also important to highlight that, in areas of South America, meteorological surface stations are scarce [19,20], which leaves important gaps in defining extreme precipitation thresholds, frequency, and patterns. Hence, products such as reanalysis and estimated satellite data are commonly used to provide such information [20], but with no previous validation with some existing stations.

Since extreme precipitation events on both a weather and climate scale can be responsible for biodiversity and economic losses, such as failures in the supply of energy and water for agriculture and navigation, it is important to study the frequency, duration, and intensity of such events. In this framework, the main objective of this study is to describe the occurrence and trends of extreme weather (wet) and climate (wet/drought) events in the Madeira, São Francisco, and Paraná river basins during the extended rainy season of South America (October to March) through observed and grid point datasets: the Climate Prediction Center (CPC) precipitation analysis and ERA5 reanalysis. We also aim to highlight the grid dataset that better represents the observations in the SAMS region. This study is part of the project "Combinando Modelagem Numérica Climática e uso de Aprendizado de Máquina na detecção de extremos climáticos em horizontes subsazonais, sazonais e multi-decadais", project 88887.688971/2022-00, funded by the Coordination for the Improvement of Higher Education Personnel (CAPES).

## 2. Materials and Methods

### 2.1. Study Area

This study focuses on three socio-economically important river basins located in the SAMS region [21–23]: the Madeira River Basin (MRB), the Paraná River Basin (PRB), and the São Francisco River Basin (SFRB) (Figure 1).

#### 2.1.1. Madeira River Basin

The MRB is mainly composed of the Madeira River, which is one of the five rivers with the highest water flow in the world, the longest and most significant tributary of the Amazon River, and one of the main Brazilian rivers [23]. With a hydrographic basin of approximately 1.25 million km$^2$, the river runs 3315 km, covering Bolivia (51%), Brazil (42%), and Peru (7%), and is the 17th longest river in the world [23,24]. Originating from the Beni, Mamoré, and Madre de Dios rivers, the river crosses Rondônia and flows into the Amazon [25,26]. The MRB is divided into three regions: Upper, Middle, and Lower Madeira (Figure 1).

Its importance ranges from the subsistence of riverside populations to the regional economy, through the use of its water in agriculture, fishing, industry, waterway transport, and energy production [23–25]. The MRB has a hydroelectric complex, which includes the Jirau and Santo Antônio plants, generating around 3500 MW of energy each, which is subsequently distributed to the Center-West and Southeast regions of Brazil [25,27,28]. The MRB is located in the Amazon biome and features a diverse range of vegetation, including areas of campinaranas, swamps, grasslands, savannas, and upland and floodplain forests [29,30]. The climate present in the basin is equatorial, being hot and humid, with an average precipitation of 1940 mm year$^{-1}$ and temperatures between 24 and 26 °C [31,32], influenced mainly by the SAMS [2,4,6].

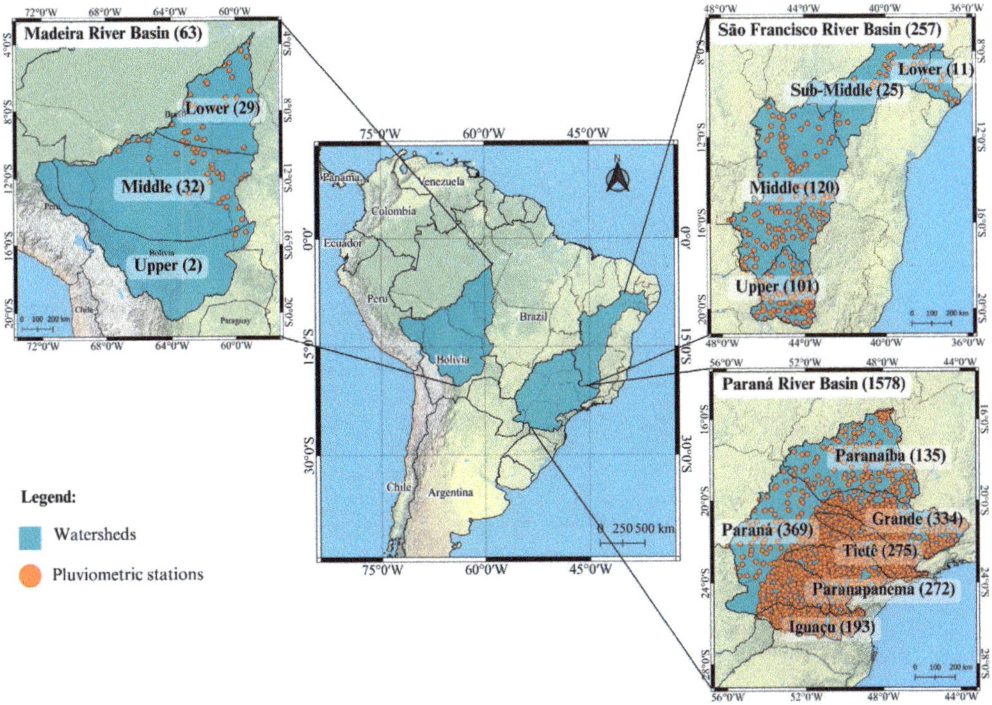


**Figure 1.** Representation of the study area with a zoom on the basins that were analyzed and the locations of the pluviometric stations used. The number of stations in each basin is also indicated in the figure in parentheses.

### 2.1.2. Paraná River Basin

The PRB occupies around 10% of the national territory, covering six Brazilian states and the Federal District, with an area of approximately 879,873 km$^2$ [22,33]. The PRB is divided into six regions: Paranaíba, Grande, Tietê, Paraná, Paranapanema, and Iguaçu (Figure 1). Located in the most economically developed region of the country, the PRB meets 32% of the national water demand, mainly for irrigation, industrial, and urban use [33,34]. With a population of around 61.4 million inhabitants, the PRB contains 36% of the urban population and produces 45% of the country's GDP [33,34]. It houses important metropolitan regions such as São Paulo, Brasília, and Curitiba [22,33,35]. Furthermore, this region includes the Itaipu Hydroelectric Plant, the largest generator of clean and renewable energy in the world [36,37].

In this basin, the predominant biomes are the Atlantic Forest and the Cerrado [38] and the climatic parameters vary spatially and temporally, affecting water availability in the region. The dominant climate is tropical, except in Paraná and Santa Catarina, where a temperate subtropical climate prevails [33,35]. The SACZ is the main system responsible for rainfall in summer [39,40].

### 2.1.3. São Francisco River Basin

The SFRB contains an important river, the São Francisco, also known as the National Integration River, which extends for 2863 km from latitude 7°00′ S to 21°00′ S, crossing six Brazilian states [22,41,42]. The SFRB covers 638,466 km$^2$, around 7.5% of the Brazilian territory, and is home to more than 14 million inhabitants [22,34]. The SFRB is divided into four regions [21], Upper, Middle, Sub-Middle, and Lower São Francisco (Figure 1), and stands out for its energy generation potential, with hydroelectric plants such as Três Marias,

Sobradinho, Paulo Afonso, Itaparica, and Xingó, and its water supply for the population, navigation, fishing, agriculture, and basic sanitation [43].

Fragments of biomes such as Cerrado, Caatinga, and Atlantic Forest are found in this basin. Furthermore, due to its territorial extension, its climate varies from hot and humid tropical to semi-arid, influenced by systems such as SACZ [39,40,44] in tUpper and Middle São Francisco and the Intertropical Convergence Zone (ITCZ) [45] in Sub-Middle and Lower São Francisco.

### 2.2. Data

Daily rainfall data registered between 1980 and 2022 from 1898 pluviometric stations of the National Water and Basic Sanitation Agency (ANA; https://www.snirh.gov.br/hidroweb/serieshistoricas, accessed on 10 June 2024) and located in the MRB, PRB, and SFRB were used in this study (Figure 1). However, only sites with at least 7300 days of valid observations (approximately 20 years of data) were considered. ANA data were classified according to consistency: level 1 included pre-analyzed data, used in this study, while level 2 comprised raw data.

Daily rainfall data from the Gauge-Based Analysis of Global Daily Precipitation (CPC) [46,47] generated by the National Oceanic and Atmospheric Administration/Climate Prediction Center (NOAA/CPC) and reanalysis data from ERA5 [48], provided by the European Center for Medium-Range Weather Forecasts (ECMWF), from 1980 to 2022, were also used. CPC analysis has a horizontal spatial resolution of 0.5° and combines data from surface observations of approximately 30,000 sites from around the world, creating a unified, high-quality dataset. ERA5 reanalysis has a 0.25° horizontal spatial resolution and combines model data and global observations to create a complete and consistent dataset, based on the laws of physics.

The flowchart below summarizes all the analyses performed (Figure 2), which will be described in more detail in the following sections.

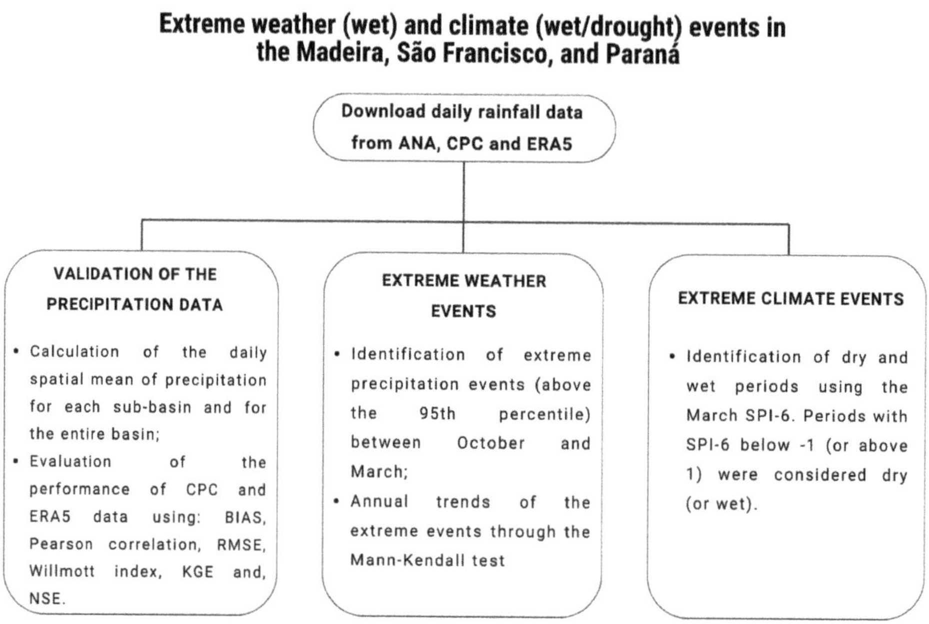


**Figure 2.** Flowchart summarizing the steps of the analyses performed in this article.

### 2.3. Validation of the Precipitation Data

Since we aimed to conduct an integrated study of each sub-basin and the basins as a whole, the precipitation spatial average for each sub-basin and the entire basin was initially calculated for each day of the study period and each dataset (CPC, ERA5, and rain-gauge stations). To delimitate the sub-basin and basin areas, the shapefiles provided by ANA were used. Then, to evaluate the performance of CPC analysis and ERA5 reanalysis in the study region, we compared the daily time series obtained with the averaged area with those from rain-gauge stations through a set of statistical measures: mean error (BIAS) [49], Pearson correlation coefficient [50,51], root mean square error (RMSE) [49,51,52], Willmott Index [53,54], Kling–Gupta efficiency coefficient (KGE) [55,56], and Nash–Sutcliffe efficiency (NSE) [57,58].

### 2.4. Extreme Weather Events

The time series (ANA, CPC, and ERA5) obtained with the methodology from Section 2.3 were used to identify extreme weather events. Daily rainfall extremes occur when the daily precipitation values exceed the 95th percentile (P95), calculated considering only rainfall values above 1 mm within the months of the extended rainy season (from October to March). The number of days per year with precipitation values over the threshold of the 95th percentile was calculated. Following this, the Sun's slope was computed and the Mann–Kendall test [51] was applied to verify the statistical significance (at 0.05 level) of the trends related to the frequency of occurrence of precipitation extremes.

### 2.5. Extreme Climate Events

Drought and wet periods were identified through the Standardized Precipitation Index (SPI), calculated considering the average time series of the entire basins for all datasets (ANA, CPC, and ERA5). The SPI is an index that only uses historical precipitation series to identify drought and wet periods [59]. It was developed by Mckee et al. [60] and has the advantage of being able to be calculated for different time scales, and it is recommended by WMO [61]. In this study, the SPI-6 from March, which refers to precipitation between October (of the previous year) to March, was analyzed. The choice of the March SPI-6 was made based on the desire to evaluate the extended rainy season of South America, in order to check seasonal drought and wet events in the basins during the rainiest months, and to verify which database (CPC or ERA5) has more similarity with observations. Periods in which the March SPI-6 was less (more) than −1 (1) were considered as drought (wet) episodes, the value at which a drought (wet) event begins, according to the SPI classification developed by Mckee et al. [60].

## 3. Results and Discussion

### 3.1. Basic Statistical Analysis

Although our main focus is on the representation of extreme weather and climate events by CPC and ERA5, here we also present the performance of these grid datasets in reproducing the basic climatology of each subbasin. For the MRB, both datasets (CPC and ERA5) indicate higher precipitation volumes in the northern portion of the basin, where the Lower Madeira (III) region is located, which includes the Madeira river hydroelectric complex, comprising the Jirau and Santo Antônio plants. Higher rainfall is also observed in the western region, near the borders of the Upper (I) and Middle Madeira (II) regions (Figure 3a,b). The southern part of the basin, particularly in the Upper Madeira (I) region, shows lower accumulated rainfall rates, consistent with Vergasta et al. [62], and this subregion presents the greatest discrepancies between the CPC and ERA5, with the CPC indicating low precipitation accumulations throughout the region, while ERA5 shows low values only in the southern part, with some areas in the north accumulating more than 2000 mm. Regarding monthly averages, while CPC tends to underestimate the ANA observations, particularly during the rainy season, ERA5 shows more subtle differences when compared to observations (Figure 3d). This pattern shifts during the driest periods, with CPC data aligning more closely with observations, while ERA5 tends to overestimate precipitation in the basin, as also observed by Liu et al. [63], who identified this same tendency for overestimation in relation to monthly precipitation in the Amazon basin region. The statistical results calculated from daily precipitation data revealed smaller bias for ERA5; however, there were no large differences in other statistical parameters computed for CPC and ERA5, respectively (Figure 3c).

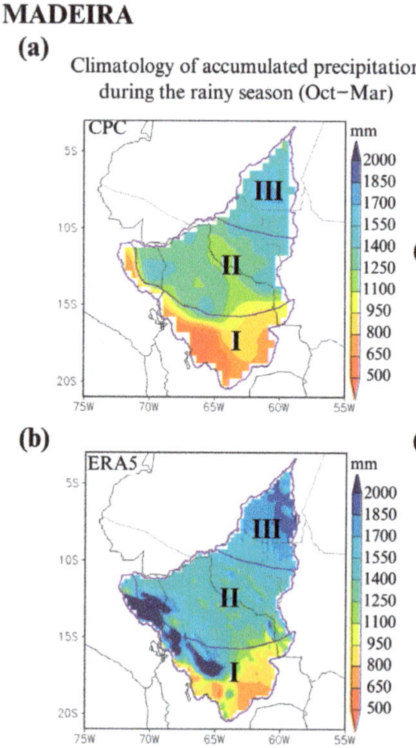


(c)

| Statistics Madeira | CPC | ERA |
|---|---|---|
| BIAS | −1.17 | 0.13 |
| Pearson | 0.78 | 0.70 |
| RMSE | 3.29 | 3.69 |
| Willmott Index | 0.86 | 0.83 |
| KGE | 0.65 | 0.70 |
| NSE | 0.55 | 0.43 |

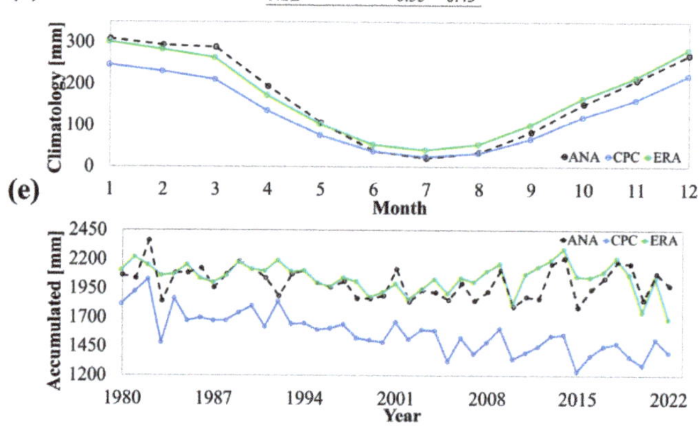


**Figure 3.** Analysis for the MRB: climatology of accumulated precipitation for the rainy season (October to March) for (**a**) CPC and (**b**) ERA5; (**c**) statistical analysis based on daily data; (**d**) monthly climatology; and (**e**) annual accumulation. In (**d**,**e**), the ERA5 database is indicated in green, the CPC is indicated in blue, and the observations are shown in black. Regions I, II and III indicate the Upper, Middle and Lower Madeira regions, respectively.

To improve understanding, especially in the Upper Madeira region, where there are few meteorological stations available, two-point comparisons were added (Figure 4). The results show that the CPC model outperforms ERA in both locations in Upper Madeira (1559000 and 1560000). CPC presents lower errors (BIAS and RMSE) and better correlations (Pearson), in addition to greater agreement with observations, reflected in metrics such as the Willmott Index, KGE, and NSE. The monthly climatology and annual accumulated precipitation curves indicate that CPC more accurately captures the seasonal and interannual variability of precipitation, while ERA presents greater deviations, especially in correlation and mean square error.

**(a)**

| Statistics Upper Madeira (1559000) | CPC | ERA |
|---|---|---|
| BIAS | −0.29 | −0.14 |
| Pearson | 0.75 | 0.29 |
| RMSE | 7.02 | 10.51 |
| Willmott Index | 0.84 | 0.45 |
| KGE | 0.63 | 0.15 |
| NSE | 0.56 | 0.02 |

**(b)**

| Statistics Upper Madeira (1560000) | CPC | ERA |
|---|---|---|
| BIAS | −0.03 | 0.37 |
| Pearson | 0.55 | 0.27 |
| RMSE | 8.16 | 9.71 |
| Willmott Index | 0.71 | 0.47 |
| KGE | 0.49 | 0.17 |
| NSE | 0.26 | −0.05 |

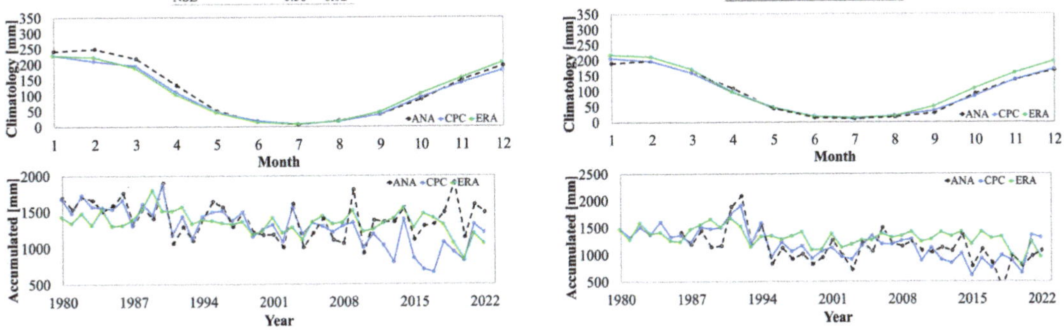


**Figure 4.** Statistical analysis based on daily data, monthly climatology, and annual accumulation for 2 stations in Upper Madeira: (**a**) 1559000 and (**b**) 1560000. The ERA5 database is indicated in green, the CPC is indicated in blue, and the observations are shown in black.

The spatial patterns of precipitation provided by CPC and ERA5 are similar in the PRB (Figure 5a,b). The highest rainfall values are concentrated in the northern (I and II) and southern (VI) extremities of the basin, as also shown by Rafee et al. [64] and Freitas et al. [12]. In the southern part of the region (VI), the Itaipu Power Plant is present—the largest global generator of clean and renewable energy [65]. Despite CPC slightly underestimating precipitation data (negative bias), the results indicate that CPC performs better than ERA5 across this entire basin (Figure 5c). This good performance, albeit with a slight underestimation of the CPC data, was also observed by Cardoso and Quadro [66], who evaluated precipitation in the southern region of Brazil, where a large part of the basin is located.

Due to its vast latitudinal extension, precipitation in the SFRB exhibits higher rainfall values concentrated in the southern sector of the basin (I and II) and lower rainfall values in the Sub-Middle (III) to Lower (IV) São Francisco regions (Figure 6a,b), as also shown by Freitas et al. [11]. The Brazilian semi-arid region is located in subregions III and IV, and the water demand here is greater than the supply, which generates socioeconomic impacts [67,68]. CPC and ERA5 indicate that the monthly and annual average have similar patterns to the observation, although both datasets underestimate precipitation values (Figure 6d,e). The statistical analysis shows that, although closely matched, CPC has superior performance (Figure 6c). Overall, both datasets exhibit high correlation and low bias, with agreement with ANA, especially notable for the CPC database. Torres et al. [69], when evaluating different databases with regard to the SFRB, also identified good performance of CPC data in this region.

In general, for the three basins, both datasets show strong correlation coefficients, low error (RMSE), and high agreement values with the ANA data, indicating good performance (Willmott Index, KGE, and NSE). However, the CPC results stand out, especially in the PRB, demonstrating superior performance compared to the other datasets.

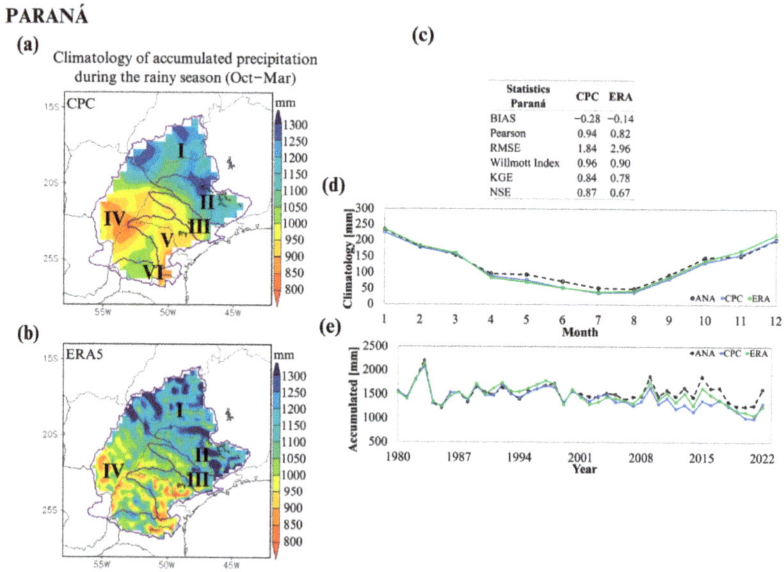


**Figure 5.** Analysis for the PRB: climatology of accumulated precipitation for the rainy season (October to March) for (**a**) CPC and (**b**) ERA5; (**c**) statistical analysis based on daily data; (**d**) monthly climatology; and (**e**) annual accumulation. In (**d**,**e**), the ERA5 database is indicated in green, the CPC is indicated in blue, and the observations are shown in black. Regions I, II, III, IV, V, and VI indicate the Paranaíba, Grande, Tietê, Paraná, Paranapanema, and Iguaçu regions.

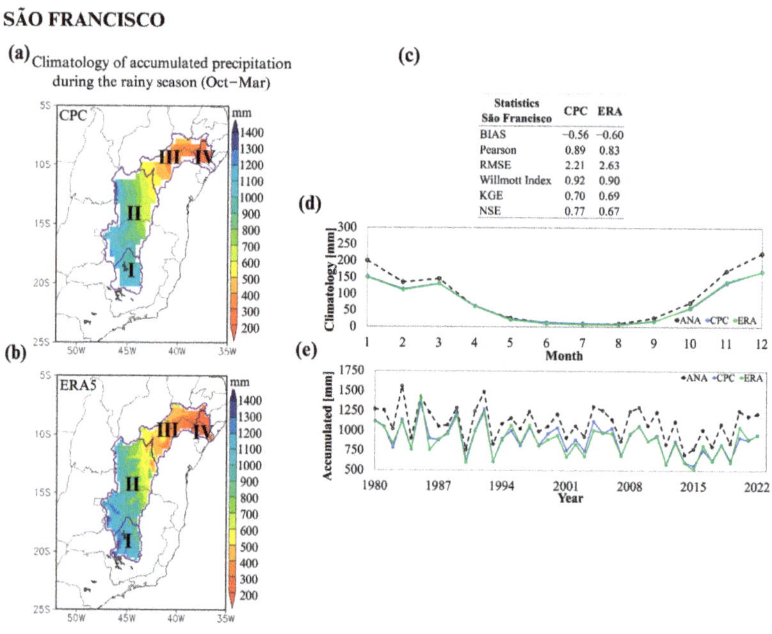


**Figure 6.** Analysis for the SFRB: climatology of accumulated precipitation for the rainy season (October to March) for (**a**) CPC and (**b**) ERA5; (**c**) statistical analysis based on daily data; (**d**) monthly

climatology; and (**e**) annual accumulation. In (**d**,**e**), the ERA5 database is indicated in green, the CPC is indicated in blue, and the observations are shown in black. Regions I, II, III, and IV indicate the Upper, Middle, Sub-Middle, and Lower São Francisco regions.

### 3.2. Extreme Daily Precipitation Events and Trend Analysis

The occurrence of extreme rainfall events was assessed by comparing daily precipitation data (averaged within each subregion of the basin) during the extended rainy season with the 95th percentile values obtained from the time series (ANA, CPC, and ERA5). Except for the Upper Madeira region, which had a limited number of stations available, all other regions showed consistent P95 values across the different datasets (Figure 7), with observations presenting a slightly higher value.

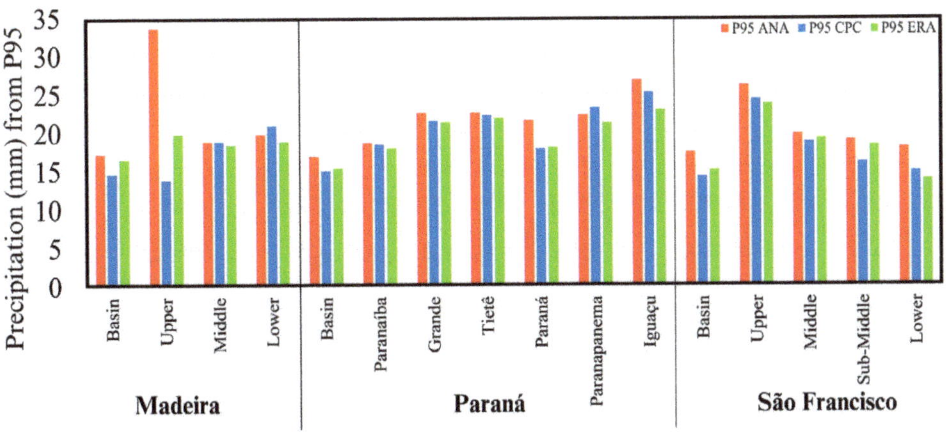


**Figure 7.** P95 values (precipitation in mm) identified by the ANA, CPC, and ERA5 datasets considering the average in the basins and for each of the subregions. The ERA5 database is indicated in green, the CPC in blue and the observations are shown in red.

The occurrence of extreme events in the MRB (Figure 8a) exhibits a similar average across the analyzed period (1980–2022) in most subregions, except for Upper Madeira, which has only two rain gauges. In the Upper Madeira region, the CPC underestimates the number of extreme events, whereas ERA5 overestimates it. For instance, the years 2007 and 2008 were marked by flood events in this region, leading to significant economic impacts and numerous fatalities [70]. During these years, both CPC and ERA5 recorded extreme precipitation events above the average (Figure 8d). However, ANA data did not indicate a trend, likely due to the limited number of rain gauges in this region. Overall, in the MRB, ANA indicates a decreasing trend in the number of days with extreme rainfall events across the entire basin, with statistical significance in almost every region except Middle Madeira (Table 1). The CPC results corroborate those obtained from ANA data, although they do not show statistical significance. On the other hand, ERA5 indicates a significant increasing trend in the number of occurrences in almost the entire basin. Espinoza et al. [71] also identified a downward trend in the annual frequency of rainy days (precipitation greater than 10 mm) in the period from 1981 to 2009 in the region where the MRB is located, more specifically in the Bolivian Amazon, in the south of the Brazilian Amazon, and in the border area between Peru, Colombia, and Brazil.

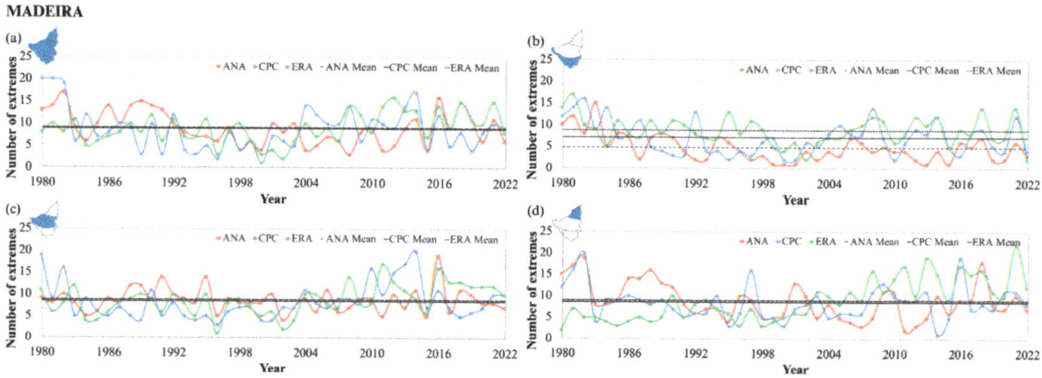


**Figure 8.** Number of days with extreme precipitation higher than P95 identified by the ANA, CPC, and ERA5 datasets in the MRB regions between 1980 and 2022: (**a**) entire basin; (**b**) Upper; (**c**) Middle; and (**d**) Lower. In some figures, the means of the datasets are quite close, which causes the black line representing the average to appear as a single line. The ERA5 database is indicated in green, the CPC is indicated in blue, and the observations are shown in red.

**Table 1.** Trends (increase represented by the letter I and decrease by D) and statistical significance obtained with Mann–Kendall analysis of the different datasets (ANA, CPC, and ERA5), considering the entire basin and the subregions, from 1980 and 2022. Significant trends occur when *p*-value is lower than 0.05.

| Region | Mann–Kendall | | | | | | | | |
|---|---|---|---|---|---|---|---|---|---|
| | ANA | | | CPC | | | ERA5 | | |
| | Trend | *p*-Value | Significant | Trend | *p*-Value | Significant | Trend | *p*-Value | Significant |
| Madeira | | | | | | | | | |
| Basin | D | 0.0444 | Yes | D | 0.9413 | No | I | 0.0189 | Yes |
| Upper | D | 0.0052 | Yes | D | 0.4058 | No | D | 0.4294 | No |
| Middle | D | 0.5604 | No | I | 0.0795 | No | I | 0.0068 | Yes |
| Lower | D | 0.0143 | Yes | D | 0.9161 | No | I | 0.0000 | Yes |
| Paraná | | | | | | | | | |
| Basin | D | 0.0579 | No | D | 0.0009 | Yes | D | 0.0686 | No |
| Paranaíba | D | 0.0840 | No | D | 0.0003 | Yes | I | 0.8580 | No |
| Grande | D | 0.0027 | Yes | D | 0.0003 | Yes | D | 0.1280 | No |
| Tietê | D | 0.0044 | Yes | D | 0.0009 | Yes | D | 0.6265 | No |
| Paraná | I | 0.0472 | Yes | I | 0.1741 | No | I | 0.8159 | No |
| Paranapanema | D | 0.3367 | No | D | 0.7502 | No | D | 0.3358 | No |
| Iguaçu | D | 0.5185 | No | D | 0.3259 | No | I | 0.5330 | No |
| São Francisco | | | | | | | | | |
| Basin | D | 0.0354 | Yes | D | 0.0668 | No | D | 0.3306 | No |
| Upper | D | 0.1456 | No | D | 0.0106 | Yes | D | 0.3067 | No |
| Middle | D | 0.0920 | No | D | 0.2510 | No | D | 0.6578 | No |
| Sub-Middle | D | 0.3559 | No | I | 0.1625 | No | D | 0.2238 | No |
| Lower | D | 0.8805 | No | I | 0.0637 | No | D | 0.3760 | No |

The average number of extreme events identified through the CPC and ERA5 in the PRB was closely aligned (Figure 9a). Additionally, the evolution in the frequency of extreme rainfall events over the years follows a consistent pattern, albeit with occasional discrepancies where ANA data were either underestimated or overestimated. Across most areas of the basin, notable peaks occurred during the years 1982/1983 and 1997/1998, marked by above-average values which resulted in economic and social losses due to floods [72]. Strong El Niño events, associated with the movement of the SACZ to the southeast of Brazil, where most of the PRB is located, were detected in both periods, as identified by Valverde and Marengo [72]. However, during the 2020/2021 period, many subregions experienced values below the average, coinciding with severe drought conditions affecting parts of the basin [12]. Overall, the observations also indicated a decreasing trend in the annual occurrence of extreme rainfall events across almost the entire PRB (Table 1). The northern and northeastern portions (Grande and Tietê) showed a statistically significant decrease (Table 1). However, in the central part of the basin, in the Paraná region, there was a significant increase in the annual occurrence of rainfall events. The CPC trends were found to be very similar to those from ANA, indicating a statistically significant decrease in the number of events both for the basin as a whole and in the northern and northeastern portions (Paranaíba, Grande, and Tietê). However, the ERA5 results showed some differences, with an increase in the number of events in more areas, but without statistical significance. The results from ANA and CPC are consistent with previous studies that analyzed the trend of extremes in the PRB region, which also indicate negative trends in the occurrence of extreme rainfall events in the northern and northeastern portions and positive trends in the central and southern parts of the basin [64,73].

**PARANÁ**

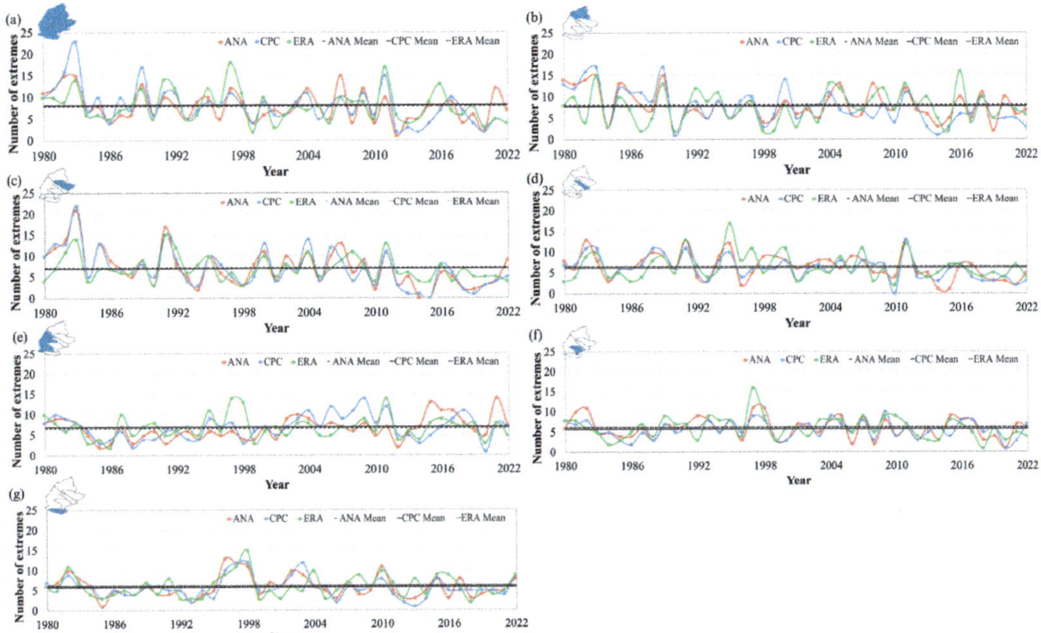


**Figure 9.** Number of days with extreme precipitation higher than P95 identified by the ANA, CPC, and ERA5 datasets in the PRB regions between 1980 and 2022: (**a**) entire basin; (**b**) Paranaíba; (**c**) Grande; (**d**) Tietê; (**e**) Paraná; (**f**) Paranapanema; (**g**) and Iguaçu. In some figures, the means of the datasets are quite close, which causes the black line representing the average to appear as a single line. The ERA5 database is indicated in green, the CPC is indicated in blue, and the observations are shown in red.

In the SFRB, the average results of extreme rainfall events were consistently similar across all datasets, with minimal variation over time (Figure 10a). The middle São Francisco region exhibited a more uniform pattern in the frequency of extreme events compared to the other subregions analyzed. For example, 1992 presented the highest number of extreme rainfall events recorded (Figure 10c), coinciding with one of the major floods in the basin [72,74]. Conversely, the SFRB presented the highest number of years without extreme events compared to the MRB and PRB. Specifically, the period 2014/2015 experienced a number of extreme events which was below the average, coinciding with one of the severest droughts in the region [11,75,76]. Overall, ANA data indicated a decreasing trend in the number of extreme rainfall events, with significant decreases across the basin as a whole (Table 1). ERA5 presented results similar to ANA, indicating a decreasing trend in events, but without statistical significance. On the contrary, CPC trends were different, with two regions (Sub-Middle and Lower) showing an increase in the number of extreme events, though not statistically significant. Additionally, the CPC for the Upper São Francisco region, located in southeastern Brazil where the river's source is found, showed a significant decrease in the number of rainfall events. Santos et al. [77] also identified in their study on the semi-arid region of Brazil—where a large part of the SFRB is concentrated—a negative trend in the frequency of extreme events in a large part of the area analyzed in the period from 2001 to 2020.

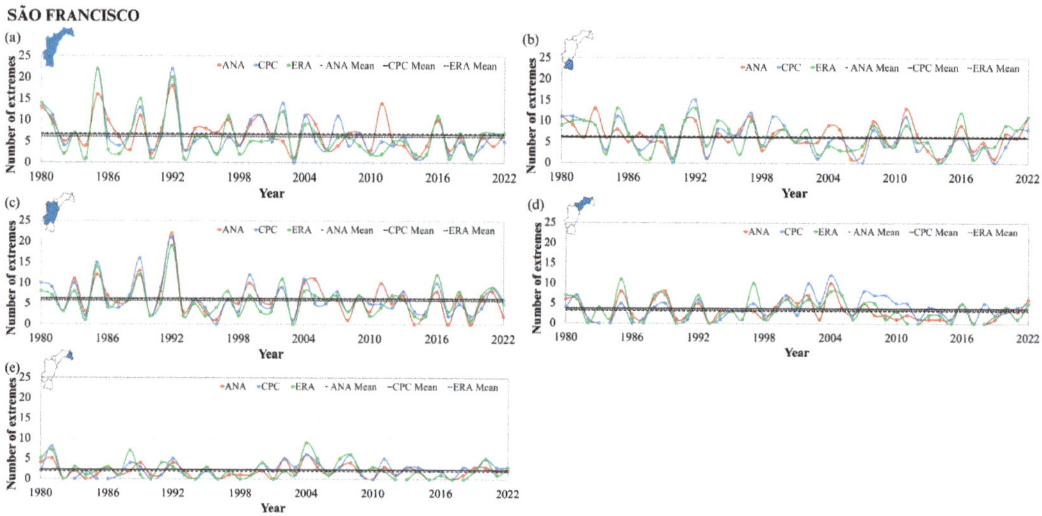


**Figure 10.** Number of days with extreme precipitation higher than P95 identified by the ANA, CPC, and ERA5 datasets in the SFRB regions between 1980 and 2022: (**a**) entire basin; (**b**) Upper; (**c**) Middle; (**d**) Sub-Middle; and (**e**) Lower. In some figures, the means of the datasets are quite close, which causes the black line representing the average to appear as a single line. The ERA5 database is indicated in green, the CPC is indicated in blue, and the observations are shown in red.

### 3.3. Extreme Climate Events

Climate-related extreme wet and drought events were identified using the SPI-6 from March, which stored the information regarding the extended rainy season. The results revealed a shift in the patterns of occurrence of extreme events across the basin, evident in both datasets (Figure 11). Above-average rainfall events have become less frequent, while drought events have become increasingly common. In the MRB, ANA recorded seven drought episodes and eight wet episodes (Figure 11a). The CPC, in turn, captured six droughts and six wet events, while ERA5 identified six droughts and only five wet periods (Figure 11b,c). Some of the wetter events recorded by ANA are associated with flood years in Porto Velho, a municipality belonging to the MRB, such as 1982, 1984, 1988 and 1997 [78], and these events were more effectively captured by the CPC data (Figure 11b). The year 2014 stands out as a period of unprecedented flooding in both the Brazilian and

Peruvian parts of the basin, which was reflected in the wet conditions observed in the SPI, especially in ERA5 (Figure 11c) [79]. Regarding drought episodes, the literature highlights the years 2005, 2010, 2016, 2020, and 2022 [9,80,81]. The year 2005 was identified as dry by ANA and ERA5, but not by CPC. However, the droughts of 2016, 2020, and 2022 were effectively captured by CPC. The year 2010 was not captured by any dataset, corroborating the observations of Molina-Carpio et al. [82], who also did not identify this period as dry. In general, the ANA and CPC datasets show a similar variation pattern, with more wet episodes at the beginning of the series and more dry episodes at the end. However, ERA5 exhibits an inverse behavior, with an increase in wet events in more recent years.

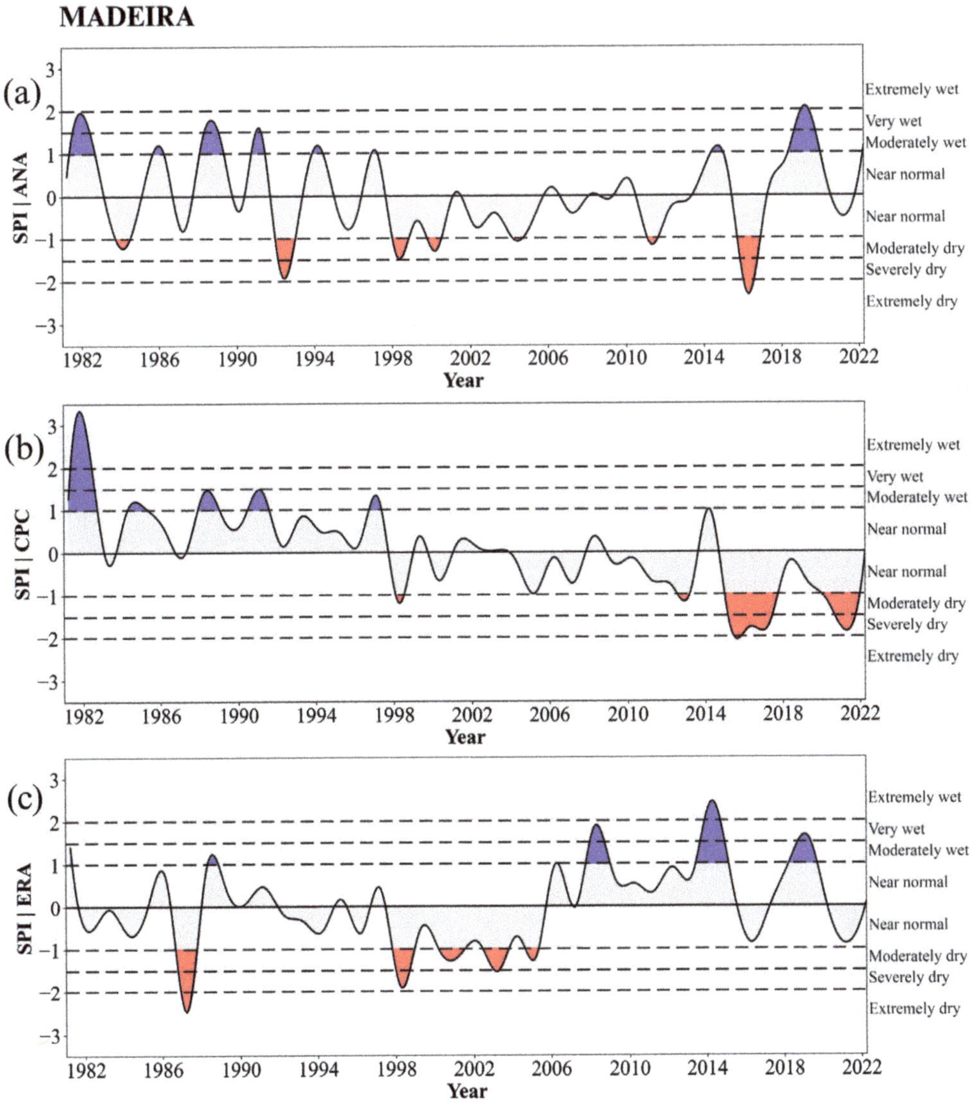


**Figure 11.** March SPI-6 for the MRB considering the datasets (**a**) ANA, (**b**) CPC, and (**c**) ERA5.

In the case of the PRB, four dry periods and eight wet periods were identified using SPI-6 calculated with ANA data (Figure 12a). Using CPC data, seven dry and five wet episodes were identified, while ERA5 indicated six dry and seven wet periods (Figure 12b,c). In general, the wettest periods occurred at the beginning of the series of all datasets, coinciding with several flood episodes in the basin, mainly between 1983/1984 and during the 1990s [83,84]. The SPI-6 results indicate that the most recent years have been drier in the basin, which is in line with previous findings in the literature [12,85]. Overall, both datasets present consistent SPI results, demonstrating similar behavior throughout the analyzed period.

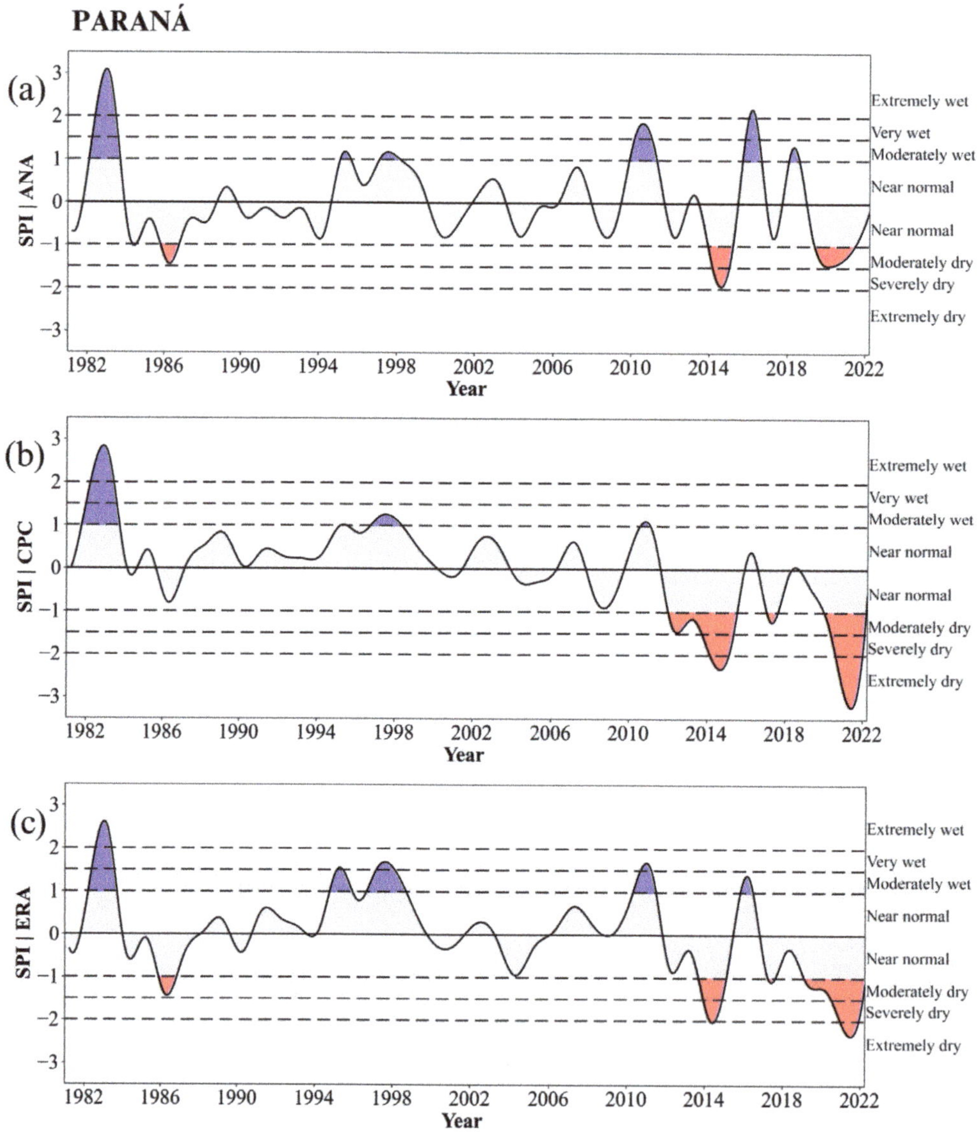


**Figure 12.** March SPI-6 for the PRB considering the datasets (**a**) ANA, (**b**) CPC, and (**c**) ERA5.

In total, the SPI-6 data from ANA indicate the occurrence of eight dry and seven wet episodes in the SFRB (Figure 13a). The CPC identified six dry and nine wet periods, while ERA5 showed six dry and eight wet periods (Figure 13b,c). As in the PRB, the wettest periods in the SFRB were recorded early in the series in both datasets, coinciding with flood episodes in the basin, such as in 1983, 1992, and 2004 [84,86]. Since 2013, the SFRB has experienced prolonged periods of below-average precipitation, as indicated by the SPI-6 of the analyzed databases, which is in line with reports in the literature [11,76]. Overall, the CPC and ERA5 data effectively identified the behavior of the SPI-6 series in the SFRB during the period from 1980 to 2022.

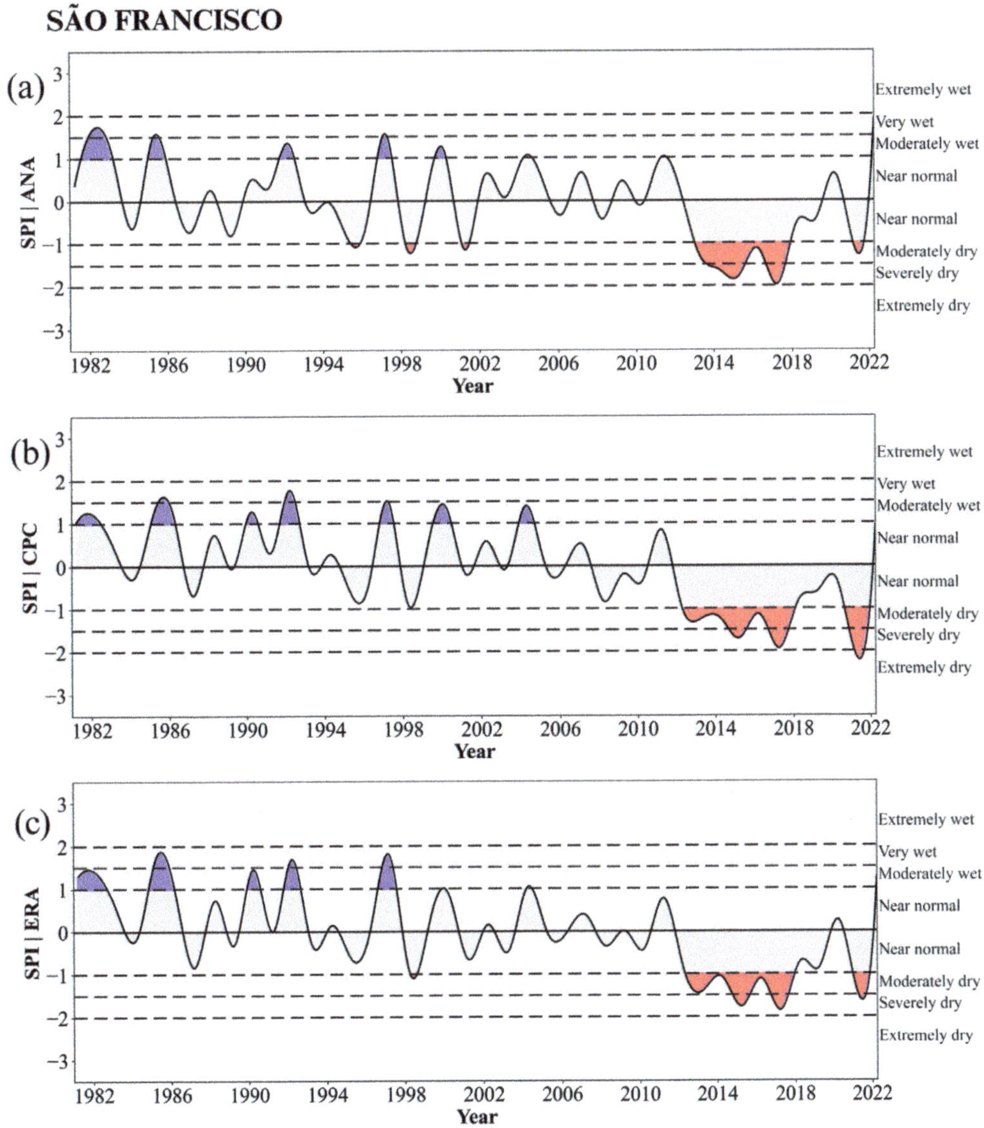


**Figure 13.** March SPI-6 for the SFRB considering the datasets (**a**) ANA, (**b**) CPC, and (**c**) ERA5.

In general, in the past, drought events in the Amazon region were mainly associated with the warm phase of the El Niño-Southern Oscillation and/or the positive phase of the North Atlantic Oscillation (NAO). The NAO affects the transport of atmospheric moisture to the Amazon, impacting annual rainfall variability [87,88]. In addition, increased deforestation and land use changes in the North and Central regions of Brazil have aggravated these effects, altering local climate dynamics and intensifying the occurrence of dry events in the study areas, especially in the MRB and PRB [89,90]. In the case of the SFRB, the tendency towards desertification is one of the main problems, affecting the hydrological cycle and making the region more arid and susceptible to extreme events [91,92].

## 4. Conclusions

In this study, an analysis of extreme weather and climate precipitation events during the rainy season was conducted in the MRB, PRB, and SFRB, which belong to the SAMS region. Additionally, the performance of CPC and ERA5 data in identifying the occurrence of these events was evaluated. Statistical analysis of the daily data revealed that the CPC set performed better in all basins analyzed.

The analysis of extreme weather rainfall events showed similar patterns for both datasets in most subregions of the basins. In general, with the exception of Upper Madeira, both datasets captured the pattern of the time series, but CPC better reproduces the extreme values as well the trend of the data. In the MRB, episodes such as the 2014 flood, which caused losses, were recorded as above-average extreme events by both datasets. In the PRB, CPC and ERA5 records of extremes were aligned, with notable peaks in 1982/1983 and 1997/1998 due to floods that resulted in economic and social losses, and below-average values in 2020/2021 due to a severe drought affecting parts of the basin. In the SFRB, the results were also consistent, with peaks in 1992 coinciding with one of the largest floods in the basin, and periods of low occurrence of extreme events during the severe drought of 2014/2015. Most of the subregions also showed a decreasing trend in the number of identified extreme weather rainfall events between 1980 and 2022.

According to the SPI-6 results, the pattern of extreme climate rainfall and drought events in the basins has changed, with a decrease in the frequency of above-average rainfall events and an increase in droughts. This change has been observed since the mid-1990s in the MRB and PRB and since 2010 in the SFRB. Overall, this change in the pattern of occurrence of extreme weather events was effectively reproduced by both datasets (CPC and ERA5). In the MRB, the extreme drought of 2010 was well represented by the SPI-6 from the datasets. In the PRB, recent SPI-6 data from both datasets indicate an intensification of droughts starting from 2020/2021. In the SFRB, the region has been experiencing prolonged periods of drought since 2013, which are also identified in both datasets. The increase in the occurrence of droughts in recent decades may have significant implications for water availability, agriculture, and biodiversity in the studied basins.

The trends observed in the study basins are expected to persist under climate change scenarios. Studies such as that by [93] indicate drier conditions in the North and Northeast regions of Brazil. In the Amazon, Llopart et al. [94] highlighted that the hydrological cycle is partially affected by the reduction in moisture convergence in the region. These studies serve as important warnings for decision-makers, who need to prepare for problems that are already underway and that may intensify in the future, such as water scarcity and increased occurrence of fires [95,96].

Figure 14 summarizes all analyses performed, indicating which dataset performed best. Overall, the CPC data presented the best statistical results for all basins. Regarding extreme precipitation events, although ERA5 performed better in percentage values, the CPC data identified precipitation extremes better, except for the MRB. For the analysis of drought events and Mann–Kendall trends, both datasets performed well. The lack of meteorological stations in the Upper Madeira region may have negatively influenced the results of the comparison between the datasets. Overall, both datasets performed well, indicating that they can be considered in the absence of station data. It is important to note the differences in the spatial resolutions of the data, with the CPC having a resolution of 0.5°, while ERA5 has a finer resolution of 0.25°. Furthermore, it is also worth mentioning some limitations of this study, in addition to highlighting the difficulties faced by Latin American researchers, which are mainly associated with the low density of meteorological stations, the quality of the data, which often presents inconsistencies and precision limitations, and the use of interpolation techniques to fill areas with little station coverage, which can result in information that does not always accurately represent real conditions [20].

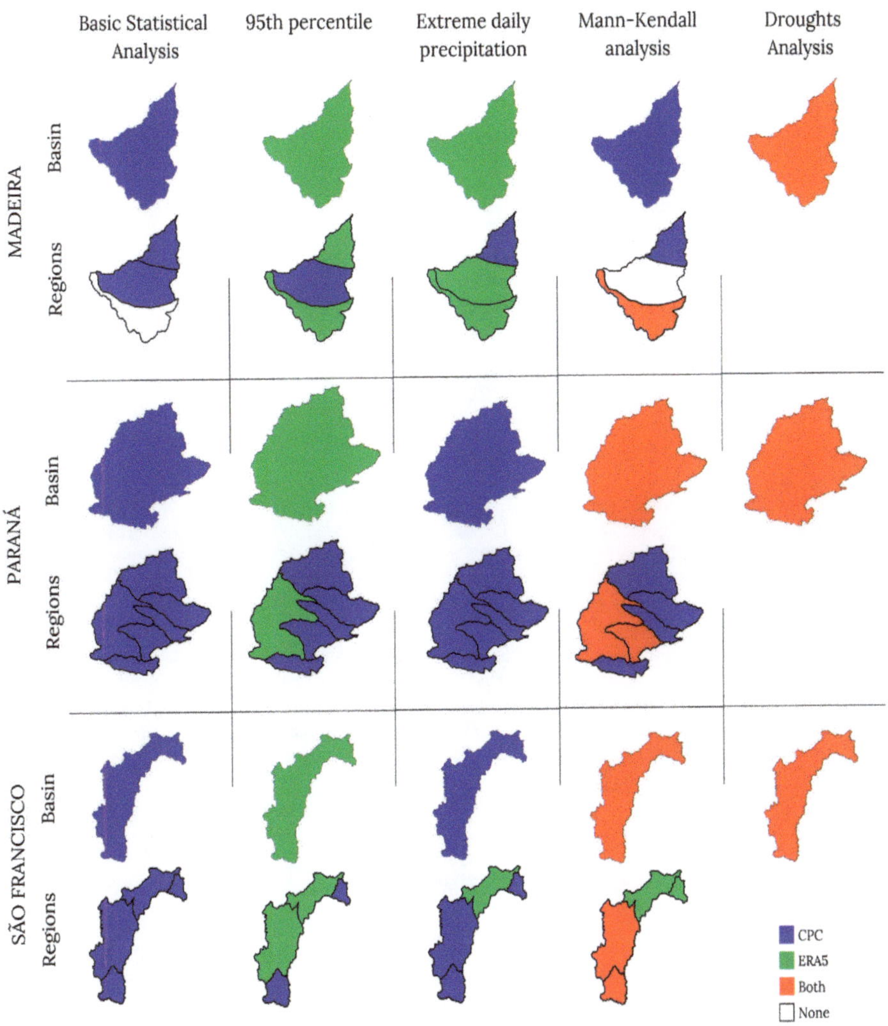


**Figure 14.** Summary of all analyses performed and which database presented the best results, considering the basin as a whole and the subregions of MRB, PRB, and SFRB.

**Author Contributions:** Conceptualization, A.A.d.F., V.S.B.C. and M.S.R.; methodology, A.A.d.F., V.S.B.C. and M.S.R.; software, A.A.d.F.; validation, A.A.d.F., V.S.B.C. and M.S.R.; formal analysis, A.A.d.F., V.S.B.C. and M.S.R.; investigation, A.A.d.F., V.S.B.C. and M.S.R.; writing—original draft preparation, A.A.d.F., V.S.B.C. and M.S.R.; writing—review and editing, A.A.d.F., V.S.B.C. and M.S.R.; visualization, A.A.d.F., V.S.B.C. and M.S.R.; supervision, A.A.d.F., V.S.B.C. and M.S.R.; project administration, A.A.d.F., V.S.B.C. and M.S.R. All authors have read and agreed to the published version of the manuscript.

**Funding:** This research was funded by the Coordination for the Improvement of Higher Education Personnel (CAPES), grant number 88887.921175/2023-00.

**Data Availability Statement:** Data are contained within the article.

**Acknowledgments:** The authors would like to thank the Coordination for the Improvement of Higher Education Personnel (CAPES). We would like to thank ECMWF, CPC, and ANA for providing the data. The authors would like to thank the editor and the anonymous reviewers.

**Conflicts of Interest:** The authors declare no conflicts of interest.

## References


1. Lenters, J.D.; Cook, K.H. Simulation and diagnosis of the regional summertime precipitation climatology of South America. *J. Clim.* **1995**, *8*, 2988–3005. [CrossRef]
2. Zhou, J.; Lau, K.M. Does a monsoon climate exist over South America? *J. Clim.* **1998**, *11*, 1020–1040. [CrossRef]
3. Jones, C. and Carvalho, L.M. Active and break phases in the South American monsoon system. *J. Clim.* **2002**, *15*, 905–914. [CrossRef]
4. Marengo, J.A.; Liebmann, B.; Grimm, A.M.; Misra, V.; Silva Dias, P.L.; Cavalcanti, I.F.A.; Carvalho, L.M.V.; Berbery, E.H.; Ambrizzi, T.; Vera, C.S.; et al. Recent developments on the South American monsoon system. *Int. J. Climatol.* **2012**, *32*, 1–21. [CrossRef]
5. Ferreira, G.W.S.; Reboita, M.S. A new look into the South America precipitation regimes: Observation and forecast. *Atmosphere* **2022**, *13*, 873. [CrossRef]
6. Reboita, M.S.; Teodoro, T.A.; Ferreira, G.W.S.; Souza, C.A. Ciclo de vida do sistema de monção da América do Sul: Clima presente e futuro. *Rev. Bras. Geogr. FÍS.* **2022**, *15*, 343–358. [CrossRef]
7. Coelho, C.A.S.; Oliveira, C.P.; Ambrizzi, T.; Reboita, M.S.; Carpenedo, C.B.; Campos, J.L.P.S.; Tomaziello, A.C.N.; Pampuch, L.A.; Custódio, M.S.; Dutra, L.M.M.; et al. The 2014 southeast Brazil austral summer drought: Regional scale mechanisms and teleconnections. *Clim. Dyn.* **2016**, *46*, 3737–3752. [CrossRef]
8. Marengo, J.A.; Williams, E.R.; Alves, L.M.; Soares, W.R.; Rodriguez, D.A. Extreme seasonal climate variations in the Amazon Basin: Droughts and floods. *Interact. Biosph. Atmos. Hum. Land Use Amaz. Basin* **2016**, *227*, 55–76. [CrossRef]
9. Marengo, J.A.; Borma, L.S.; Rodríguez, D.A.; Pinho, P.; Soares, W.R.; Alves, L.M. Recent extremes of drought and flooding in Amazonia: Vulnerabilities and human adaptation. *Am. J. Clim. Change* **2013**, *2*, 87–96. [CrossRef]
10. Santos, E.B.; Freitas, E.D.; Rafee, S.A.A.; Fujita, T.; Rudke, A.P.; Martins, L.D.; Martins, J.A. Spatio-temporal variability of wet and drought events in the Paraná River basin—Brazil and its association with the El Niño—Southern Oscillation phenomenon. *Int. J. Climatol.* **2021**, *41*, 4879–4897. [CrossRef]
11. Freitas, A.A.; Drumond, A.; Carvalho, V.S.B.; Reboita, M.S.; Silva, B.C.; Uvo, C.B. Drought assessment in São Francisco river basin, Brazil: Characterization through SPI and associated anomalous climate patterns. *Atmosphere* **2022**, *13*, 41. [CrossRef]
12. Freitas, A.A.; Reboita, M.S.; Carvalho, V.S.B.; Drumond, A.; Ferraz, S.E.T.; Silva, B.C.; Rocha, R.P. Atmospheric and oceanic patterns associated with extreme drought events over the Paraná Hydrographic Region, Brazil. *Climate* **2023**, *11*, 12. [CrossRef]
13. IPCC. Annex II: Glossary. In *Climate Change 2022: Impacts, Adaptation and Vulnerability. Contribution of Working Group II to the Sixth Assessment Report of the Intergovernmental Panel on Climate Change*; Pörtner, H.-O., Roberts, D.C., Tignor, M., Poloczanska, S.E., Mintenbeck, K., Alegría, A., Craig, M., Langsdorf, S., Löschke, S., Möller, V., et al., Eds.; Cambridge University Press: Cambridge, UK; New York, NY, USA, 2022; pp. 2897–2930.
14. Silva, V.B.; Kousky, V.E. The South American monsoon system: Climatology and variability. *Mod. Climatol.* **2012**, *123*, 152. Available online: https://www.whoi.edu/cms/files/silva_kousky_217705.pdf (accessed on 17 June 2024).
15. Vu, T.M.; Mishra, A.K. Nonstationary frequency analysis of the recent extreme precipitation events in the United States. *J. Hydrol.* **2019**, *575*, 999–1010. [CrossRef]
16. Van Lanen, H.A.J.; Laaha, G.; Kingston, D.G.; Gauster, T.; Ionita, M.; Vidal, J.P.; Vlnas, R.; Tallaksen, L.M.; Stahl, K.; Hannaford, J.; et al. Hydrology needed to manage droughts: The 2015 European case. *Hydrol. Process.* **2016**, *30*, 3097–3104. [CrossRef]
17. Marengo, J.A.; Tomasella, J.; Alves, L.M.; Soares, W.R.; Rodriguez, D.A. The drought of 2010 in the context of historical droughts in the Amazon region. *Geophys. Res. Lett.* **2011**, *38*, 1–5. [CrossRef]
18. Lima, L.S.; Silva, F.E.O.E.; Dorio Anastácio, P.R.; Kolanski, M.M.D.P.; Pires Pereira, A.C.; Menezes, M.S.R.; Macedo, M.N. Severe droughts reduce river navigability and isolate communities in the Brazilian Amazon. *Commun. Earth Environ.* **2024**, *5*, 370. [CrossRef]
19. Lagos-Zúniga, M.; Balmaceda-Huarte, R.; Regoto, P.; Torrez, L.; Olmo, M.; Lyra, A.; Pareja-Quispe, D.; Bettolli, M.L. Extreme indices of temperature and precipitation in South America: Trends and intercomparison of regional climate models. *Clim. Dyn.* **2022**, *62*, 4541–4562. [CrossRef]
20. Cavazos, T.; Bettolli, M.L.; Campbell, D.; Sánchez Rodríguez, R.A.; Mycoo, M.; Arias, P.A.; Rivera, J.; Simões Reboita, M.; Gulizia, C.; Hidalgo, H.G.; et al. Challenges for climate change adaptation in Latin America and the Caribbean region. *Front. Clim.* **2024**, *6*, 1392033. [CrossRef]

21. ANA. *Projeto de Gerenciamento Integrado das Atividades Desenvolvidas em Terra na Bacia do Rio São Francisco: Programa de ações estratégicas para o gerenciamento integrado da Bacia do Rio São Francisco e da sua Zona Costeira*; PAE: Brasília, Brazil, 2004; p. 50.
22. ANA. *Conjuntura dos recursos hídricos no Brasil: Regiões hidrográficas brasileiras—Edição Especial*; ANA: Brasília, Brazil, 2015; p. 164.
23. ANA. Madeira. 2024. Available online: https://www.gov.br/ana/pt-br/sala-de-situacao/rio-madeira/saiba-mais (accessed on 10 June 2024).
24. CEMADEN. 13/04/2018—Previsão de vazão para bacia do Rio Madeira. 2018. Available online: http://www2.cemaden.gov.br/13042018-previsao-de-vazao-para-bacia-do-rio-madeira/ (accessed on 10 June 2024).
25. ANA. *Plano estratégico de recursos hídricos dos afluentes da margem direita do rio Amazonas: Resumo executivo/Agência Nacional de Águas*; ANA: Brasília, Brazil, 2012; p. 146.
26. Santo Antônio Energia. Saiba Mais Rio Madeira. 2022. Available online: https://www.santoantonioenergia.com.br/wp-content/uploads/2022/08/02_rio_madeira.pdf (accessed on 10 June 2024).
27. Eletrobras Furnas. Usina de Santo Antônio. 2024. Available online: https://www.furnas.com.br/subsecao/134/usina-de-santo-antonio?culture=pt (accessed on 10 June 2024).
28. JIRAU. A estrutura da usina hidrelétrica Jirau. 2024. Available online: https://www.jirauenergia.com.br/conheca-a-uhe/ (accessed on 10 June 2024).
29. INPA. *Biodiversidade do médio Madeira: Bases científicas para propostas de conservação*; INPA: Manaus, Brazil, 2007; p. 239.
30. Perigolo, N.A.; Medeiros, M.B.; Simon, M.F. Vegetation types of the upper Madeira River in Rondônia, Brazil. *Brittonia* **2017**, *69*, 423–446. [CrossRef]
31. Muniz, L.S. Análise dos padrões fluviométricos da Bacia do Rio Madeira-Brasil. Master's Thesis, Universidade Federal do Amazonas, Manaus, Brazil, 2013; p. 146. Available online: https://tede.ufam.edu.br/handle/tede/3961 (accessed on 10 June 2024).
32. Ribeiro Neto, A.; Silva, R.C.V.; Collischonn, W.; Tucci, C.E.M. Simulação na bacia Amazônica com dados limitados: Rio Madeira. *RBRH* **2008**, *13*, 47–58. [CrossRef]
33. MMA. *Caderno da Região Hidrográfica do Paraná/Ministério do Meio Ambiente*; Secretaria de Recursos Hídricos; MMA: Brasília, Brazil, 2006; p. 122.
34. SNIRH. Atlas Geográfico digital de recursos hídricos do Brasil. 2013. Available online: https://portal1.snirh.gov.br/atlasrh2013/ (accessed on 17 July 2024).
35. ANA. *Atlas Brasil: Abastecimento urbano de água: Panorama nacional*; ANA: Brasília, Brazil, 2010; p. 72.
36. Itaipu Binacional. *Assegurar o acesso confiável, sustentável, moderno e a preço acessível à energia para todas e todos/Itaipu Binacional*; Central Hidrelétrica de Itaipu: Paraná, Brazil, 2019; p. 56. Available online: https://www.itaipu.gov.br/sites/default/files/af_df/Estudo_de_caso_Itaipu_ODS_7.pdf (accessed on 17 June 2024).
37. Itaipu Binacional. *Relatório Anual Itaipu Binacional 2020*; Central Hidrelétrica de Itaipu: Paraná, Brazil, 2020; p. 132. Available online: https://www.itaipu.gov.br/sites/default/files/af_df/Relatorio_Anual_2020_v2.pdf (accessed on 17 June 2024).
38. Rudke, A.P.; Xavier, A.C.F.; Martins, L.D.; Freitas, E.D.; Uvo, C.B.; Hallak, R.; Souza, R.A.F.; Andreoli, R.V.; Almeida Albuquerque, T.T.; Martins, J.A. Landscape changes over 30 years of intense economic activity in the upper Paraná River basin. *Ecol. Informat.* **2022**, *72*, 101882. [CrossRef]
39. Reboita, M.S.; Marietto, D.M.G.; Souza, A.; Barbosa, M. Caracterização atmosférica quando da ocorrência de eventos extremos de chuva na região sul de Minas Gerais. *Rev. Bras. Climatol.* **2017**, *21*, 20–37. [CrossRef]
40. Escobar, G.C.J.; Reboita, M.S. Relationship between daily atmospheric circulation patterns and South Atlantic Convergence Zone (SACZ) events. *AtmÓSfera* **2022**, *35*, 1–25. [CrossRef]
41. CODEVASF. *Plano Nascente: Plano de preservação e recuperação de nascentes da bacia do rio São Francisco*; IABS: Brasília, Brazil, 2015; p. 129.
42. CBHSF. A Bacia. 2024. Available online: https://cbhsaofrancisco.org.br/a-bacia/ (accessed on 8 July 2024).
43. IBGE. *Vetores Estruturantes da Dimensão Socioeconômica da Bacia Hidrográfica do Rio São Francisco*; IBGE: Rio de Janeiro, Brazil, 2009; p. 174.
44. Silva, D.F.; Brito, J.I.B. Variabilidade do vento na bacia hidrográfica do rio São Francisco durante a ocorrência da ZCAS. *AmbiÊNcia* **2008**, *4*, 221–235. Available online: https://revistas.unicentro.br/index.php/ambiencia/article/view/164 (accessed on 17 June 2024).
45. Utida, G.; Cruz, F.W.; Etourneau, J.; Bouloubassi, I.; Schefuß, E.; Vuille, M.; Novello, V.F.; Prado, L.F.; Sifeddine, A.; Klein, V.; et al. Tropical South Atlantic influence on Northeastern Brazil precipitation and ITCZ displacement during the past 2300 years. *Sci. Rep.* **2019**, *9*, 1698. [CrossRef]
46. Chen, M.; Shi, W.; Xie, P.; Silva, V.B.; Kousky, V.E.; Higgins, R.W.; Janowiak, J.E. Assessing objective techniques for gauge-based analyses of global daily precipitation. *J. Geophys. Res. Atmos.* **2008**, *113*, D04110. [CrossRef]
47. Sun, Q.; Miao, C.; Duan, Q.; Ashouri, H.; Sorooshian, S.; Hsu, K.L. A review of global precipitation data sets: Data sources, estimation, and intercomparisons. *Rev. Geophys.* **2018**, *56*, 79–107. [CrossRef]
48. Hersbach, H.; Bell, B.; Berrisford, P.; Hirahara, S.; Horányi, A.; Muñoz-Sabater, J.; Nicolas, J.; Peubey, C.; Radu, R.; Schepers, D.; et al. The ERA5 global reanalysis. *Q. J. R. Meteorol. Soc.* **2020**, *146*, 1999–2049. [CrossRef]
49. Montgomery, D.C.; Jennings, C.L.; Kulahci, M. *Introduction to Time Series Analysis and Forecasting*; John Wiley & Sons: Hoboken, NJ, USA, 2015; p. 671.

50. Hinkle, D.E.; Wiersma, W.; Jurs, S.G. *Applied Statistics for the Behavioral Sciences*; Houghton Mifflin: Boston, MA, USA, 2003; Volume 663, p. 748.
51. Wilks, D.S. *Statistical Methods in the Atmospheric Sciences*, 4th ed.; Elsevier: Amsterdam, The Netherlands, 2019; p. 807.
52. Hallak, R.; Pereira Filho, A.J. Metodologia para análise de desempenho de simulações de sistemas convectivos na região metropolitana de São Paulo com o modelo ARPS: Sensibilidade a variações com os esquemas de advecção e assimilação de dados. *Rev. Bras. Meteorol.* **2011**, *26*, 591–608. [CrossRef]
53. Willmott, C.J. On the validation of models. *Phys. Geogr.* **1981**, *2*, 184–194. [CrossRef]
54. Willmott, C.J.; Robeson, S.M.; Matsuura, K. A refined index of model performance. *Int. J. Climatol.* **2012**, *32*, 2088–2094. [CrossRef]
55. Gupta, H.V.; Kling, H.; Yilmaz, K.K.; Martinez, G.F. Decomposition of the mean squared error and NSE performance criteria: Implications for improving hydrological modelling. *J. Hydrol.* **2009**, *377*, 80–91. [CrossRef]
56. Kling, H.; Fuchs, M.; Paulin, M. Runoff conditions in the upper Danube basin under an ensemble of climate change scenarios. *J. Hydrol.* **2012**, *424*, 264–277. [CrossRef]
57. Nash, J.E.; Sutcliffe, J.V. River flow forecasting through conceptual models part I—A discussion of principles. *J. Hydrol.* **1970**, *10*, 282–290. [CrossRef]
58. Moriasi, D.N.; Arnold, J.G.; Van Liew, M.W.; Bingner, R.L.; Harmel, R.D.; Veith, T.L. Model evaluation guidelines for systematic quantification of accuracy in watershed simulations. *Trans. ASABE* **2007**, *50*, 885–900. [CrossRef]
59. Fernandes, D.S.; Heinemann, A.B.; Paz, R.L.; Amorim, A.O.; Cardoso, A.S. *Índices Para a Quantificação da Seca*; Embrapa Arroz e Feijão-Documentos: Colombo, Brazil, 2009; p. 48. Available online: http://www.infoteca.cnptia.embrapa.br/infoteca/handle/doc/663874 (accessed on 3 July 2024).
60. Mckee, T.B.; Doesken, N.J.; Kleist, J. The relationship of drought frequency and duration to time scales. In Proceedings of the 8th Conference on Applied Climatology, Anaheim, CA, USA, 17–22 January 1993; pp. 179–183.
61. WMO. Standardized Precipitation Index. In *User Guide*; WMO: Geneva, Switzerland, 2012; p. 24.
62. Vergasta, L.A.; Correia, F.W.S.; Chou, S.C.; Nobre, P.; Lyra, A.D.A.; Gomes, W.D.B.; Capistrano, V.; Veiga, J.A.P. Avaliação do Balanço de água na Bacia do Rio Madeira Simulado Pelo Modelo Regional Climático Eta e o Modelo Hidrológico de Grandes Bacias MGB. *Rev. Bras. Meteorol.* **2021**, *36*, 153–169. [CrossRef]
63. Liu, R.; Zhang, X.; Wang, W.; Wang, Y.; Liu, H.; Ma, M.; Tang, G. Global-scale ERA5 product precipitation and temperature evaluation. *Ecol. Indic.* **2024**, *166*, 112481. [CrossRef]
64. Rafee, S.A.A.; Freitas, E.D.; Martins, J.A.; Martins, L.D.; Domingues, L.M.; Nascimento, J.M.; Machado, C.B.; Santos, E.B.; Rudke, A.P.; Fujita, T.; et al. Spatial trends of extreme precipitation events in the Paraná river basin. *J. Appl. Meteorol. Climatol.* **2020**, *59*, 443–454. [CrossRef]
65. Itaipu Binacional. Revista Itaipu Binacional. 2017. Available online: https://www.itaipu.gov.br/sites/default/files/af_df/1702_022_atualizacao_revista_IB_2017_digital_b.pdf (accessed on 17 June 2024).
66. Cardoso, C.S.; Quadro, M.F.L. Análise comparativa de dados de precipitação gerados pelo "Climate Prediction Center–CPC" versus dados observados para o Sul do Brasil. *Rev. Bras. Geogr. FÍS.* **2017**, *10*, 1180–1198. [CrossRef]
67. Correia, R.C.; Kiill, L.H.P.; Moura, M.S.B.; Cunha, T.J.F.; Jesus Júnior, L.A.; Araújo, J.L.P. *A Região Semiárida Brasileira*; Embrapa: Brasília, Brazil, 2011; p. 28. Available online: https://www.embrapa.br/busca-de-publicacoes/-/publicacao/916891/a-regiao-semiarida-brasileira (accessed on 17 June 2024).
68. Marengo, J.A.; Alves, L.M.; Beserra, E.A.; Lacerda, F.F. Variabilidade e mudanças climáticas no semiárido brasileiro. *Rec. HÍDricos Regiões ÁRidas SemiÁRidas* **2011**, *1*, 385–422. Available online: https://www.ccst.inpe.br/publicacao/variabilidade-e-mudancas-climaticas-no-semiarido-brasileiro/ (accessed on 3 July 2024).
69. Torres, F.L.R.; Souza Ferreira, G.W.; Kuki, C.A.C.; Vasconcellos, B.T.C.; Freitas, A.A.; Nascimento Silva, P.; Souza, C.A.; Reboita, M.S. Validação de diferentes bases de dados de precipitação nas bacias hidrográficas do Sapucaí e São Francisco. *Rev. Bras. Climatol.* **2020**, *27*, 368–404. [CrossRef]
70. Ovando, A.; Tomasella, J.; Rodriguez, D.A.; Martinez, J.M.; Siqueira-Junior, J.L.; Pinto, G.L.N.; Passy, P.; Vauchel, P.; Noriega, L.; von Randow, C. Extreme flood events in the Bolivian Amazon wetlands. *J. Hydrol. Reg. Stud.* **2016**, *5*, 293–308. [CrossRef]
71. Espinoza, J.C.; Ronchail, J.; Marengo, J.A.; Segura, H. Contrasting North–South changes in Amazon wet-day and dry-day frequency and related atmospheric features (1981–2017). *Clim. Dyn.* **2019**, *52*, 5413–5430. [CrossRef]
72. Valverde, M.C.; Marengo, J.A. Extreme rainfall indices in the hydrographic basins of Brazil. *Open J. Mod. Hydrol.* **2014**, *4*, 10–26. [CrossRef]
73. Cerón, W.L.; Kayano, M.T.; Andreoli, R.V.; Avila-Diaz, A.; Ayes, I.; Freitas, E.D.; Martins, J.A.; Souza, R.A. Recent intensification of extreme precipitation events in the La Plata Basin in Southern South America (1981–2018). *Atmos. Res.* **2021**, *249*, 105299. [CrossRef]
74. CBHSF. *Plano de recursos hídricos da bacia hidrográfica do rio São Francisco*; CBHSF: Bahia, Brazil, 2004; p. 337.
75. Marengo, J.A.; Cunha, A.P.; Alves, L.M. A seca de 2012-15 no semiárido do Nordeste do Brasil no contexto histórico. *Rev. ClimanÁLise* **2016**, *3*, 49–54. Available online: http://climanalise.cptec.inpe.br/~rclimanl/revista/pdf/30anos/marengoetal.pdf (accessed on 3 July 2024).
76. Cunha, A.P.M.A.; Zeri, M.; Deusdará Leal, K.; Costa, L.; Cuartas, L.A.; Marengo, J.A.; Tomasella, J.; Vieira, R.M.; Barbosa, A.A.; Cunningham, C.; et al. Extreme drought events over Brazil from 2011 to 2019. *Atmosphere* **2019**, *10*, 642. [CrossRef]

77. Santos, A.L.M.; Gonçalves, W.A.; Andrade, L.D.M.B.; Rodrigues, D.T.; Batista, F.F.; Lima, G.C.; Silva, C.M.S. Space–Time Characterization of Extreme Precipitation Indices for the Semiarid Region of Brazil. *Climate* **2024**, *12*, 43. [CrossRef]
78. Bezerra, H.C. *Monitoramento Hidrológico*; Boletim No 2—25 January 2012; CPRM—Serviço Geológico do Brasil: Brasília, Brazil, 2012; p. 11.
79. Espinoza, J.C.; Marengo, J.A.; Ronchail, J.; Carpio, J.M.; Flores, L.N.; Guyot, J.L. The extreme 2014 flood in south-western Amazon basin: The role of tropical-subtropical South Atlantic SST gradient. *Environ. Res. Lett.* **2014**, *9*, 124007. [CrossRef]
80. Tomasella, J.; Borma, L.S.; Marengo, J.A.; Rodriguez, D.A.; Cuartas, L.A.; Nobre, C.A.; Prado, M.C. The droughts of 1996–1997 and 2004–2005 in Amazonia: Hydrological response in the river main-stem. *Hydrol. Process.* **2011**, *25*, 1228–1242. [CrossRef]
81. Laureanti, N.C.; Tavares, P.D.S.; Tavares, M.; Rodrigues, D.C.; Gomes, J.L.; Chou, S.C.; Correia, F.W.S. Extreme Seasonal Droughts and Floods in the Madeira River Basin, Brazil: Diagnosis, Causes, and Trends. *Climate* **2024**, *12*, 111. [CrossRef]
82. Molina-Carpio, J.; Espinoza, J.C.; Vauchel, P.; Ronchail, J.; Gutierrez Caloir, B.; Guyot, J.L.; Noriega, L. Hydroclimatology of the Upper Madeira River basin: Spatio-temporal variability and trends. *Hydrol. Sci. J.* **2017**, *62*, 911–927. [CrossRef]
83. Itaipu Binacional. Chuvas provocam vazão histórica no encontro dos rios Iguaçu e Paraná. 2013. Available online: https://www.itaipu.gov.br/sala-de-imprensa/noticia/chuvas-provocam-vazao-historica-no-encontro-dos-rios-iguacu-e-parana#:~:text=Houve%20registro%20de%20enchentes%20em,foram%20quatro%20acima%20desse%20%C3%ADndice (accessed on 3 July 2024).
84. Fleischmann, A.S.; Siqueira, V.A.; Wongchuig-Correa, S.; Collischonn, W.; Paiva, R.C.D.D. The great 1983 floods in South American large rivers: A continental hydrological modelling approach. *Hydrol. Sci. J.* **2020**, *65*, 1358–1373. [CrossRef]
85. Cumplido, M.A.; Inocente, M.C.; Medeiros, T.P.; Oliveira, G.S.; Marengo, J.A. Secas e crises hídricas no Sudeste do Brasil: Um histórico comparativo entre os eventos de 2001, 2014 e 2021 com enfoque na bacia do rio Paraná. *Rev. Bras. Climatol.* **2023**, *32*, 129–153. [CrossRef]
86. CBHSF. RP1A —Diagnóstico da Dimensão Técnica e Institucional. In *Plano de Recursos Hídricos da Bacia Hidrográfica do rio São Francisco (2016–2025)*; CBHSF: São Paulo, Brazil, 2015; p. 318. Available online: https://2017.cbhsaofrancisco.org.br/wp-content/uploads/2015/08/V7_usos_diag_v2.pdf (accessed on 8 July 2024).
87. Marengo, J.A.; Espinoza, J.C. Extreme seasonal droughts and floods in Amazonia: Causes, trends and impacts. *Int. J. Climatol.* **2015**, *36*. [CrossRef]
88. Marengo, J.A.; Cunha, A.P.; Espinoza, J.C.; Fu, R.; Schöngart, J.; Jimenez, J.C.; Costa, M.C.; Ribeiro, J.M.; Wongchuig, S.; Zhao, S. The Drought of Amazonia in 2023–2024. *Am. J. Clim. Change* **2024**, *13*, 567–597. [CrossRef]
89. Sorí, R.; Marengo, J.A.; Nieto, R.; Drumond, A.; Gimeno, L. The atmospheric branch of the hydrological cycle over the Negro and Madeira river basins in the Amazon Region. *Water* **2018**, *10*, 738. [CrossRef]
90. Naumann, G.; Podesta, G.; Marengo, J.; Luterbacher, J.; Bavera, D.; Arias Muñoz, C.; Barbosa, P.; Cammalleri, C.; Chamorro, L.; Cuartas, A.; et al. *The 2019-2021 Extreme Drought Episode in La Plata Basin*; Publications Office of the European Union: Luxembourg, 2021.
91. Vieira, R.M.D.S.P.; Sestini, M.F.; Tomasella, J.; Marchezini, V.; Pereira, G.R.; Barbosa, A.A.; Santos, F.C.; Rodriguez, D.A.; do Nascimento, F.R.; Santana, M.O.; et al. Characterizing spatio-temporal patterns of social vulnerability to droughts, degradation and desertification in the Brazilian northeast. *Environ. Sustain. Indic.* **2020**, *5*, 100016. [CrossRef]
92. Vieira, R.M.D.S.P.; Tomasella, J.; Cunha, A.P.M.D.A.; Barbosa, A.A.; Pompeu, J.; Ferreira, Y.; Santos, F.C.; Alves, L.M.; Ometto, J. Socio-environmental vulnerability to drought conditions and land degradation: An assessment in two northeastern brazilian river basins. *Sustainability* **2023**, *15*, 8029. [CrossRef]
93. Reboita, M.S.; Kuki, C.A.C.; Marrafon, V.H.; Souza, C.A.; Ferreira, G.W.S.; Teodoro, T.; Lima, J.W.M. South America climate change revealed through climate indices projected by GCMs and Eta-RCM ensembles. *Clim. Dyn.* **2022**, *58*, 459–485. [CrossRef]
94. Llopart, M.; Domingues, L.M.; Torma, C.; Giorgi, F.; Rocha, R.P.; Ambrizzi, T.; Reboita, M.S.; Alves, L.M.; Coppola, L.M.; da Silva, M.L.; et al. Assessing changes in the atmospheric water budget as drivers for precipitation change over two CORDEX-CORE domains. *Clim. Dyn.* **2021**, *57*, 1615–1628. [CrossRef]
95. Jones, M.W.; Veraverbeke, S.; Andela, N.; Doerr, S.H.; Kolden, C.; Mataveli, G.; Pettinari, M.L.; Le Quéré, C.; Rosan, T.M.; van der Werf, G.R.; et al. Global rise in forest fire emissions linked to climate change in the extratropics. *Science* **2024**, *386*, eadl5889. [CrossRef]
96. Vicente-Serrano, S.M.; Tramblay, Y.; Murphy, C.; Ocampo-Melgar, A.; Guan, H.; Spinoni, J.; Marengo, J.; Ghosh, S.; Yang, Y.; Cheval, S.; et al. First Issue of Water Scarcity and Drought. *Water Scarcity Drought* **2024**, *1*, 1–5. Available online: https://www.sciltp.com/journals/wsd/article/view/479 (accessed on 3 July 2024).

climate

MDPI

Article

# Spatio-Temporal Trends of Precipitation and Temperature Extremes across the North-East Region of Côte d'Ivoire over the Period 1981–2020

**Kouamé Donald Kouman [1,2,3,*], Amos T. Kabo-bah [2], Boyossoro Hélène Kouadio [3] and Komlavi Akpoti [4]**

[1] Regional Centre for Energy and Environmental Sustainability, University of Energy and Natural Resources, Sunyani P.O. Box 214, Ghana
[2] Civil and Environmental Engineering Department, University of Energy and Natural Resources, Sunyani P.O. Box 214, Ghana; amos.kabobah@uenr.edu.gh
[3] UFR-Science de la Terre et des Ressources Minières (STRM), Université Félix Houphouët-Boigny, Abidjan 22 BP 582, Côte d'Ivoire; helene.kouadio@curat-edu.org
[4] International Water Management Institute (IWMI), Cantonments Accra, Greater Accra PMB CT 112, Ghana; k.akpoti@cgiar.org
* Correspondence: donald.kouman.stu@uenr.edu.gh

**Abstract:** The northeast region of Côte d'Ivoire, where agriculture is the main economic activity, is potentially vulnerable to extreme climatic conditions. This study aims to make a comprehensive spatio-temporal analysis of trends in extreme indices related to precipitation and temperature for the Zanzan region of Côte d'Ivoire over the period of 1981–2020. The statistical significance of the calculated trends was assessed using the non-parametric Mann–Kendall test, while Sen's slope estimation was used to define the amount of change. For extreme precipitations, the results showed a decreasing trend in annual total precipitations estimated at 112.37 mm and in daily precipitations intensity indices. Furthermore, the consecutive dry days' index showed an increasing trend estimated at 18.67 days. Unlike the trends in precipitation extremes, which showed statistically non-significant trends, the trends in temperature extremes were mostly significant over the entire study area. The cold spells indices all show decreasing trends, while the warm spells show increasing trends. Drawing inferences from the results, it becomes clear that the study area may be threatened by food insecurity and water scarcity. The results are aimed to support climate adaptation efforts and policy intervention in the region.

**Keywords:** extreme climate; spatio-temporal analysis; extreme indices; trend analysis

## 1. Introduction

The achievement of sustainable development goals (SDGs), especially those related to ensuring environmental sustainability, reduction of poverty and hunger, and access to clean water among others are under severe threat from climate change [1,2]. This is important, particularly for Sub-Saharan Africa (SSA), where agriculture is the primary source of income and livelihood for many people. With nearly 19% of the region's gross output produced in the agricultural sector, SSA is the most agriculturally dependent region in the world after South Asia [3]. However, there is a high dependence of crops on weather events, with 97% of agricultural land being rainfed [4], and only 1% of cultivated land being irrigated [5,6]. For several decades climate change has strongly affected this part of Africa [7]. This manifestation of climate change in SSA is perceived through the increase in temperature, the presence of storms and extreme precipitation [8], the increase in land salinity, the worsening of weather conditions, and tornadoes that result in the ruin of ecosystems [9]. In its third assessment report, the Intergovernmental Panel on Climate Change (IPCC) stated that the number of hydro-meteorological disasters has increased significantly through heavy rainfall, severe landslides, increased incidence of floods, and severe frequency of droughts in different parts of Africa [10].

The change in rainfall and temperature over the past decades has caused significant vulnerabilities for food security. In maize and rice farming, for example, rainfall and temperature are important factors for yield, water availability, and the potential of the land to support crop production [11], and excesses or reductions in the value of these parameters produce negative effects [12]. Yields of common crops, such as cassava, sorghum, maize, and millet, are currently declining and these declines may be further accentuated in the future under the influence of climate change [13] in SSA. Studies conducted in several SSA countries show that these low agricultural productivity levels are an important threat to human health and nutrition, especially for children [14,15]. In 2006, among the 39 countries in the world that expressed a need for external food aid to cover the consumption needs of their population, 25 were in sub-Saharan Africa [16]. Food insecurity remains a major threat for SSA. With an already precarious economy, economic performance in SSA could be reduced by climate change [17] considering the significant contribution of agriculture to the gross domestic product (GDP). For example, in one of its reports, the World Bank stated that 17% of GDP in SSA comes from agriculture, and this proportion even exceeded 50% in some countries in 2005 [18]. In addition, the provision of ecosystem services and the functioning of ecosystems are also affected by climate change and therefore constitute a considerable concern for economic development and human societies [19]. In terms of health, climate change is leading to the modifications in certain disease vectors, involving changes in the transmissibility of some infectious diseases in SSA [20]. The increase in desertification [21], the threat to the existence of large wetlands that cover 9.01% of the landmass with a large population of poor people dependent on them for their livelihoods [22], and natural disasters, such as heat waves and droughts, coastal, river and rainfall floods that threaten already vulnerable urban areas [23] are also noted.

West Africa, as part of the global climate context of SSA, is heavily reliant on rainfed agriculture [24], high seasonal climate variability, and weak economic and institutional capacity to respond to climate variability and change, the region is perceived to be highly vulnerable to climate change [25]. Several studies on which this statement is based have looked extensively at the impacts of climate extreme events [26–28]. A climate extreme is described as a weather or climate variable having a value far from its average value and for simplicity, this term is used to refer to both weather and climate extremes [29]. Rapid climate change is occurring in West Africa, with widespread warming, more frequent climate extremes, and the reappearance of monsoonal rains in some areas [25]. Under both high and moderate emission scenarios, the existing climate trends are likely to persist into the twenty-first century [30]. Heat waves, floods, and droughts, among other events of climate extreme, are predicted to have a large impact in developing nations, especially in the rural areas [31]. However, understanding how climate extremes change at the global, regional, and local levels is a major first step in planning appropriate adaptation measures, as variations in climate extremes have significant socioeconomic impacts [32,33]. Extreme climate events are localized weather events, both in spatial and temporal terms, that cause significant damage to agriculture, livestock, and natural resources [34]. Indeed, rising temperatures coupled with increased variability in precipitation will lead to dysfunctional agricultural seasons, disruption of crop life cycles, and deterioration in agricultural production [35]. According to multiple research studies, cereal production in West Africa is expected to drop about 10% by 2050 [36,37] due to extreme temperature and rainfall. Food insecurity is still an issue in West Africa, where the need for food is anticipated to grow fivefold by 2050 as the region's population doubles [38]. Agriculture is therefore directly affected by both temperature and rainfall extremes and is the most vulnerable and exposed economic sector in West Africa. Climate extremes have a significant impact on the current economic development of African countries, particularly in areas of traditional rainfed agriculture and pastoralism, and on water resources at all levels. This factor could therefore worsen poverty in Africa where economic development is already precarious. The 2013 report on Geography of Poverty, Disasters and Climate Extremes in 2030 mapped the places where the poorest people were likely to live. It found that many regions with concentrated poor populations would also experience an increasing number of extreme climate events. According to the report, by 2030, up to 325 million extremely poor people will be living in the 49 countries most at risk, the majority of which are located in South Asia and sub-Saharan Africa [39].

Since its independence, Côte d'Ivoire has considered agriculture as its key economic strength. Agriculture currently employs 46% of the working population and contributes 21.2% of the country's GDP [40]. Furthermore, Côte d'Ivoire's agriculture is predominantly rainfed, with only 0.2% of the country's cultivated land currently equipped for irrigation [41]. Concerns about the effects of climate change are therefore high, especially as climate change has already led to warmer temperatures, greater meteorological variability, shifting precipitation patterns and more extreme weather conditions, and increased drought occurrence [42]. In Côte d'Ivoire, rainfall deficits and extreme temperatures have already been highlighted in the central part of the country [43,44], with high values noticed for the southern [45] and the northern parts of the country [46]. The risks to rainfed agriculture related to the new rainfall conditions are disruption of the agricultural cycle, seed losses, and yield reduction [41]. A global food security survey revealed that 12.6% of the rural population in Côte d'Ivoire are food insecure, with 2.5% being severely food insecure and 10.1% being moderately food insecure [40]. One of the reasons for this vulnerability to food insecurity is the occurrence of drought and climate extremes [47]. Communities in the northern part of Côte d'Ivoire have long suffered from recurring climate extremes that have limited their ability to improve agricultural activities [48]. The Zanzan region in the northeastern part of Côte d'Ivoire appears to be one of the least rainfed regions in the country [49]. However, the Zanzan region represents an economic potential for the Côte d'Ivoire. In fact, in the Zanzan region, agriculture is the main source of income for the population, and it occupies more than 90% of the population. In terms of food crops, this region is a potential supplier of food products, such as yams, cassava, and corn, for Côte d'Ivoire. Its main commercial crop, cashew nuts, contributes essentially to the Ivorian GDP. For such a region, climate variability through, for example, rainfall disruption, which is a constraint to agricultural development, would considerably influence socio-economic activities. Indeed, considered as one of the driving forces in the practice of rainfed agriculture, previous studies have shown that, to some extent, the climate by its extreme has played a role in the modification of past and present cropping systems in this region [50]. The study by Dje et al. [51] in the region showed periods of disruption in the rainy season. These previous studies have been limited to the study of climate variability without quantifying these climate disturbances in the region. According to WMO [52,53], the ability to handle the risks related to extreme climate events is critical to the long-term viability of agricultural development and the economic life of the local population. Therefore, the quantification of climate extremes for better understanding of their evolution in the region is essential for long-term socio-economic development. In addition, with the advent of climate change and variability that have disrupted socio-economic patterns in many parts of the world, studies on climate aspects will support local authorities in immediate decision making and planning for the welfare of their populations. The current study's main objective is to map the spatio-temporal evolution of climatic extremes in the Zanzan region of northeastern Côte d'Ivoire from 1981 to 2020 through the indices generated by the WMO Expert Team on Climate Change Detection Monitoring and Indices (ETCCDMI).

## 2. Materials and Methods

### 2.1. Study Area

Côte d'Ivoire is a West African country currently divided into 31 regions including the Zanzan region. The Zanzan region is in the northeast of Côte d'Ivoire between latitudes 7° and 10° North, and longitudes 4°30′ and 2°30′ West (Figure 1). This region, whose regional capital is the city of Bondoukou, is dominated by a humid tropical Sudanese-

Guinean climate. The main climate has four seasons, two of which are rainy seasons from March to June and from September to October. These rainy seasons are interspersed with two dry seasons which run from November to February and from July to August. The annual rainfall varies between 800 and 1200 mm with an average temperature of about 28 °C [54]. The Zanzan region has a surface area of approximately 38,000 km$^2$. Based on the 2014 national population and housing census, its population is estimated at 934,352 inhabitants [55]. The region's economy is mainly based on agriculture, livestock production, trade (food crops, handicrafts, etc.), and services. Cashew nuts are the main cash crop in the region with a contribution of 22% of national production [49]. The cashew industry constitutes 8.8% of the GDP of Côte d'Ivoire with a production of 700,000 tons [56]. In addition to this crop, cocoa and coffee are produced in small quantities in the southern part of the Zanzan region.

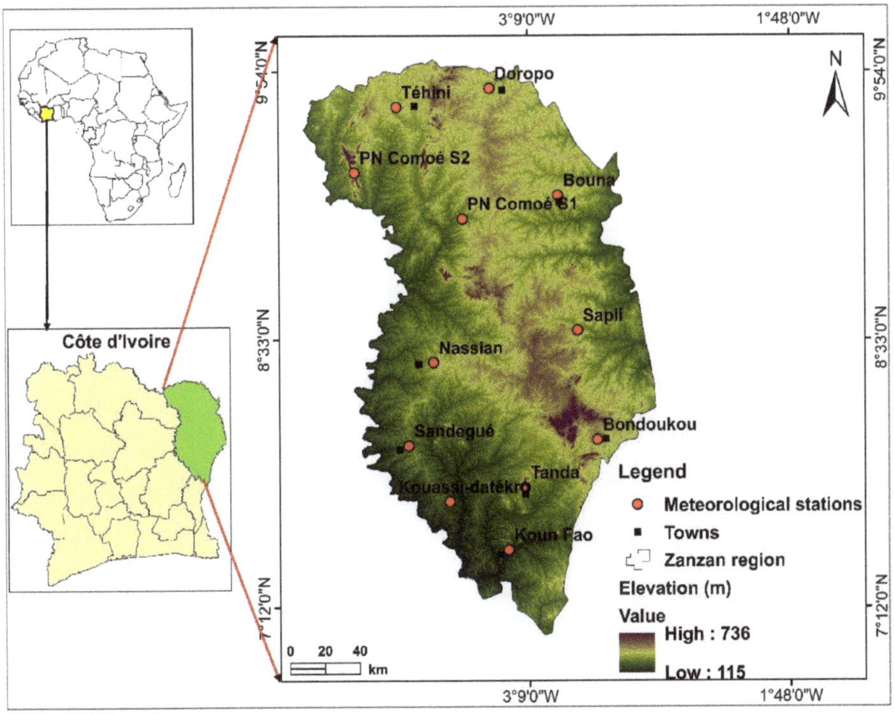


**Figure 1.** Geographical location of the Zanzan region and spatial representation of the meteorological stations of the study.

*2.2. Data*

The present research is based on a climatic database, spatially covering the entire Zanzan region. The data are a 40-year (1981 to 2020) time series of daily precipitation and temperature data from 12 measuring stations (Figure 1) taken from the Côte d'Ivoire meteorological station (SODEXAM). The percentage of missing data was used to verify data quality. The percentages of missing data for minimum temperature (Tmin), maximum temperature (Tmax), and precipitation ranged from 0 to 6%, 0 to 7%, and 0 to 11%, respectively (Table 1). To fill in the missing data, the method of using a gridded meteorological database, based on reanalysis and satellite data was used in this study, especially the data from NASA Prediction of Worldwide Energy Resource (POWER). NASA POWER data are a suitable alternative when ground-based data are missing or of doubtful quality [57]. These data based on reanalysis models developed by NASA's Global Modeling and Assimilation

Office (GMAO) [58] can be accessed and downloaded via the POWER website homepage at http://power.larc.nasa.gov (accessed on 11 March 2021). NASA POWER data have long been used and validated in the US in several studies [59]. Before using the NASA POWER data to fill in the missing data from the ground stations (SODEXAM), correlation tests were conducted between the different data sets using Pearson's correlation coefficient ($r$) for the twelve stations (Table 2). The positive correlations closer to 1 illustrated by the results in Table 2 show a good correlation between the different data sets. It is at the end of the different tests that the POWER data were used to fill our missing data. Other studies have found a strong link between ground data and NASA POWER data, such as Bai et al. [60] in China, Bender and Sentelhas [61] in Brazil, and Aboelkhair et al. [62] in Egypt, or have used NASA POWER data to fill the gaps in their ground data, such Berhane et al. [63] in Ethiopia and West Africa [64].

**Table 1.** Geolocation of the stations with the percentage of missing data.

| Station Name | Elevation (m) | Longitude ($^0$W) | Latitude ($^0$N) | Percentage (%) of Missing Data | | |
|---|---|---|---|---|---|---|
| | | | | $T_{min}$ | $T_{max}$ | Precip. |
| Bondoukou | 286.62 | −2.79 | 8.04 | 0.3 | 0.6 | 2 |
| Sapli | 283.55 | −2.89 | 8.59 | 4.5 | 4.4 | 6 |
| Bouna | 288.64 | −2.99 | 9.27 | 3 | 0 | 3.4 |
| Doropo | 315.91 | −3.34 | 9.81 | 4 | 5 | 7 |
| Kouassi-datékro | 211.45 | −3.56 | 7.73 | 0.6 | 2.3 | 4.1 |
| Koun Fao | 224.41 | −3.25 | 7.48 | 0 | 0 | 0.3 |
| Nassian | 241.9 | −3.64 | 8.43 | 6 | 7 | 11 |
| PN Comoé S1 | 311.65 | −3.49 | 9.15 | 0 | 0 | 0 |
| PN Comoé S2 | 284.6 | −4.04 | 9.38 | 0 | 0 | 0 |
| Sandegué | 242.1 | −3.77 | 8.01 | 3.1 | 4.6 | 8 |
| Tanda | 263.51 | −3.16 | 7.80 | 2 | 2.2 | 4 |
| Téhini | 324.98 | −3.83 | 9.72 | 0 | 0 | 0.2 |

**Table 2.** Pearson's correlation coefficient ($r$) between the ground data (SODEXAM) and NASA POWER for the twelve stations.

| Station Name | Coefficient, $r$ | | |
|---|---|---|---|
| | $T_{min}$ | $T_{max}$ | Precip. |
| Bondoukou | 0.98 | 0.97 | 0.86 |
| Sapli | 0.96 | 0.96 | 0.95 |
| Bouna | 0.97 | 0.97 | 0.96 |
| Doropo | 0.98 | 0.99 | 0.98 |
| Kouassi-datékro | 0.99 | 0.98 | 0.88 |
| Koun Fao | 0.94 | 0.93 | 0.87 |
| Nassian | 0.96 | 0.96 | 0.95 |
| PN Comoé S1 | 0.95 | 0.96 | 0.97 |
| PN Comoé S2 | 0.91 | 0.95 | 0.99 |
| Sandegué | 0.97 | 0.97 | 0.88 |
| Tanda | 0.98 | 0.96 | 0.94 |
| Téhini | 0.96 | 0.97 | 0.96 |

### 2.3. Methods

#### 2.3.1. Homogeneity Test of the Data Used

Before calculating indices related to climate extremes, it is necessary to check the homogeneity of the daily input data, look for points of change in the time series, assess whether they are natural or not, and then rectify them. Homogeneity refers to a series' consistency through time, and it is an evident necessity for reliable climatic time series analysis [65]. There is presently no mechanism to homogenize the data in ClimPACT2, the software used to calculate the indices related to climatic extremes, except for simple quality control. The software application *RHtest* of the "Canadian Meteorological Service" and adopted by the ETCCDMI [32,66], was used in the current research work to correct the homogeneity of the daily data. This software is run in the R statistical and programming environment. A double regression model is used in this process [67].

### 2.3.2. Analysis of Climate Extremes Indices

For the characterization of the intensity, frequency, and duration of extreme climate events, indices have been developed and recommended by the Expert Team on Sector-specific Climate Indices (ET-SCI) and the Expert Team on Climate Change Detection Monitoring and Indices (ETCCDMI), both from the WMO [65]. The ETCCDMI indices are integrated into the RClimDex software [68]. Those of ET-SCI, which already contain the ETCCDMI indices with additional indices to be used in sectoral applications based on specialists from the health, water, and agriculture sectors, are integrated into ClimPACT2 software [65]. ClimPACT2 software is an update of ClimPACT, which was based on ETCCDMI's RClimDex software. RClimDex and ClimPACT2 are written in R, statistical computing, and graphic language and environment. A series of 27 indices including 11 precipitation and 16 temperature-related indices were defined by ETCCDMI [65,69]. Climate indices are generated values that can be used to reflect the state of variations in the climate system. They allow statistical study and comparison of time series, extremes, trends, and means [70]. The use of indices to detect climate change has the advantage of allowing for a simple comparison of trends across different climate zones and areas [71]. In this study, the indices were calculated using the ClimPACT2 Master program after a quality control and homogeneity test of the daily input data, but only the indices deemed relevant and applicable to Côte d'Ivoire were selected for a detailed analysis (Table 3). The quality control was performed by the software before the calculation of the climate indices in order to check for fundamental input mistakes, such as intentionally exaggerated results or rounding bias, and missing data. The indices listed in Table 2 were chosen for this study because they have already been applied several times in other studies in West Africa [27,72–74] and Côte d'Ivoire [71,75,76] for the detection of change in climate extremes. These indices are therefore considered relevant and adaptable to the climatic context of the Zanzan region for the detection of changes in precipitation and temperature extremes.

### 2.3.3. Trend Analysis

After verification of data quality and coherency, the next step was to calculate the climate indices by ClimPACT2 software. Then a statistical analysis of the trends was performed. There are several methods of statistical trend analysis that other similar studies have already used, where we can mention among others the linear trend analysis [77], the linear trend analysis associated with the *t*-test [75,78], the non-parametric Kendall's tau test with linear least squares trend [33,79], the Sen's slope estimator with non-parametric Mann–Kendall test [63,74,80,81]. The non-parametric Mann–Kendall test has been utilized in various research works since it makes no assumptions regarding data distribution or trend linearity [82]. Unlike parametric trend tests, which demand that the data be both normally distributed and independent, non-parametric trend tests only require independent data [28]. The Sen's slope estimator and non-parametric Mann–Kendall (MK) test have been used in this study to assess the statistical significance of the trends and to accurately determine the magnitude of the trends in climate extremes indices due to their robustness in addition to their insensitivity to outliers in the times series. The MK test was performed using the MAKESENS software. The null hypothesis ($H_0$) in the calculation of this test asserts that there is no trend in climate extremes over time, whereas the alternative hypothesis

($H_1$) indicates that there is a monotonic decreasing or increasing trend. The mathematical equations used to calculate the Mann–Kendall test are as follows:

$$S = \sum_{k=1}^{n-1} \sum_{j=k+1}^{n} sgn(X_j - X_k) \tag{1}$$

**Table 3.** Indices of daily precipitation and temperature extremes and their descriptions as defined by ETCCDMI.

| Indices | Names | Definitions | Units |
|---|---|---|---|
| Precipitations | | | |
| PRCPTOT | Annual total precipitations PR | Sum of daily PR ≥ 1.0 mm | mm |
| SDII | Daily precipitations (PR) intensity | Annual total PR divided by the number of wet days (when total PR ≥ 1.0 mm) | mm/d |
| CDD | Consecutive dry days | Maximum number of consecutive dry days (when PR < 1.0 mm) | day |
| CWD | Consecutive wet days | Maximum annual number of consecutive wet days (when PR ≥ 1.0 mm) | day |
| R99p | Total annual PR from very heavy rain days | Annual sum of daily PR > 99th percentile | mm |
| R99pTOT | Contribution from extremely wet days | $100 \times R99p/PRCPTOT$ | % |
| Temperatures | | | |
| TX10p | Number of cool days | Percentage of days when TX < 10th percentile | % |
| TN10P | Number of cold nights | Percentage of days when TN < 10th percentile | % |
| TX90p | Number of hot days | Percentage of days when TX > 90th percentile | % |
| TN90p | Number of warm nights | Percentage of days when TN > 90th percentile | % |
| WSDI | Warm spell duration indicator | Annual number of days contributing to events where 6 or more consecutive days experience TX > 90th percentile | day |
| CSDI | Cold spell duration indicator | Annual number of days contributing to events where 6 or more consecutive days experience TN < 10th percentile | day |

In a data set of length $n$, $X_j$ and $X_k$ are consecutive data values.

$$sgn(X_j - X_k) = \begin{cases} 1, \ if \ X_j > X_k \\ 0, \ if \ X_j = X_k \\ -1, \ if \ X_j < X_k \end{cases} \tag{2}$$

The behavior of the S statistic is approximately the same for a normal distribution when $n \geq 8$, so the test is performed with a normal distribution with E(S) = 0 with the variance as follows:

$$\text{Var}(S_{mk}) = \frac{n(n-1)(2n+5) - \sum_{p=1}^{q} t_p(t_p - 1)(2t_p + 5)}{18} \tag{3}$$

Here $t_p$ is the number of input values inside the $p$-th affiliated group and $q$ is the number of affiliated groups in the data set. The test statistic $Z_{mk}$ is calculated using the values of S and Var ($S_{mk}$) as follows:

$$Z_{mk} = \begin{cases} \frac{S-1}{\sqrt{Var(S_{mk})}}, \ if \ S > 0 \\ 0, \ if \ S = 0 \\ \frac{S+1}{\sqrt{Var(S_{mk})}}, \ if \ S < 0 \end{cases} \tag{4}$$

$Z_{mk}$ is a statistically significant trend detection tool. A positive (negative) value of $Z_{mk}$ indicates the upward (downward) trend. To check for a monotonic upward or downward trend (two-tailed test) at the $\alpha$ level of significance, $H_0$ is rejected if the absolute value of $Z_{mk} > Z_{1-\frac{\alpha}{2}}$ where $Z_{1-\frac{\alpha}{2}}$ is acquired from the usual normal cumulative distribution tables. The trend outcomes in this study were evaluated at the 5% level of significance, with the null hypothesis of no trend being rejected if $|Z_{mk}| > 1.96$. The estimator of Sen [83] which is defined by Equations (5) and (6), was used to calculate the real slope of an existing trend.

$$Q_{i=\frac{x_j-x_k}{j-k}\ for\ all\ j>k} \quad (5)$$

There will be $N = n(n-1)/2$ slope $Q_i$ estimations if the time series has $n$ $x_j$ values. The median of these N values of $Q_i$ is Sen's slope estimator.

$$Q_{med} = \begin{cases} Q_{[(N+1)/2]}, if\ N\ is\ odd \\ \frac{1}{2}\left(Q_{[N/2]} + Q_{[(N+2)/2]}\right),\ if\ N\ is\ even \end{cases} \quad (6)$$

#### 2.3.4. Spatial Interpolation of Climate Indices

To produce a continuous surface, spatial distribution or interpolation improves the representation of a surface and forecasts values for other uncharted territories [84]. It is a mathematical approach or function for estimating values in situations when there are no measurable values [85]. Interpolation methods (e.g., spline, inverse distance weighting (IDW), Kriging, etc.) are numerous and vary greatly in complexity and effectiveness. The Kriging approach was employed in this study to interpolate the extreme climate indices from the ArcGIS software for spatial analysis. Kriging is a geostatistical method that provides unbiased estimates with minimum variance taking into account the spatial relationship between the data points [86]. The advantage of using geostatistical interpolation approaches over conventional techniques for spatial estimation of weather data has been reported by several authors [87–89]. One of the main advantages of Kriging over simpler techniques, such as IDW, besides providing a measure of estimation uncertainty (Kriging variance), is that it can use correlated dense secondary variables to improve the prediction of a sparsely sampled primary variable [86]. Considering the topography as a secondary variable, the prediction of meteorological data can be improved by using Kriging. In this study, among the interpolation methods, Kriging gives the best representativeness of our results.

## 3. Results

### 3.1. Analysis of the Zanzan Region's Spatio-Temporal Evolution of Extreme Precipitation Trends

#### 3.1.1. Total Annual Precipitation Index (PRCPTOT)

Table 4 shows both the Sen slope and Mann–Kendall statistical test for all of the extreme rainfall indices studied for the 12 stations. In the Zanzan region, an examination of cumulative annual precipitations PRCPTOT shows a general decreasing trend. This is observed in 10 of the study stations, excluding the Doropo and Téhini sites, which show increasing trends. This downward trend is however statistically non-significant at the 5% level of significance. Over a decade, the largest decreases are estimated at 45.1 mm/decade at Kouassi-datékro, 41.5 mm/decade at Sandegué, and 38.2 mm/decade at Bondoukou. Over the period of 1981–2020, the annual total precipitations decreased by an average of 112.35 mm over the entire study area. The total annual precipitation index is shown in Figure 2 as a spatial distribution per decade. From 1981 to 2020, the cumulative annual rainfall ranged from 837.69 to 1404.3 mm. In the first decade (i.e., 1981 to 1990), the cumulative precipitation was between 947.1 and 1404.3 mm and from 1991 to 2000, the value was between 837.69 and 1276.9 mm, representing a decrease of 127.4 mm. For the 2001–2010 decade, the total annual precipitation varied between 852.14 and 1145.3 mm, for a decrease of 131.6 mm compared to the previous decade. From 2011 to 2020, the total precipitation was between 895.56 and 1327.2 mm, corresponding to an increase of 181.9 mm. Compared to the last two decades, the 2011–2020 decade shows a slight return of precipitation, but less important than the precipitation of the 1981–1990 decade. The analysis of Figure 2 indicates that the largest overall precipitation was found in the Zanzan region's south, while the lowest was found in the study area's northern part.

**Table 4.** Mann–Kendall (Z) trend test statistics and Sen's slope (S) of extreme precipitation indices from 1981 to 2020 for the twelve stations.

| Station Name | PRCPTOT | | SDII | | CWD | | CDD | | R99p | | R99pTOT | |
|---|---|---|---|---|---|---|---|---|---|---|---|---|
| | Z | S | Z | S | Z | S | Z | S | Z | S | Z | S |
| Bondoukou | −1.11 | −3.82 | −0.66 | −0.01 | **−3.00** * | −0.77 | 0.83 | 0.27 | 1.12 | 0.26 | 1.31 | 0.05 |
| Sapli | −1.20 | −3.28 | −0.41 | −0.01 | **−2.87** * | −0.68 | 1.29 | 0.41 | 0.88 | 0.17 | 1.14 | 0.04 |
| Bouna | −0.78 | −1.88 | −0.20 | −0.00 | −1.85 | −0.44 | 0.90 | 0.35 | 0.77 | 0.00 | 1.06 | 0.01 |
| Doropo | 0.64 | 1.23 | 1.08 | 0.01 | −1.39 | −0.33 | **2.18** * | 0.89 | 0.85 | 0.09 | 1.11 | 0.03 |
| Kouassi-datékro | −1.27 | −4.51 | −0.94 | −0.03 | **−2.74** * | −0.58 | 0.00 | 0.00 | 0.52 | 0.01 | 0.68 | 0.01 |
| Koun Fao | −0.38 | −1.22 | −0.36 | −0.01 | −1.89 | −0.52 | −0.11 | 0.00 | 0.85 | 0.05 | 0.94 | 0.02 |
| Nassian | −1.36 | −3.75 | −1.08 | −0.01 | −1.80 | −0.29 | 1.29 | 0.40 | 0.56 | 0.00 | 0.56 | 0.00 |
| PN Comoé S1 | −0.62 | −1.34 | −0.36 | −0.01 | −1.29 | −0.22 | 1.33 | 0.44 | 0.31 | 0.00 | 0.54 | 0.00 |
| PN Comoé S2 | −0.78 | −2.26 | −0.64 | −0.01 | −1.12 | −0.22 | 1.78 | 0.65 | 0.21 | 0.00 | 0.42 | 0.00 |
| Sandegué | −1.57 | −4.15 | −0.90 | −0.02 | −1.17 | −0.22 | 0.99 | 0.35 | 0.34 | 0.00 | 0.34 | 0.00 |
| Tanda | −0.69 | −2.60 | −0.64 | −0.01 | −1.94 | −0.57 | −0.03 | 0.00 | 1.06 | 0.23 | 1.04 | 0.05 |
| Téhini | 0.22 | 0.49 | 0.85 | 0.01 | −1.07 | −0.22 | 1.39 | 0.55 | 1.06 | 0.61 | 1.32 | 0.08 |

* Trends that are statistically significant at the 5% level; positive (+)/negative (−) values imply an increasing/decreasing trend respectively.

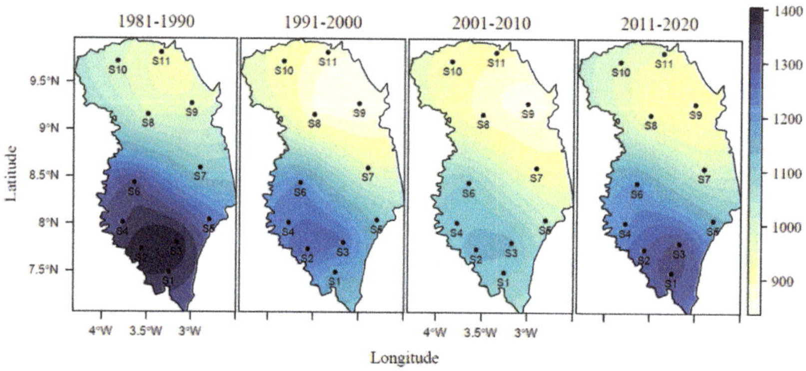


**Figure 2.** Spatial distribution of annual total precipitations (PRCPTOT) (mm) per decade in the Zanzan region. Stations (S1 = Koun Fao, S2 = Kouassi-datékro, S3 = Tanda, S4 = Sandegué, S5 = Bondoukou, S6 = Nassian, S7 = Sapli, S8 = PN Comoé S1, S9 = Bouna, S10 = Téhini, S11 = Doropo).

### 3.1.2. Daily Precipitation Intensity Index (SDII)

The Mann–Kendall test of trends in the daily precipitation intensity index (Table 4) reveals general declining trends at most of the research sites, but are not significant at the 5% level of significance. The daily precipitation intensity in the studied area decreased by 0.47 mm/d on average across the study period. The largest drops were observed in Kouassi-datékro and Sandegué stations, with estimates of 1.17 mm/d and 0.78 mm/d, respectively. Figure 3 displays the spatial distribution of the index per decade. From 1981 to 1990, the daily rainfall intensity in the Zanzan region varied between 5.34 and 7.22 mm/d, and over the 1991–2000 decade this variation was between 4.82 and 6.59 mm/d, corresponding to a decrease of 0.63 mm/d. From 2001 to 2010, the values of the index were between 4.81 and 6.15 mm/d, which implies a decrease of 0.09 mm/d, and from 2011 to 2020 the variation was between 5.26 and 6.73 mm/d, which is an increase of 0.58 mm/d. From 1981 to 2020, the southwestern part of the Zanzan region experienced the highest daily precipitation intensity, while the north-eastern half experienced the lowest.

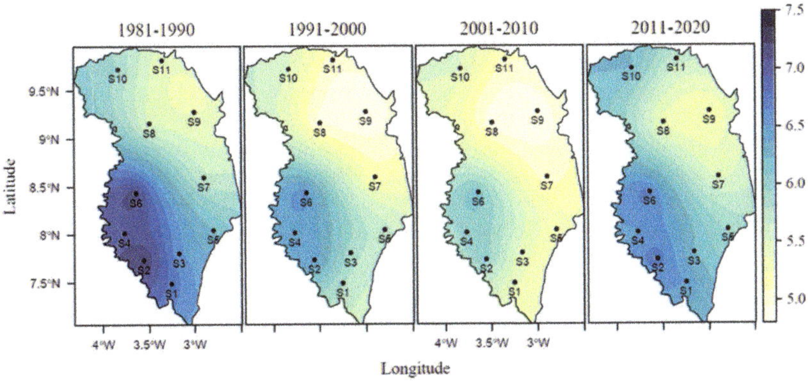


**Figure 3.** Spatial distribution of the daily precipitation intensity (SDII) (mm/d) per decade in the Zanzan region. Stations (S1 = Koun Fao, S2 = Kouassi-datékro, S3 = Tanda, S4 = Sandegué, S5 = Bondoukou, S6 = Nassian, S7 = Sapli, S8 = PN Comoé S1, S9 = Bouna, S10 = Téhini, S11 = Doropo).

### 3.1.3. Index of Consecutive Wet Days (CWD)

Figure 4 and Table 4 show the outputs of the Mann–Kendall statistical test of the data series and the geographical extent of the consecutive wet day's index in the Zanzan region, respectively. All stations in the study area have a general dropping trend in consecutive wet days, and three of these stations have a significant declining trend at the 5% significance level, according to the descriptive statistics. Over the study period of 1981–2020, these significant decreases were estimated at 30.03 days at Bondoukou, 26.52 days at Sapli, and 22.62 days at Kouassi-datékro. Over the whole study area, the total average decline was 16.45 days. The spatial distribution of consecutive wet days per decade shows a variation from 37.7 to 69.8 days over the period of 1981–1990 and from 1991 to 2000 the index varied from 32.1 to 51.99, indicating a decrease of 17.81 days. From 2001 to 2010 the evolution of the index indicates a decrease of 3.29 days while from 2011 to 2020 a decrease of 5.9 days was observed. The greatest reductions are seen in the southeast of the region under consideration for the entire study period. The duration of rainy seasons in a certain location was determined by the sequence of rainy days CWD.

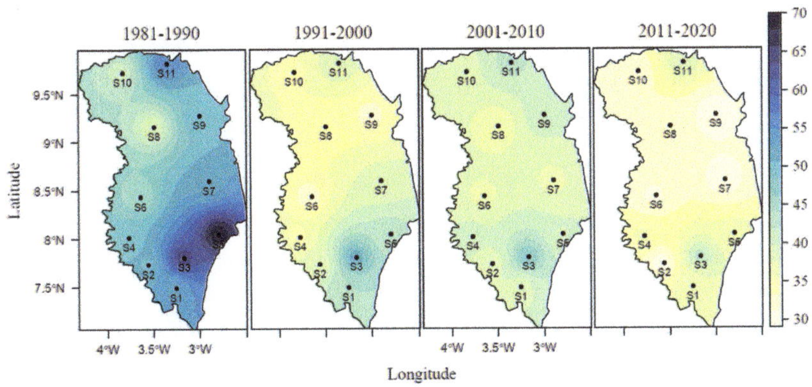


**Figure 4.** Spatial distribution of the consecutive wet days (CWD) (days) per decade in the Zanzan region. Stations (S1 = Koun Fao, S2 = Kouassi-datékro, S3 = Tanda, S4 = Sandegué, S5 = Bondoukou, S6 = Nassian, S7 = Sapli, S8 = PN Comoé S1, S9 = Bouna, S10 = Téhini, S11 = Doropo).

### 3.1.4. Index of Consecutive Dry Days (CDD)

In the Zanzan region, the trends in the time series of the consecutive dry days' index assessed using the Mann–Kendall test (Table 4) showed an increase in 10 of the stations tested, except for Koun Fao and Tanda, which showed non-significant declining trends. The almost general upward trend in consecutive dry days in the study area was however non-significant at the 5% significance level for most stations except for the Doropo station, which presented a significant increasing trend. In Doropo the increase was 0.89 days/year, 8.9 days per decade, and 34.71 days over the entire study period. On average, the increase was estimated at 18.67 days over the study period. Figure 5 presents the index of the consecutive dry days' spatial patterns per decade. Over the 1981–1990 decade, the consecutive dry days were between 50.4 and 82.4 days, and from 1991 to 2000, the index was between 52.1 and 98.8 days, indicating an increase of 16.4 days. Over 2001 and 2010, the variation in consecutive dry days was estimated between 55.1 and 90.6 days, which is a decrease of 8.2 days, and from 2011 to 2020, it was estimated between 51.2 and 102 days, corresponding to an increase of 11.4 days. When comparing the first decade of the study (i.e., 1981–1990), to the last decade (i.e., 2011–2020), there was an increase of 19.6 consecutive dry days. The high values of the index were found in the northern part of the study area, in places such as Doropo, Téhini, and Bouna, and the lowest values were found in the southern part of the Zanzan region, such as Tanda and Koun Fao.

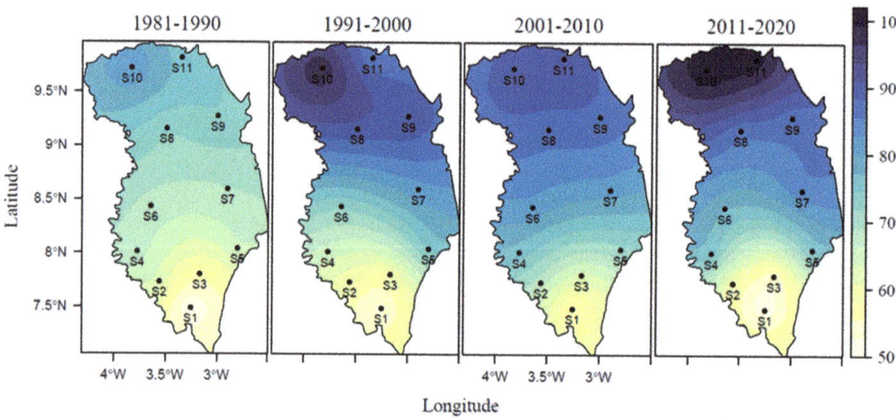


**Figure 5.** Spatial distribution of the consecutive dry days (CDD) (days) per decade in the Zanzan region. Stations (S1 = Koun Fao, S2 = Kouassi-datékro, S3 = Tanda, S4 = Sandegué, S5 = Bondoukou, S6 = Nassian, S7 = Sapli, S8 = PN Comoé S1, S9 = Bouna, S10 = Téhini, S11 = Doropo).

### 3.1.5. Index of Total Annual Precipitations from Very Heavy Rain Days (R99p)

Table 4 shows the findings of the Mann–Kendall statistical test, which demonstrated a general trend in the Zanzan region of increasing total yearly precipitation from very heavy rain days. At any of the sites studied, the general rise was non-significance at the 5% threshold of significance. Over the study period, the strong increases were estimated at 23.79 mm in Téhini, 10.14 mm in Bondoukou, and 8.97 mm in Tanda. Over the entire research area, the overall average was estimated at 4.61 mm. Figure 6 illustrates the spatial distribution of the total annual precipitations from very heavy rain days. Over the 1981–1990 decade, the evolution of extreme rainy days ranged from 37.4 to 98.5 mm, and from 1991 to 2000, it ranged from 34.1 to 66.9 mm, which represents a decrease of 31.6 mm. From 2001 to 2010 the index varied between 26.4 and 37.4 mm, for a decrease of 29. mm, and from 2011 to 2020 the variation was between 91.1 and 130 mm for an increase of 92.6 mm. Compared to the decades of 1991–2000 and 2001–2010, the last decade of the study witnessed a return to a significant quantity of extreme rainy days. These extremely wet days were more pronounced in the south of the Zanzan region across the study decades. The highest daily water intakes relative to the 99th percentile threshold are total annual precipitations from very heavy rain days (R99p).

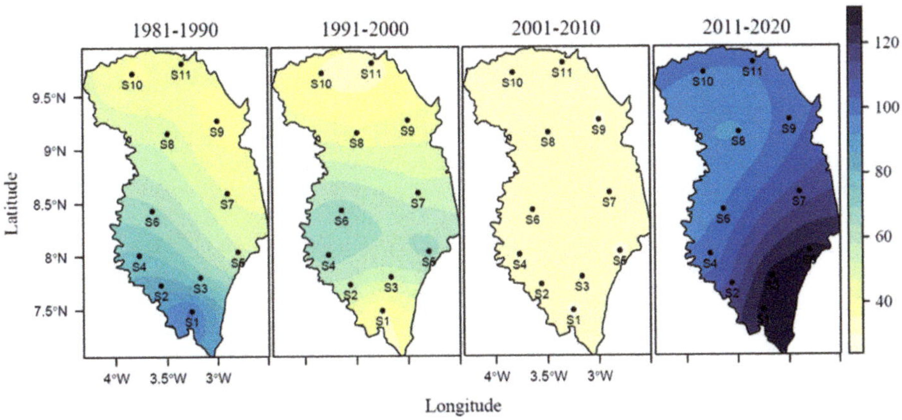


**Figure 6.** Spatial distribution of the total annual precipitation from very heavy rain days (R99p) (mm) per decade in the Zanzan region. Stations (S1 = Koun Fao, S2 = Kouassi-datékro, S3 = Tanda, S4 = Sandegué, S5 = Bondoukou, S6 = Nassian, S7 = Sapli, S8 = PN Comoé S1, S9 = Bouna, S10 = Téhini, S11 = Doropo).

### 3.1.6. Contribution of Extremely Wet Days (R99pTOT)

The contribution of extremely wet days analyzed using the Mann–Kendall test displays a general increasing trend that was non-significant at the 5% level of significance over the entire study area. The most significant increases were in the locality of Téhini estimated at 3.12%, Bondoukou estimated at 1.95% and Tanda estimated at 1.95% of days. On average, this increase was estimated at 1.41% of days from 1981 to 2020 for the entire region. The spatial distribution per decade of the contribution of extremely wet days illustrated in Figure 7 indicates a variation between 3.87% and 6.18% from 1981 to 1990, and from 1991 to 2000 the contribution was between 3.25% and 4.65%, hence a decrease of 1.53%. During the period 2001–2010, these values were between 2.82% and 3.09%, corresponding to a decrease of 1.56%, and from 2011 to 2020, the contribution of extremely wet days was between 7.58% and 10.08%, representing an increase of 6.99%. The last decade thus witnessed an important contribution of extremely wet days located in the east of the study area.

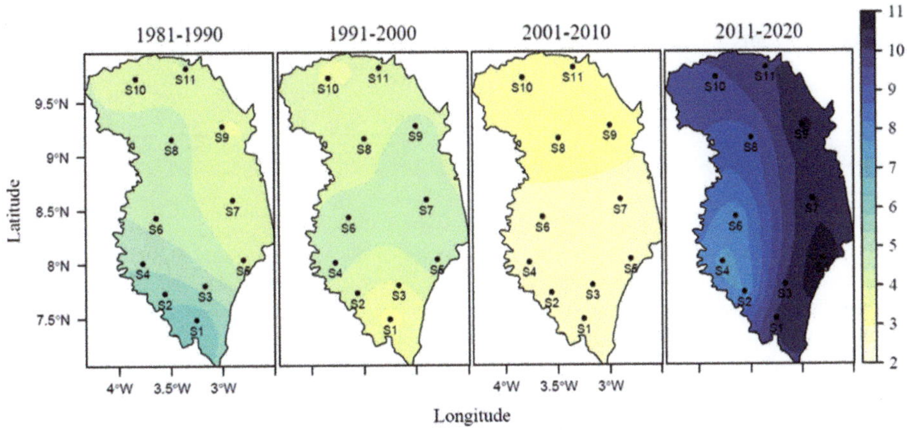


**Figure 7.** Spatial distribution of the contribution of extremely wet days (R99pTOT) (%) per decade in the Zanzan region. Stations (S1 = Koun Fao, S2 = Kouassi-datékro, S3 = Tanda, S4 = Sandegué, S5 = Bondoukou, S6 = Nassian, S7 = Sapli, S8 = PN Comoé S1, S9 = Bouna, S10 = Téhini, S11 = Doropo).

### 3.2. *Analysis of the Spatio-Temporal Evolution of Extreme Temperature Trends in the Zanzan Region*

#### 3.2.1. Evolution of the Index of the Number of Cool Days (TX10p)

The Mann–Kendall statistical test and the Sen slope of all the indices of temperature extremes assessed for the 12 study sites are shown in Table 5. The analysis of the number of cool days indicates a general declining trend in the Zanzan region for all the stations studied. Within the general declining trend, eight of the studied stations showed significant trends at the level of 5%. The values of these declines over the 1981–2020 study period for the localities were estimated at 14.54% for Sapli, 9.98% for Bondoukou, 9.08% for Sandegué, 8.85% for Nassian, and 5.96% of days for Tanda sites. Generally, the average decline in the study area was 6.58% of days. To better appreciate this decreasing trend, Figure 8 illustrates the spatial distribution of the number of cool days per decade. From 1981 to 2020, the number of cool days was between 5.61% and 18.53% of days. In the first decade (i.e., 1981 to 1990), the number was between 15.57% and 18.53% and from 1991 to 2000 the values were between 6.92% and 10.51%, which means a decrease of 8.02%. Over the decade 2001 to 2010, the number of cool days ranged between 6.68% and 8.25%, representing a decrease of 2.26% compared to the previous decade. From 2011 to 2020 the number of days varied between 5.61% and 10.35%, corresponding to an increase of 2.1%. Compared to the last two decades, the decade 2011–2020 showed a slight return of cool days, but less than the number of cool days in the decade 1981–1990. The analysis of Figure 8 revealed that the high percentages of cool days in the decades 1981–1990 and 1991–2000 were concentrated in the center of the Zanzan region, but in the decades 2001–2010 and 2011–2020, they were concentrated in the study area's northern and southern parts.

**Table 5.** Mann–Kendall (Z) trend test statistics and Sen's slope (S) of extreme precipitation indices from 1981 to 2020 for the twelve stations.

| Station Name | TX10P | | TN10P | | TX90P | | TN90P | | WSDI | | CSDI | |
|---|---|---|---|---|---|---|---|---|---|---|---|---|
| | Z | S | Z | S | Z | S | Z | S | Z | S | Z | S |
| Bondoukou | **−3.40 *** | −0.26 | **−4.84 *** | −0.37 | 1.43 | 0.12 | **4.31 *** | 0.38 | 0.82 | 0.17 | **−4.45 *** | −1.00 |
| Sapli | **−4.54 *** | −0.37 | **−2.90 *** | −0.20 | **3.72 *** | 0.34 | **1.67** | 0.16 | **2.77 *** | 0.62 | **−2.39 *** | −0.40 |
| Bouna | **−1.99 *** | −0.15 | **−3.81 *** | −0.29 | 1.55 | 0.15 | **3.23 *** | 0.34 | 1.53 | 0.34 | **−3.42 *** | −0.62 |
| Doropo | −0.17 | −0.02 | **−2.67 *** | −0.22 | 0.79 | 0.07 | **2.33 *** | 0.19 | 0.77 | 0.17 | **−2.57 *** | −0.55 |
| Kouassi-datékro | **−2.50 *** | −0.15 | **−4.03 *** | −0.31 | 1.15 | 0.11 | **3.95 *** | 0.35 | 0.83 | 0.21 | **−3.59 *** | −0.78 |
| Koun Fao | −1.39 | −0.11 | **−3.62 *** | −0.31 | 0.23 | 0.03 | **3.55 *** | 0.30 | 0.13 | 0.00 | **−3.72 *** | −0.67 |
| Nassian | **−3.25 *** | −0.23 | **−3.97 *** | −0.32 | 1.90 | 0.17 | **4.23 *** | 0.35 | 1.73 | 0.30 | **−3.92 *** | −0.77 |
| PN Comoé S1 | −1.48 | −0.13 | **−3.30 *** | −0.26 | 1.46 | 0.12 | **2.90 *** | 0.27 | 1.35 | 0.24 | **−3.05 *** | −0.56 |
| PN Comoé S2 | **−2.13 *** | −0.16 | **−3.50 *** | −0.29 | 1.35 | 0.15 | **2.69 *** | 0.25 | 1.34 | 0.29 | **−3.21 *** | −0.67 |
| Sandegué | **−3.32 *** | −0.23 | **−3.93 *** | −0.32 | **2.13 *** | 0.19 | **4.23 *** | 0.37 | **2.08 *** | 0.37 | **−3.82 *** | −0.77 |
| Tanda | **−2.50 *** | −0.15 | **−3.86 *** | −0.32 | 0.86 | 0.08 | **3.97 *** | 0.36 | 0.44 | 0.11 | **−3.33 *** | −0.76 |
| Téhini | −0.90 | −0.07 | **−2.95 *** | −0.44 | 0.79 | 0.08 | **2.60 *** | 0.23 | 0.72 | 0.17 | **−2.82 *** | −0.50 |

* Trends that are statistically significant at the 5% level; positive (+)/negative (−) values imply an increasing/decreasing trend, respectively.

#### 3.2.2. Changes in the Index of the Number of Cold Nights (TN10p)

The Mann–Kendall test of trends in the index of the number of cold nights (Table 5) showed a general declining trend across all stations in the research area, which was significant at the 5% statistical significance. Over the research period from 1981 to 2020, the decline in the number of cold nights in the Zanzan region was estimated at an average of 11.83% of nights. The strongest decreases of cold nights were observed in Bondoukou (14.23%), Nassian (12.36%), Sandegué (12.48%), and Tanda (12.55%). Figure 9 shows the spatial distribution of the index per decade. From 1981 to 1990, the number of cold nights in the Zanzan region varied between 16.69% and 18.73%, and over the decade 1991–2000, the variation was between 8.79% and 11.52%, which represents a decrease of 7.21% of cold nights. From 2001 to 2010 the values of the index were between 6.18% and 7.18%, meaning a decrease of 4.34% and from 2011 to 2020, the variation was between 6.04% and 8.18%, indicating an insignificant increase of 0.99% of cold nights. From 1981 to 1990 and 1991 to 2000, the highest percentages of cold nights in the Zanzan region were recorded in the southeast and southwest of the region, respectively. From 2001 to 2010 and from 2011 to 2020, high percentages were seen in the northern part of Zanzan region.

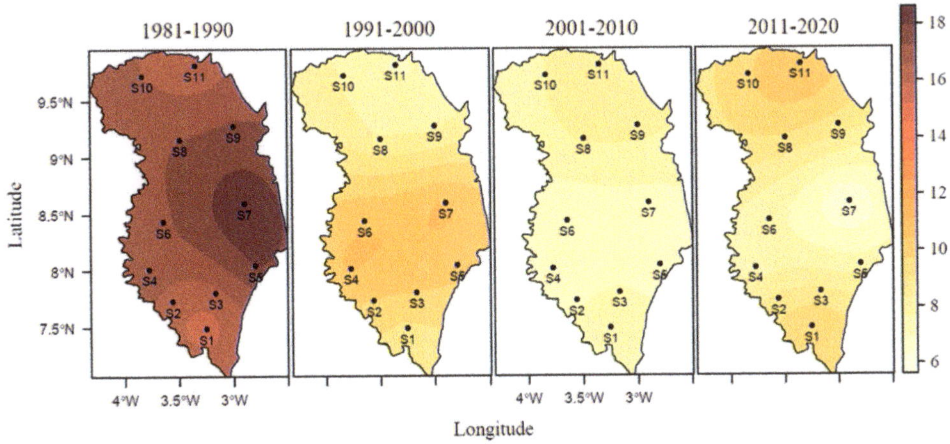


**Figure 8.** Spatial distribution of number of cool days (TX10p) (%) per decade in Zanzan region. Stations (S1 = Koun Fao, S2 = Kouassi-datékro, S3 = Tanda, S4 = Sandegué, S5 = Bondoukou, S6 = Nassian, S7 = Sapli, S8 = PN Comoé S1, S9 = Bouna, S10 = Téhini, S11 = Doropo).

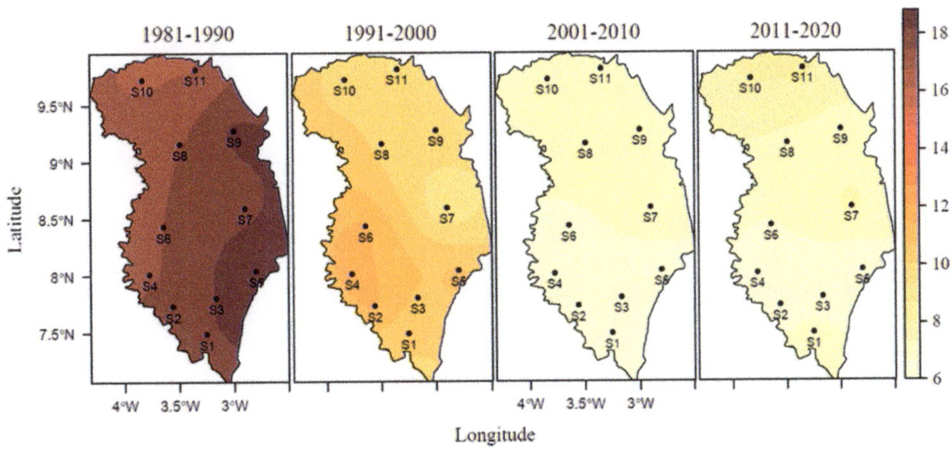


**Figure 9.** Spatial distribution of number of cold nights (TN10p) (%) per decade in Zanzan region. Stations (S1 = Koun Fao, S2 = Kouassi-datékro, S3 = Tanda, S4 = Sandegué, S5 = Bondoukou, S6 = Nassian, S7 = Sapli, S8 = PN Comoé S1, S9 = Bouna, S10 = Téhini, S11 = Doropo).

#### 3.2.3. Evolution of the Index of Cold Spell Duration Indicator (CSDI)

The Mann–Kendall test demonstrated a general declining trend in the index of cold spell duration indicator (Table 5) for the entire research area. At the 5% significance level, the general decreasing trends were significant at all study sites. Over the research period of 1981–2020, the largest significant declines were in Bondoukou estimated at 39 days, Kouassi-datékro estimated at 30.57 days, Nassian estimated at 29.99 days, and Sandegué estimated at 29.95 days. On average, this decline was estimated at 26.15 days from 1981 to 2020 for the entire region. The spatial distribution per decade of the cold spell duration indicator shown in Figure 10 indicates that from 1981 to 1990, the variation was between 36.65 and 46.97 days, and from 1991 to 2000 the variation was between 14.43 and 25.8 days, representing a decrease of 21.17 days. During the period 2001 to 2010, the values were between 10.21 and 14.1 days, which was a decrease of 11.7 days, and from 2011 to 2020, the variations were between 11.8 and 19.59 days, implying an increase of 5.49 days. In comparison with the preceding two decades, the cold spell duration indicator in the north of the region under consideration has made a slight return in the last decade.

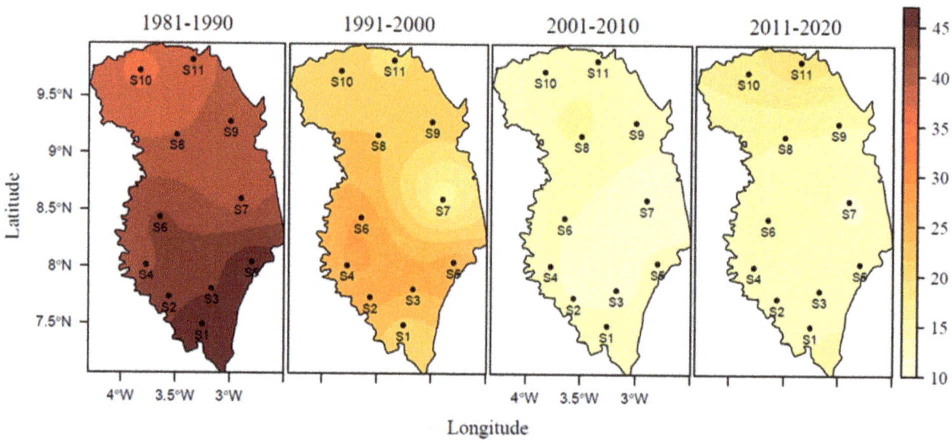


**Figure 10.** Spatial distribution of cold spell duration indicator (CSDI) (%) per decade in Zanzan region. Stations (S1 = Koun Fao, S2 = Kouassi-datékro, S3 = Tanda, S4 = Sandegué, S5 = Bondoukou, S6 = Nassian, S7 = Sapli, S8 = PN Comoé S1, S9 = Bouna, S10 = Téhini, S11 = Doropo).

#### 3.2.4. Changes in the Number of Hot Days' Index (TX90p)

Table 5 and Figure 11 give the outputs of the Mann–Kendall statistical test and the spatial distribution of the number of hot days' index in the Zanzan region, respectively. In all of the stations within the research area, the statistical analysis provides an overall increase in the number of hot days, and two of these stations show significant increasing trends at the 5% significance level. Over the study period of 1981–2020, the significant increases were estimated at 13.18% of days at Sapli, and 7.44% of days at Sandegué. The overall average increasing trend was 5.19% of days across the entire research area. The spatial distribution of the number of hot days per decade showed a variation from 5.74% to 9.11% over the period of 1981–1990, and from 1991 to 2000, the index varied from 8.07% to 12.3%, which represents an increase of 3.19% of days. The highest index values over the last two decades have been observed in the northern and southern parts of the study area. From 2001 to 2010 the variation of the index was estimated between 12.2% and 15.4%, meaning an increase of 3.1% and from 2011 to 2020, the variation was between 6.72% and 14.5%, meaning a small decrease of 0.9% of days. The decade 2001–2010 appears to be the hottest in the spatial analysis of the number of relatively hot days, and this significant heating was more pronounced in the south-western part of the Zanzan region.

#### 3.2.5. Evolution of the Index of the Number of Warm Nights (TN90p)

The trends in the temporal series of the index of the number of warm nights analyzed through the Mann–Kendall test (Table 5) in the Zanzan region revealed a general increase over all the stations studied. Except for the Sapli station, which indicated a non-significant rising trend, the general rising trend in the number of warm nights in the study region was significant at the 5% level of statistical significance for 11 of the study stations. Geographically over the study period, these increases were estimated at 14.89% in Bondoukou, 14.27% in Sandegué, 13.68% in Nassian, and 13.80% of nights in Tanda. On average in the study area, this increase was estimated at 11.50% of nights. The spatial distribution of the warm nights' index per decade is presented in Figure 12. Over the decade 1981–1990, the number of warm nights was between 4.84% and 7.77% and from 1991 to 2000 it varied between 8.1% and 10.25%, which is an increase of 2.48%. From 2001 to 2010, the variation in the number of warm nights was estimated between 11.19% and 14.73%, representing an increase of 4.48%, and from 2011 to 2020, the variation was between 10.44% and 14.56%, indicating a small decrease of 0.17% in warm nights. The comparison between the first decade of this study (i.e., 1981–1990) and the last decade (i.e., 2011–2020) indicated an increase in the number of warm nights by 5.39%. For the first two decades of the study period, the highest values of the index were found in the north of the area under study, in places such as Doropo, Téhini, and Bouna. For the decades of 2001–2010 and 2011–2020, the high values were found in the southeast and southwest of the Zanzan region respectively in locations such as Tanda and Bondoukou.

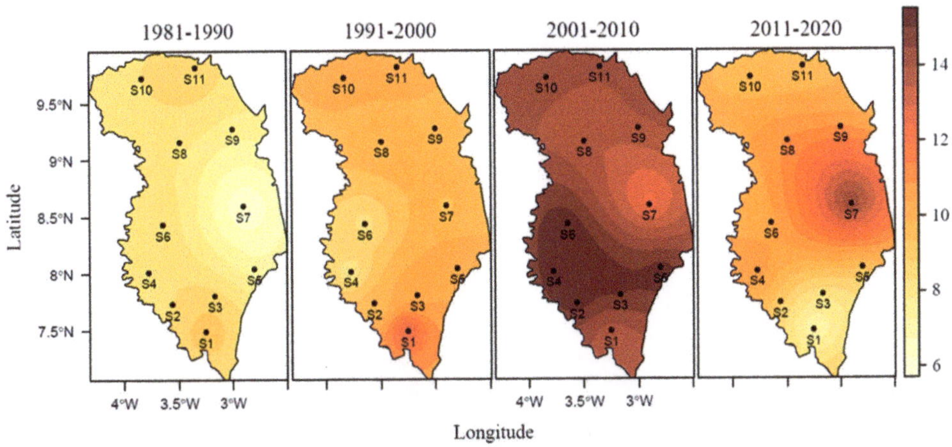


**Figure 11.** Spatial distribution of number of hot days (TX90p) (%) per decade in Zanzan region. Stations (S1 = Koun Fao, S2 = Kouassi-datékro, S3 = Tanda, S4 = Sandegué, S5 = Bondoukou, S6 = Nassian, S7 = Sapli, S8 = PN Comoé S1, S9 = Bouna, S10 = Téhini, S11 = Doropo).

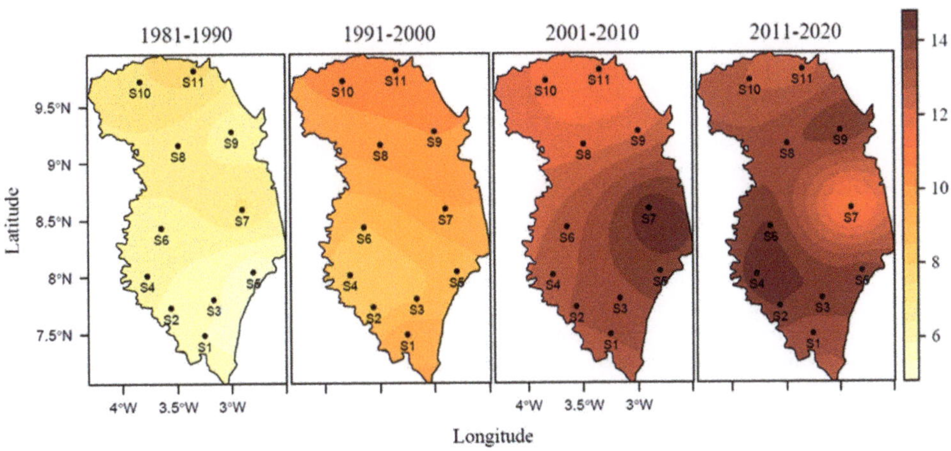


**Figure 12.** Spatial distribution of number of warm nights (TN90p) (%) per decade in Zanzan region. (S1 = Koun Fao, S2 = Kouassi-datékro, S3 = Tanda, S4 = Sandegué, S5 = Bondoukou, S6 = Nassian, S7 = Sapli, S8 = PN Comoé S1, S9 = Bouna, S10 = Téhini, S11 = Doropo).

3.2.6. Evolution of the Warm Spell Duration Indicator (WSDI)

The data in Table 5 on the outputs of the Mann–Kendall statistical test showed a general increasing trend of the warm spell duration indicator in the Zanzan region. The general rising trend was nevertheless non-significant at the 5% significance level, except for the Sapli and Sandegué stations, which displayed significant rising trends. Over the study period, these significant increases were estimated at 24.29 days in Sapli, 14.39 days in Sandegué, and the overall average increase was estimated at 9.71 days over the entire study area. Figure 13 illustrates the spatial pattern of the warm spell duration indicator. Over the decade 1981–1990, the evolution of the warm spell duration indicator was between 13.5 and 26.7 days, and from 1991 to 2000, the evolution was between 16.8 and 31.3 days, meaning an increase of 4.6 days. Over the first decade, the strong warm spells were more intense in the southern and northern part of the research area, and over the second decade 1991–2000, the strong warm spells were more noticeable in the southern part of the region. From 2001 to 2010 the warm spell duration indicator varied between 25.3 and 40.1 days, which corresponds to an increase of 8.8 days, and from 2011 to 2020, the variation was between 14.9 and 29.8 days, which corresponds to a decrease of 10.3 days. Compared to other decades, the decade 2001–2010 stands out as the warmest with longer warm spells that cover almost the entire Zanzan region.

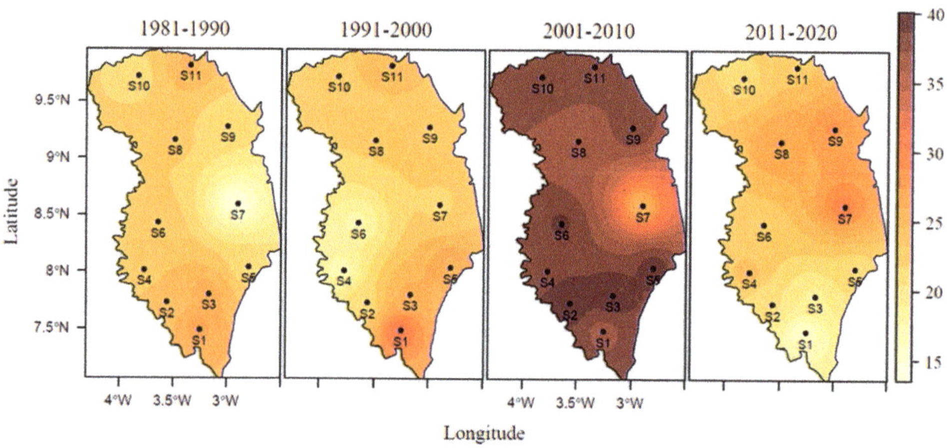


**Figure 13.** Spatial distribution of warm spell duration indicator (WSDI) (days) per decade in Zanzan region. Stations (S1 = Koun Fao, S2 = Kouassi-datékro, S3 = Tanda, S4 = Sandegué, S5 = Bondoukou, S6 = Nassian, S7 = Sapli, S8 = PN Comoé S1, S9 = Bouna, S10 = Téhini, S11 = Doropo).

## 4. Discussion

### 4.1. Extreme Precipitations

This study carried out in the Zanzan region showed a decrease in total annual precipitation (PRCPTOT) at most of the stations studied (10 of the stations), with an average decrease of 112 mm over the research period. Furthermore, a generalized decrease in daily precipitation intensity SDII was seen across the entire study area, with the example of the localities in the southeast of the region, namely, Kouassi-datékro and Sandegué with respective decreases estimated at 1.17 mm/d and 0.78 mm/d. The decrease of the PRCPTOT index was also observed in the spatial distribution following the decades even if in the last decade we notice a slight return of precipitations. The primary influence of the decrease in precipitation was observed through the low intensity of rainfall received by the study area, as the daily precipitation intensity was also decreasing even though the decrease seemed insignificant. The increasing trend in daily precipitation intensity over the last decade had also led to a slight overall increase in rainfall. The general decrease in rainfall in the study area followed the general trend of decreasing rainfall that is observed throughout Côte d'Ivoire and varies according to the regions of the country. Indeed, for the PRCPTOT index, Allechy et al. [76] in the central-western part of Côte d'Ivoire noted a decrease of 788 mm between 1984 and 2013, Atcheremi et al. [71] in the south-western part of Côte d'Ivoire revealed a decrease of 85 mm in Daloa and 493 mm in Sassandra as well as a decrease of 159 mm according to Balliet et al. [75] in Gagnoa. In the forested south of Côte d'Ivoire, N'guessan et al. [90] determined a decrease in PRCPTOT of 20 mm followed by a decrease of 0.5 mm/d in rainfall intensities over two decades. The generalized decline in rainfall, which had even created a major drought wave starting in the 1970s, as revealed in the study conducted by Paturel et al. [91], continued over the following decades in Côte d'Ivoire, particularly in the Zanzan region, even though a slight return of rainfall was observed over the last decade. However, this decline in total annual precipitation and daily precipitation intensity is more than problematic for agricultural development, especially for the main cash crop in the Zanzan region, cashew nuts. For acceptable cashew nut production, rainfall between 800 and 1800 mm is required to be considered sufficient [92], whereas in the Zanzan region this rainfall, which varies between 800 and 1400 mm, is decreasing. A future study of the impact of rainfall variations on agricultural products, especially cashew nuts, will provide information on the real economic threat in the Zanzan region. The observation of rainfall disruption is also shown elsewhere in other studies. In the West African sub-region, studies such as that of New et al. [33] showed that the regions as a whole experienced a decrease in total annual precipitation and the number of rainy days. Thus, Hountondji et al. [72] in Benin found a decrease in total annual precipitation PRCPTOT for 95% of the stations studied and stability of the daily precipitation intensity SDII. This decrease in rainfall extends even across Africa as Vondou et al. [74], reported that the cumulative annual rainfall and the daily precipitation intensity SDII are decreasing in northern Cameroon. Unlike the decrease in the daily precipitation intensity SDII found in our study area, Gebrechorkos et al. [81] in their work found an increase in the SDII index in Kenya.

The CWD index for consecutive wet days was declining by an average of 16.45 days across all of the stations analyzed in the Zanzan region, while the CDD index for consecutive dry days was rising by an average of 18.67 days throughout most of the stations studied, according to the findings of this study. The CDD index is a marker of the presence of drought in a region. Its increase in a locality should be a concern, especially in an area such as the Zanzan region where the source of income of the population is based on agriculture. It is also a factor in the threat to water resources. These indicators have been studied in other regions of Côte d'Ivoire. In the southwest of Côte d'Ivoire, contrary to our result, Atcheremi et al. [71] observed an increase in consecutive wet days in Daloa, but in Gagnoa in the same area, Balliet et al. [75] indicated that the CWD was decreasing while the CDD was increasing. Furthermore, in the west-central area of Côte d'Ivoire, Allechi et al. [76] demonstrated that the CWD index has decreased while the CDD index has increased by 0.812 days/year. The occurrence of dry sequences within the rainy season thus seems to be widespread in Côte d'Ivoire, which is a cause for concern in an essentially agricultural country. For a proper practice of rainfed agriculture, when during the rainy season there are sequences of dry days, this can cause the drying of the young plant at the primary stage of its growth, especially for crops such as maize and yam, which are cultivated in the Zanzan region. This impact may be minimal on cashew trees because they have more or less short-drought resistant characteristics. In the West African sub-region, this observation of the evolution of these indices seems to be the same. The study in Ghana by Larbi et al. [93] showed rising trends in CDD and a declining trend in the CWD index, and Ozer et al. [94] identified the same results in Niger. These rainfall pattern parallels would be worldwide in scope, as they have been confirmed in several parts of the Pacific [31], and Donat et al. [95] detected that the CDD index was trending toward arid conditions across South Africa, East Asia, and South America in their analytical research of precipitation and temperature extremes since the turn of the 20th century. The R99p and R99pTOT indices in this study were slightly increased in the Zanzan region. The average increase in total annual precipitation from very heavy rain days was 1.41% of days. The spatial distribution clearly showed this evolution over the decades. The last decade, 2011–2020, showed a significant increase for both indices. A rise in the R99p and R99pTOT indices is a very important water contribution. For rainfed agriculture, these high water inputs can contribute to the development of the plant, but only if the plant has already exceeded the primary growth period, at the risk of being destroyed by these heavy rains, as could be the case for maize or cashew. These indices are sometimes the cause of flooding and landslides. Balliet et al. [75] identified a rise in R99p index at Gagnoa in Côte d'Ivoire. These patterns of change in these indices are similar and sometimes different in other parts of Africa. Hountondji et al. [72] found more stable conditions of the R99p and R99pTOT index in Benin. In the Central Sahel, Panthou et al. [73] reported that the occurrence of extreme rainfall was increasing as well as in the West African Sahel [96]. In contrast to the increase in R99p index found in this study, Berhane et al. [63] found in their study a significant decrease of R99p in Ethiopia in the Tigray region. With the advent of climate change, some areas of the world will receive more rainfall, and sometimes the scarcity of rainfall in some areas will increase, hence the similarity and differences in the results of this study compared to others. In the Zanzan region, these high rainfall amounts have certainly influenced the slight return of rainfall observed in the decade 2011–2020.

### 4.2. Temperature Extremes

A considerable drop in the trends of the cold spells, such as the index of the number of cold nights TN10p, the index of the number of cool days TX10p, and the cold spell duration indicator CSDI, was noted at all research stations in the Zanzan region in the north-east of Côte d'Ivoire. On the other hand, the warm spells, such as the index of the number of hot days TX90p, the index of warm spell duration indicator WSDI, and the index of the number of warm nights TN90p increased in all the stations studied and were more significant in the stations such as Bondoukou, Sapli, and Sandegué. Regarding the spatial distribution per decade, the cold and warm spells have evolved in contrast. The first decade (i.e., 1981–1990) showed a large number of cool days and this has evolved in a decreasing way until the last decade (i.e., 2011–2020) which showed a very low number of cool days. For the warm spells, the evolution started with a low number of hot days before displaying a high number of hot days over the last two decades. However, the decade 2001–2010 has shown to be the warmest in terms of days and the last two decades in terms of nights in the study area. The humid climate of the Zanzan region has been transformed into a hot climate. In the climatic context of Côte d'Ivoire, Atcheremi et al. [71] found similar results concerning the increase in the WSDI index and the decrease in a CSDI index in the southwest of Côte d'Ivoire, precisely in Daloa, Gagnoa, and Sassandra. The various trends observed related to the warming of the study area are under the influence of climate change. For the Zanzan region, this situation is a concern. The shortening of the rainy seasons and the warming of the study area will have an influence on water resources in the region and will greatly impact agriculture, which is essentially rainfed in this area. For cashew farming, for example, the presence of high temperatures during the flowering period can be very problematic. The threshold temperature should not be above 32 °C in order not to compromise the flowering and fruiting of the cashew tree. Indeed, beyond this threshold value, the temperature can provoke scald and flower abortion phenomena and significantly reduce the fruiting of the cashew tree [97]. The lack of cold nights in favor of the warming could impact the health of the population as shown in other studies, namely, Ambouta et al. [98] in Niger and Abatan et al. [26] in Nigeria. The climate context related to global warming is a sub-regional observation. Barry et al. [27] found in West Africa that warm nights and hot days are becoming more common, whereas cold nights and cool days are becoming less common. Furthermore, Chaney et al. [99] concluded that in Sub-Saharan Africa, there is a more rapid rise in warm climatic occurrences and a decline in cold spells. This warming is even spreading across Africa and around the world. In a study conducted in the "Greater

Horn of Africa" in countries such as Burundi, Sudan, Rwanda, Djibouti, Tanzania, Uganda, and Ethiopia, Omondi et al. [100] stated that there was a rise in extreme heat (TX90P, WSDI) while cold spells decreased. In South Africa [101] and Morocco [102,103], the scholars concluded that there was an increase in warm extreme events TX90p and WSDI while cold extreme conditions TX10p and CSDI declined. Popov et al. [104] evaluated the extreme temperature indices in Bosnia and Herzegovina and concluded that there was a significant rise in warm temperature (TN90p, TX90p, WSDI) and similar observations were made in Israel and Palestine by Salameh et al. [105], as well as in China by Zhou et al. [106]. The decade 2001–2010 was found to be warmer than other decades in this study. This increase in warmer conditions during the 2001–2010 decade according to the WMO [107] and Coumou and Rahmstorf [108] studies was felt globally in several regions of the world. According to the IPCC [31], the temperature change follows natural and anthropogenic factors, especially in the causal relationships between the emissions of greenhouse gas (GHG). For the IPCC, continued GHG emissions at the current rate or higher are expected to increase warming and profoundly alter the climate system in the 21st century. It should be noted that the decrease in rainfall and the temperature rise are also locally linked to the regression of forests following the ongoing deforestation in Côte d'Ivoire. Indeed, according to Yao [109], the Ivorian forest has declined from 13 million hectares in colonial times to less than two million hectares currently.

## 5. Conclusions

The current study, whose major objective was to spatiotemporally assess the climate change in precipitation and temperature extremes in the Zanzan region using ETCCDMI indices, has produced a summary of the region's extreme climatic trends. Daily observations of minimum and maximum temperatures, as well as daily rainfall, were used to calculate the indices. For the extreme precipitation trends, the stations studied showed indices that have mostly non-significant statistical trends. Indeed, the results showed a decline in annual total precipitation PRCPTOT, daily precipitation intensity SDII, and consecutive wet days CWD, indicating a decrease in rainfall in the region. These decreases were estimated to be an average over the study period at 112.35 mm for the PRCPTOT index and 16.45 days for the CWD index. Following that, the CDD (consecutive dry days) was increased, as well as the indices R99p and R99pTOT (total annual precipitation from very heavy rain days and contribution from extremely wet days, respectively). The latter indices can lead to flooding and sometimes landslides. The increase, for example, in the CDD index, was estimated at 18.67 days over the study period. The indices related to extreme temperatures were much more statistically significant than the indices related to extreme precipitation. For these temperature extremes, we observed a decline in the cold spells TX10p, TN10p, CSDI and a rise in the warm spells TX90p, TN90p, and WSDI. We can state that the decreases in the cold spells were estimated on average over the study period at 6.58% of the days for the TX10p index and 26.15 days for the CSDI index. As for the warm spells, the increases were estimated on average at 5.19% of days for the TX90p index and 9.71 days for the WSDI index. Decreased rainfall and increased temperatures in this agricultural region of Côte d'Ivoire may influence agricultural production through low yields and could also threaten water resources. Furthermore, the socio-economic situation of this region would be even more alarming if there is an intensification of the currently observed trends in the future regarding the practice of crops, such as cashew, which would provoke a vulnerable health situation for the population. The knowledge of these different trends in temperature and rainfall extremes for this region is of high importance in the measure of sustainable agricultural and economic planning by the government authorities. Measures to encourage the population to adopt crops that are more resistant to the current climatic conditions, to think about better planning of the agricultural calendar according to the rainfall conditions, and consider forest management can immediately be set up by the authorities of the Zanzan region.

**Author Contributions:** Conceptualization, K.D.K., B.H.K. and K.A.; methodology, K.D.K.; software, K.D.K. and K.A.; validation, K.D.K., A.T.K.-b., B.H.K. and K.A.; formal analysis, K.D.K.; investigation, K.D.K.; resources, K.D.K. and A.T.K.-b.; data curation, K.D.K. and B.H.K.; writing—original draft preparation, K.D.K.; writing—review and editing, K.D.K. and K.A.; visualization, A.T.K.-b., B.H.K. and K.A.; supervision, A.T.K.-b., B.H.K. and K.A.; project administration, A.T.K.-b., K.A. and K.D.K.; funding acquisition, A.T.K.-b., K.A. and K.D.K. All authors have read and agreed to the published version of the manuscript.

**Funding:** This research received no external funding.

**Institutional Review Board Statement:** Not applicable.

**Informed Consent Statement:** Not applicable.

**Data Availability Statement:** Some of the data analyzed in this study are a re-analysis of existing data, which are freely available at locations cited in the reference section and the remaining data are available from the corresponding author on reasonable request.

**Acknowledgments:** The authors would like to acknowledge the Regional Centre for Energy and Environmental Sustainability (RCEES) for their investment in this work.

**Conflicts of Interest:** The authors declare no conflict of interest.

## References

1. Orellana, M. Climate Change and the Millennium Development Goals: The Right to Development, International Cooperation and the Clean Development Mechanism. *Sur.-Int. J. Hum. Rights* **2010**, *7*, 144.
2. Akpoti, K.; Kabo-bah, A.T.; Zwart, S.J. Agricultural Land Suitability Analysis: State-of-the-Art and Outlooks for Integration of Climate Change Analysis. *Agric. Syst.* **2019**, *173*, 172–208. [CrossRef]
3. World Bank World Development Indicators. 2006. Available online: Https://Openknowledge.Worldbank.Org/Handle/10986/8151 (accessed on 8 June 2021).
4. Rockström, J.; Folke, C.; Gordon, L.; Hatibu, N.; Jewitt, G.; Penning de Vries, F.; Rwehumbiza, F.; Sally, H.; Savenije, H.; Schulze, R. A Watershed Approach to Upgrade Rainfed Agriculture in Water Scarce Regions through Water System Innovations: An Integrated Research Initiative on Water for Food and Rural Livelihoods in Balance with Ecosystem Functions. *Phys. Chem. Earth* **2004**, *29*, 1109–1118. [CrossRef]
5. Dixon, J.; Gulliver, A.; Gibbon, D. Farming Systems and Poverty: Improving Farmers' Livelihoods in a Changing World. In Farming Systems and Poverty: Improving Farmers' Livelihoods in a Changing World. Food and Agriculture Organization and the World Bank. 2001. Available online: Https://Www.Fao.Org/3/Ac349e/Ac349e (accessed on 8 June 2021).
6. Akpoti, K.; Higginbottom, T.P.; Foster, T.; Adhikari, R.; Zwart, S.J. Mapping Land Suitability for Informal, Small-Scale Irrigation Development Using Spatial Modelling and Machine Learning in the Upper East Region, Ghana. *Sci. Total Environ.* **2022**, *803*, 149959. [CrossRef]
7. Waha, K.; Müller, C.; Rolinski, S. Separate and Combined Effects of Temperature and Precipitation Change on Maize Yields in Sub-Saharan Africa for Mid-to Late-21st Century. *Glob. Planet. Chang.* **2013**, *106*, 1–12. [CrossRef]
8. Siabi, E.K.; Kabobah, A.T.; Akpoti, K.; Anornu, G.K.; Amo-Boateng, M.; Nyantakyi, E.K. Statistical Downscaling of Global Circulation Models to Assess Future Climate Changes in the Black Volta Basin of Ghana. *Environ. Chall.* **2021**, *5*, 100299. [CrossRef]
9. Ibe, G.O.; Amikuzuno, J. Climate Change in Sub-Saharan Africa: A Menace to Agricultural Productivity and Ecological Protection. *J. Appl. Sci. Environ. Manag.* **2019**, *23*, 329. [CrossRef]
10. IPCC. *The Scientific Basis, Summary for Policy Makers—Contribution of Working Group I to the Third Assessment Report of the Intergovermental Panel on Climate Change*; Houghton, J.T., Ding, Y., Griggs, D.J., Noguer, M., van der Linden, P.J., Dai, X., Maskell, K., Eds.; IPCC: Geneva, Switzerland, 2001.
11. Akpoti, K.; Dossou-Yovo, E.R.; Zwart, S.J.; Kiepe, P. The Potential for Expansion of Irrigated Rice under Alternate Wetting and Drying in Burkina Faso. *Agric. Water Manag.* **2021**, *247*, 106758. [CrossRef]
12. Shoko, R.R.; Chaminuka, P.; Belete, A. Maize yield sensitivity to climate variability in south africa: Application of the ardl-ecm approach. *J. Agribus. Rural Dev.* **2019**, *2016*, 363–371. [CrossRef]
13. Blanc, E. The Impact of Climate Change on Crop Yields in Sub-Saharan Africa. *Am. J. Clim. Change* **2012**, *1*, 1–13. [CrossRef]
14. Grace, K.; Nagle, N.N.; Husak, G. Can Small-Scale Agricultural Production Improve Children's Health? Examining Stunting Vulnerability among Very Young Children in Mali, West Africa. *Ann. Am. Assoc. Geogr.* **2016**, *106*, 722–737. [CrossRef]
15. Belesova, K.; Gasparrini, A.; Sié, A.; Sauerborn, R.; Wilkinson, P. Household Cereal Crop Harvest and Children's Nutritional Status in Rural Burkina Faso. *Environ. Health A Glob. Access Sci. Source* **2017**, *16*, 65. [CrossRef] [PubMed]
16. FAO. *The State of Food and Agriculture. Food Aid for Food Security?* FAO: Rome, Italy, 2006; ISBN 9789251056004.
17. Alagidede, P.; Adu, G.; Frimpong, P.B. The Effect of Climate Change on Economic Growth: Evidence from Sub-Saharan Africa. *Environ. Econ. Policy Stud.* **2016**, *18*, 417–436. [CrossRef]

18. World Bank. World Bank. World Development Indicators 2007. In *World Development Indicators*; World Bank: Washington, DC, USA, 2007; ISBN 0821369598.
19. Müller, C. Climate Change Impact on Sub-Saharan Africa? An Overview and Analysis of Scenarios and Models. (DIE Discussion Paper, 3/2009). In *Bonn: Deutsches Institut Für Entwicklungspolitik GGmbH*; DIE: Bonn, Germany, 2009; ISBN 9783889854513.
20. Chen, H.; Githeko, A.K.; Zhou, G.; Githure, J.I.; Yan, G. New Records of Anopheles Arabiensis Breeding on the Mount Kenya Highlands Indicate Indigenous Malaria Transmission. *Malar. J.* **2006**, *5*, 17. [CrossRef]
21. Reich, P.; Numbe, S.; Almaraz, R.; Eswaran, H. Land Resources Stresses and Desertification in Africa. *AgroSci* **2001**, *2*, 1–10. [CrossRef]
22. Mitchell, S.A. The Status of Wetlands, Threats and the Predicted Effect of Global Climate Change: The Situation in Sub-Saharan Africa. *Aquat. Sci.* **2013**, *75*, 95–112. [CrossRef]
23. Pelling, M.; Wisner, B. (Eds.) Urbanisation, Human Security and Disaster Risk in Africa. In *Disaster Risk Reduction—Cases from Urban Africa*; Earthscan: London, UK, 2009; pp. 3–16.
24. Akpoti, K.; Kabo-bah, A.T.; Dossou-Yovo, E.R.; Groen, T.A.; Zwart, S.J. Mapping Suitability for Rice Production in Inland Valley Landscapes in Benin and Togo Using Environmental Niche Modeling. *Sci. Total Environ.* **2020**, *709*, 136165. [CrossRef]
25. Sultan, B.; Gaetani, M. Agriculture in West Africa in the Twenty-First Century: Climate Change and Impacts Scenarios, and Potential for Adaptation. *Front. Plant Sci.* **2016**, *7*, 1262. [CrossRef]
26. Abatan, A.A.; Abiodun, B.J.; Lawalc, K.A.; Gutowski, W.J. Trends in Extreme Temperature over Nigeria from Percentile-Based Threshold Indices. *Int. J. Climatol.* **2016**, *36*, 2527–2540. [CrossRef]
27. Barry, A.A.; Caesar, J.; Klein Tank, A.M.G.; Aguilar, E.; McSweeney, C.; Cyrille, A.M.; Nikiema, M.P.; Narcisse, K.B.; Sima, F.; Stafford, G.; et al. West Africa Climate Extremes and Climate Change Indices. *Int. J. Climatol.* **2018**, *38*, e921–e938. [CrossRef]
28. Kpanou, M.; Laux, P.; Brou, T.; Vissin, E.; Camberlin, P.; Roucou, P. Spatial Patterns and Trends of Extreme Rainfall over the Southern Coastal Belt of West Africa. *Theor. Appl. Climatol.* **2021**, *143*, 473–487. [CrossRef]
29. IPCC. *Managing the Risks of Extreme Events and Disasters to Advance Climate Change Adaptation. A Special Report of Working Groups I and II of the Intergovernmental Panel on Climate Change*; Field, C.B., V. Barros, T.F., Stocker, D., Qin, D.J., Dokken, K.L., Ebi, M.D., Eds.; IPCC: Geneva, Switzerland, 2012; ISBN 9781139177245.
30. Roudier, P.; Sultan, B.; Quirion, P.; Berg, A. The Impact of Future Climate Change on West African Crop Yields: What Does the Recent Literature Say? *Glob. Environ. Chang.* **2011**, *21*, 1073–1083. [CrossRef]
31. IPCC. *Climate Change 2007: Impacts, Adaptation and Vulnerability. Contribution of Working Group II to the Fourth Assessment Report of the Intergovernmental Panel on Climate Change*; Parry, M.L., Canziani, O.F., Palutikof, J.P., van der Linden, P.J., Hanson, C.E., Eds.; Cambridge University Press: Cambridge, UK, 2007; ISBN 9780521880107.
32. Aguilar, E.; Barry, A.A.; Brunet, M.; Ekang, L.; Fernandes, A.; Massoukina, M.; Mbah, J.; Mhanda, A.; do Nascimento, D.J.; Peterson, T.C.; et al. Changes in Temperature and Precipitation Extremes in Western Central Africa, Guinea Conakry, and Zimbabwe, 1955–2006. *J. Geophys. Res. Atmos.* **2009**, *114*, 1–11. [CrossRef]
33. New, M.; Hewitson, B.; Stephenson, D.B.; Tsiga, A.; Kruger, A.; Manhique, A.; Gomez, B.; Coelho, C.A.S.; Masisi, D.N.; Kululanga, E.; et al. Evidence of Trends in Daily Climate Extremes over Southern and West Africa. *J. Geophys. Res. Atmos.* **2006**, *111*, 1–11. [CrossRef]
34. Karimou, B.M.; Ambouta, K.; Sarr, B.; Tychon, B. *Analyse Des Phénomènes Climtiques Extrêmes Dans Le Sud-Est Du Niger*; Publications de l'Association Internationale de Climatologie: Liège, Belgique, 2015.
35. Cameron, C.; Norrington-Davies, G.; De Agulhas, V.t.V.; Mitchell, T. Climate and Development Knowledge Network Managing Climate Extremes and Disasters in Africa: Lessons from the SREX Report; 2012. Available online: www.Cdkn.Org/Srex (accessed on 12 June 2021).
36. Schlenker, W.; Lobell, D.B. Robust Negative Impacts of Climate Change on African Agriculture. *Environ. Res. Lett.* **2010**, *5*, 014010. [CrossRef]
37. Sultan, B.; Guan, K.; Kouressy, M.; Biasutti, M.; Piani, C.; Hammer, G.L.; McLean, G.; Lobell, D.B. Robust Features of Future Climate Change Impacts on Sorghum Yields in West Africa. *Environ. Res. Lett.* **2014**, *9*. [CrossRef]
38. Rockström, J.; Falkenmark, M. Increase Water Harvesting in Africa: Meeting Global Food Needs Requires Strategies for Storing Rainwater and Retaining Soil Moisture to Bridge Dry Spells. *Nature* **2015**, *519*, 283–285. [CrossRef]
39. Wilkinson, E.; Peters, K. *Extrêmes Climatiques et Réduction de La Pauvreté Par La Résilience: Le Développement Conçu Dans l'incertitude*; Overseas Development Institute: Londres, UK, 2016; Volume 2015.
40. FAO. *Climate-Smart Agriculture in Côte d ' Ivoire. Climate-Smart Agriculture in Côte d'Ivoire*; FAO: Rome, Italy, 2018.
41. Diomande, M.; Dongo, K.; Koné, B.; Cissé, G.; Biémi, J.; Bonfoh, B. Vulnérabilité de l'agriculture Pluviale Au Changement de Régime Pluviométrique et Adaptation Des Communautés Rurales Du «V-Baoulé» En Côte d'Ivoire. *14e Colloq. Int. en Éval. Environ.* **2009**, *11*, 8–16.
42. DJE KOUAKOU BERNARD. Document de Stratégie Du Programme National Changement Climatique (2015–2020). In *84 Pages + Annexes*; Ministère de l'Enivironndment, de la Salubrité Urbaine et du Développement Durable (MESUDD): Abidjan, Côte d'Ivoire, 2014.
43. Kassin, K.E.; Doffangui, K.; Kouamé, B.; Yoro, R.G.; Assa, A. Variabilité Pluviométrique et Perspectives Pour La Replantation Cacaoyère Dans Le Centre Ouest de La Côte d'Ivoire. *J. Appl. Biosci.* **2008**, *12*, 633–641.

44. Guessan, K.; Kouassi, A.; Gnaboa, R.; Traoré, K.; Houenou, P. Analyse de phenomènes hydrologiques dans un bassin versant urbanisé: Cas de la ville de yamoussoukro (centre de la cote d ' ivoire). *Larhyss J.* **2014**, *17*, 135–154.
45. Ernest, A.K.; Blaise, K.Y.; Michel, K.A.; Gbombélé, S.; Nagnin, S.; Jean, B. Étude de La Variabilité Hydroclimatique et de Ses Conséquences Sur Les Ressources En Eau Du Sud Forestier et Agricole de La Côte d ' Ivoire: Cas de La Région d ' Abidjan-Agboville. *Int. J. Pure Appl. Biosci.* **2013**, *1*, 30–50.
46. Dekoula, C.; Brou, K.; N'goran, k.E.; Yao, F.; Ehounou, J.; Soro, N. Impact De La Variabilité Pluviométrique Sur La Saison Culturale Dans La Zone De Production Cotonnière En Côte D'. *Eur. Sci. J.* **2018**, *14*, 143–156. [CrossRef]
47. MINAGRI (Ministère de l'Agriculture). Rapport Sur La Dynamique de La Consommation Alimentaire En Côte d'Ivoire. In *98 Pages + Annexes*; MINAGRI: Abidjan, Côte d'Ivoire, 2011.
48. WFP (World Food Programme). *WFP Côte d'Ivoire Country Brief August 2019*; WFP: Rome, Italy, 2019.
49. *CI-Energies Plan Cadre de Gestion Environnementale et Sociale (PCGES) Du Projet Electrification de 1088 Localités de La Côte d'Ivoire*; Rapport Final: Abidjan, Côte d'Ivoire, 2019; pp. 1–169.
50. Dabissi, N. *Changements Hydroclimatiques et Transformations de l'agriculture: L'exemple Des Paysanneries de l'Est de La Côte d'Ivoire*; Universite Paris 1 Pantheon-Sorbonne: Paris, France, 2011.
51. Dje, K.B.; Ochou, A.D.; Kouadio, Z.A.; Alama, K. Impacts of El Nino on Rainfall and Its AgrocltmaticIncidences in the Zanzan Region: North-East of Côte d'Ivoire. *Geosci. Res.* **2018**, *3*, 9–20. [CrossRef]
52. WMO. *Guidelines on Analysis of Extremes in a Changing Climate in Support of Informed Decisions for Adaptation*; WMO-TD No. 1500: Geneva, Switzerland, 2009; pp. 361–375.
53. WMO. *Inter-Regional Workshop on Indices and Early Warning Systems for Drought*; World Meteorological Organization: Lincoln, NE, USA, 2009.
54. Brou, Y.T. La Variabilité Climatique En Côte d'Ivoire: Entre Perceptions Sociales et Réponses Agricoles. *Cah. Agric.* **2005**, *14*, 533–540.
55. *INS Recensement Général de La Population et de l'habitat (RGPH) 2014*; Répertoire Des Localités: Abidjan, Côte d'Ivoire, 2015.
56. Sinan, A.; Abou, N.K. Impacts Socio-Economiques De La Culture De L'anacarde Dans La Sous-Prefecture D'odienne (Côte d'Ivoire). *Eur. Sci. J. ESJ* **2016**, *12*, 369. [CrossRef]
57. Westberg, D.J.; Stackhouse, P.W.; Crawley, D.B.; Hoell, J.M.; Chandler, W.S.; Zhang, T. An Analysis of NASA's Merra Meteorological Data to Supplement Observational Data for Calculation of Climatic Design Conditions. *ASHRAE Trans.* **2013**, *119*, 210–221.
58. Rienecker, M.M.; Suarez, M.J.; Gelaro, R.; Todling, R.; Bacmeister, J.; Liu, E.; Bosilovich, M.G.; Schubert, S.D.; Takacs, L.; Kim, G.K.; et al. MERRA: NASA's Modern-Era Retrospective Analysis for Research and Applications. *J. Clim.* **2011**, *24*, 3624–3648. [CrossRef]
59. Chandler, W.S.; Hoell, J.M.; Westberg, D.; Whitlock, C.H.; Zhang, T., Jr.; Stackhouse, P.W. *NASA's Prediction of Worldwide Energy Resource (POWER) Web Services*; American Meteorological Society 91st Annual Meeting: Seattle, WA, USA, 2011.
60. Bai, J.; Chen, X.; Dobermann, A.; Yang, H.; Cassman, K.G.; Zhang, F. Evaluation of Nasa Satellite-and Model-Derived Weather Data for Simulation of Maize Yield Potential in China. *Agron. J.* **2010**, *102*, 9–16. [CrossRef]
61. Bender, F.D.; Sentelhas, P.C. Solar Radiation Models and Gridded Databases to Fill Gaps in Weather Series and to Project Climate Change in Brazil. *Adv. Meteorol.* **2018**, *2018*, 6204382. [CrossRef]
62. Aboelkhair, H.; Morsy, M.; El Afandi, G. Assessment of Agroclimatology NASA POWER Reanalysis Datasets for Temperature Types and Relative Humidity at 2 m against Ground Observations over Egypt. *Adv. Space Res.* **2019**, *64*, 129–142. [CrossRef]
63. Berhane, A.; Hadgu, G.; Worku, W.; Abrha, B. Trends in Extreme Temperature and Rainfall Indices in the Semi-Arid Areas of Western Tigray, Ethiopia. *Environ. Syst. Res.* **2020**, *9*, 3. [CrossRef]
64. Akpoti, K.; Antwi, E.; Kabo-bah, A. Impacts of Rainfall Variability, Land Use and Land Cover Change on Stream Flow of the Black Volta Basin, West Africa. *Hydrology* **2016**, *3*, 26. [CrossRef]
65. Alexander, L.V.; Zhang, X.; Peterson, T.C.; Caesar, J.; Gleason, B.; Klein Tank, A.M.G.; Haylock, M.; Collins, D.; Trewin, B.; Rahimzadeh, F.; et al. Global Observed Changes in Daily Climate Extremes of Temperature and Precipitation. *J. Geophys. Res. Atmos.* **2006**, *111*, 1–22. [CrossRef]
66. Aguilar, E.; Peterson, T.C.; Obando, P.R.; Frutos, R.; Retana, J.A.; Solera, M.; Soley, J.; García, I.G.; Araujo, R.M.; Santos, A.R.; et al. Changes in Precipitation and Temperature Extremes in Central America and Northern South America, 1961–2003. *J. Geophys. Res. Atmos.* **2005**, *110*, 1–15. [CrossRef]
67. Wang, X.L. Comments on "Detection of Undocumented Change Points: A Revision of the Two Phase Regression Model". *J. Clim.* **2003**, *16*, 3383–3385. [CrossRef]
68. Zhang, X.; Yang, F. *RClimDex (1.0)—User Manual, Climate Research Branch Environment Canada Downsview*; Climate Research Branch Environment Canada: Toronto, ON, Canada, 2004.
69. Zhang, X.; Alexander, L.; Hegerl, G.C.; Jones, P.; Tank, A.K.; Peterson, T.C.; Trewin, B.; Zwiers, F.W. Indices for Monitoring Changes in Extremes Based on Daily Temperature and Precipitation Data. *Wiley Interdiscip. Rev. Clim. Change* **2011**, *2*, 851–870. [CrossRef]
70. Bedoum, A.; Bouka Biona, C.; Jean Pierre, B.; Adoum, I.; Mbiake, R.; Baohoutou, L. Évolution Des Indices Des Extrêmes Climatiques En République Du Tchad de 1960 à 2008. *Atmos.-Ocean* **2017**, *55*, 42–56. [CrossRef]

71. Atcheremi, N.D.; Jourda, J.P.; Saley, M.B.; Kouame, K.J.; Balliet, R. Study of the evolution of the rainfall extremes and temperature extremes in davo river basin (south- western of cote d'ivoire) from some indices of the software rclimdex. *Larhyss J.* **2018**, *36*, 99–117.
72. Hountondji, Y.C.; De Longueville, F.; Ozer, P. Trends in Extreme Rainfall Events in Benin (West Africa), 1960–2000. In Proceedings of the 1st International Conference on Energy, Environment and Climate Change, Ho Chi Minh City, Vietnam, 26–27 August 2011.
73. Panthou, G.; Vischel, T.; Lebel, T. Recent Trends in the Regime of Extreme Rainfall in the Central Sahel. *Int. J. Climatol.* **2014**, *34*, 3998–4006. [CrossRef]
74. Vondou, D.A.; Guenang, G.M.; Kamsu-Tamo, P.H. Trends and Interannual Variability of Extreme Rainfall Indices over Cameroon. *Sustainability* **2021**, *13*, 6803. [CrossRef]
75. Balliet, R.; Saley, M.B.; Anowa Eba, E.L.; Sorokoby, M.V.; N'Guessan Bi, H.V.; N'Dri, A.O.; Djè, B.K.; Biémi, J. Évolution Des Extrêmes Pluviométriques Dans La Région Du Gôh (Centre-Ouest De La Côte d'Ivoire). *Eur. Sci. J. ESJ* **2016**, *12*, 74. [CrossRef]
76. Allechy, F.B.; Youan Ta, M.; N'Guessan Bi, V.H.; Yapi, F.A.; Koné, A.B.; Kouadio, A. Trend of Extreme Precipitations Indices in West-Central Côte d'Ivoire: Case of the Lobo Watershed. *Eur. J. Eng. Res. Sci.* **2020**, *5*, 1281–1287. [CrossRef]
77. Kruger, A.C.; Shongwe, S. Temperature Trends in South Africa: 1960–2003. *Int. J. Climatol.* **2004**, *24*, 1929–1945. [CrossRef]
78. Kruger, A.C.; Rautenbach, H.; Mbatha, S.; Ngwenya, S.; Makgole, T.E. Historical and Projected Trends in near Surface Temperature Indices for 22 Locations in South Africa. *S. Afr. J. Sci.* **2019**, *115*, 1–9. [CrossRef]
79. Croitoru, A.E.; Chiotoroiu, B.C.; Ivanova Todorova, V.; Torică, V. Changes in Precipitation Extremes on the Black Sea Western Coast. *Glob. Planet. Chang.* **2013**, *102*, 10–19. [CrossRef]
80. Kabo-Bah, A.T.; Diji, C.J.; Nokoe, K.; Mulugetta, Y.; Obeng-Ofori, D.; Akpoti, K. Multiyear Rainfall and Temperature Trends in the Volta River Basin and Their Potential Impact on Hydropower Generation in Ghana. *Climate* **2016**, *4*, 49. [CrossRef]
81. Gebrechorkos, S.H.; Hülsmann, S.; Bernhofer, C. Changes in Temperature and Precipitation Extremes in Ethiopia, Kenya, and Tanzania. *Int. J. Climatol.* **2019**, *39*, 18–30. [CrossRef]
82. Davis, C.L.; Hoffan, M.T.; Roberts, W. Recent Trends in Climate of Namaqualand, a Megadiverse Arid Region of South Africa. *S. Afr. J. Sci.* **2015**, *112*, 3/4.
83. Sen, P.K. Journal of the American Statistical Estimates of the Regression Coefficient Based on Kendall's Tau. *J. Am. Stat. Assoc.* **1968**, *63*, 1379–1389. [CrossRef]
84. Sterling, D.L. A Comparison of Spatial Interpolation Techniques for Determining Shoaling Rates of the Atlantic Ocean Channe. Master's Thesis, Virginia Polytechnic Institute and State University, Blacksburg, VA, USA, 2003.
85. Kyriakidis, P.C.; Goodchild, M.F. On the Prediction Error Variance of Three Common Spatial Interpolation Schemes. *Int. J. Geogr. Inf. Sci.* **2006**, *20*, 823–855. [CrossRef]
86. Delbari, M.; Afrasiab, P.; Jahani, S. Spatial Interpolation of Monthly and Annual Rainfall in Northeast of Iran. *Meteorol. Atmos. Phys. J.* **2013**, *122*, 103–113. [CrossRef]
87. Goovaerts, P. Geostatistical Approaches for Incorporating Elevation into the Spatial Interpolation of Rainfall. *J. Hydrol.* **2000**, *228*, 113–129. [CrossRef]
88. Tsintikidis, D.; Georgakakos, K.P.; Sperfslage, J.A.; Smith, D.E.; Carpenter, T.M. Precipitation Uncertainty, Raingauge Network Design within Folsom Lake Watershed. *ASCE J. Hydrol. Eng.* **2002**, *7*, 175–184. [CrossRef]
89. Afanasev, I.; Volkova, T.; Elizaryev, A.; Longobardi, A. Analysis of Interpolation Methods to Map the Long-Term Annual Precipitation Spatial Variability for the Republic of Bashkortostan, Russian Federation. *WSEAS Trans. Environ. Dev.* **2014**, *10*, 405–416.
90. N'guessan, H.B.V.; Arona, D.; Adjakpa, T.; Kouadio, B.H.; Kone, B.; Kouame, K.; Yapi, A.; Koaudio, A. Apport de la télédétection à l ' analyse spatio-temporelle de l' évolution des extrêmes pluviométriques dans le district d' abidjan au sud de la cote d ' ivoire. Mélanges en hommages aux Professeurs Houssou C. S., Houndagba C. *J. Thomas* **2015**, 3, 364–378.
91. Paturel, J.E.; Servat, E.; Dela, M.O. Analyse de Séries Pluviométriques de Longue Durée En Afrique de l'Ouest et Centrale Non Sahélienne Dans Un Contexte de Variabilité Climatique. *Hydrol. Sci. Sci. Hydrol.* **1998**, *43*, 937–946. [CrossRef]
92. Lacroix, E.J. *Les Anacardiers, Les Noix de Cajou et La Filière Anacarde à Bassila et Au Bénin. Projet Restauration Des Ressources Forestières de Bassila*; GFA Terra Systems: Hamburg, Allemagne, 2003.
93. Larbi, I.; Hountondji, F.C.C.; Annor, T.; Agyare, W.A.; Gathenya, J.M.; Amuzu, J. Spatio-Temporal Trend Analysis of Rainfall and Temperature Extremes in the Vea Catchment, Ghana. *Climate* **2018**, *6*, 87. [CrossRef]
94. Ozer, P.; Manzo, O.L.; Tidjani, A.D.; Djaby, B.; Longueville, F.D.E. Evolution Récente Des Extrêmes Pluviométriques Au Niger (1950–2014). *Geo-Eco-Trop.* **2017**, *41*, 375–383.
95. Donat, M.G.; Alexander, L.V.; Yang, H.; Durre, I.; Vose, R.; Dunn, R.J.H.; Willett, K.M.; Aguilar, E.; Brunet, M.; Caesar, J.; et al. Updated Analyses of Temperature and Precipitation Extreme Indices since the Beginning of the Twentieth Century: The HadEX2 Dataset. *J. Geophys. Res. Atmos.* **2013**, *118*, 2098–2118. [CrossRef]
96. Mouhamed, L.; Traore, S.B.; Alhassane, A.; Sarr, B. Evolution of Some Observed Climate Extremes in the West African Sahel. *Weather Clim. Extremes* **2013**, *1*, 19–25. [CrossRef]
97. Vaz Milheiro, A.; Neves, E. *Manual Do Cajueiro. Cultivar–Associação de Técnicos de Culturas Tropicais*; Cultivar-Associação de Técnicos de Culturas Tropicais: Porto, Portugal, 1994; p. 204.
98. Ambouta, K.; Sarr, B.; Tychon, B. *Analyse Des Phénomènes Climatiques Extrêmes Dans Le Sud-Est Du Niger*; XXVIIIe Colloque de l'Association Internationale de Climatologie: Liège, Belgium, 2015; pp. 537–542.

99. Chaney, N.W.; Sheffield, J.; Villarini, G.; Wood, E.F. Development of a High-Resolution Gridded Daily Meteorological Dataset over Sub-Saharan Africa: Spatial Analysis of Trends in Climate Extremes. *J. Clim.* **2014**, *27*, 5815–5835. [CrossRef]
100. Omondi, P.A.o.; Awange, J.L.; Forootan, E.; Ogallo, L.A.; Barakiza, R.; Girmaw, G.B.; Fesseha, I.; Kululetera, V.; Kilembe, C.; Mbati, M.M.; et al. Changes in Temperature and Precipitation Extremes over the Greater Horn of Africa Region from 1961 to 2010. *Int. J. Climatol.* **2014**, *34*, 1262–1277. [CrossRef]
101. van der Walt, A.J.; Fitchett, J.M. Exploring Extreme Warm Temperature Trends in South Africa: 1960–2016. *Theor. Appl. Climatol.* **2021**, *143*, 1341–1360. [CrossRef]
102. Filahi, S.; Tanarhte, M.; Mouhir, L.; El Morhit, M.; Tramblay, Y. Trends in Indices of Daily Temperature and Precipitations Extremes in Morocco. *Theor. Appl. Climatol.* **2016**, *124*, 959–972. [CrossRef]
103. Aoubouazza, M.; Rajel, R.; Essafi, R. Impact Des Phénomènes Climatiques Extrêmes Sur Les Ressources En Eau et l ' Agriculture Au Maroc. *Rev. Mar. Sci. Agron. Vét.* **2019**, *7*, 223–232.
104. Popov, T.; Gnjato, S.; Trbić, G.; Ivanišević, M. Recent Trends in Extreme Temperature Indices in Bosnia and Herzegovina. *Carpathian J. Earth Environ. Sci.* **2018**, *13*, 211–224. [CrossRef]
105. Salameh, A.A.M.; Gámiz-Fortis, S.R.; Castro-Díez, Y.; Abu Hammad, A.; Esteban-Parra, M.J. Spatio-Temporal Analysis for Extreme Temperature Indices over the Levant Region. *Int. J. Climatol.* **2019**, *39*, 5556–5582. [CrossRef]
106. Zhou, J.; Huang, J.; Zhao, X.; Lei, L.; Shi, W.; Wang, L.; Wei, W.; Liu, C.; Zhu, G.; Yang, X. Changes of Extreme Temperature and Its Influencing Factors in Shiyang River Basin, Northwest China. *Atmosphere* **2020**, *11*, 1171. [CrossRef]
107. WMO. *The Global Climate 2001–2010: A Decade of Climtate Extreme*; World Meteorological Organisation WMO-No. 1103: Geneva, Switzerland, 2010.
108. Coumou, D.; Rahmstorf, S. A Decade of Weather Extremes. *Nat. Clim. Chang.* **2012**, *2*, 491–496. [CrossRef]
109. YAO, K.T. *Hydrodynamisme Dans Les Aquifères de Socle Cristallin et Cristallophyllien Du Sud-Ouest de La Côte d'Ivoire: Cas Du Département de Soubré. Apport de La Télédétection, de La Géomorphologie et de l'hydrogéochimie. Thèse Unique. UFR STRM*; Université de Coco: Abidjan, Côte d'Ivoire, 2009.

MDPI AG
Grosspeteranlage 5
4052 Basel
Switzerland
Tel.: +41 61 683 77 34

*Climate* Editorial Office
E-mail: climate@mdpi.com
www.mdpi.com/journal/climate

www.ingramcontent.com/pod-product-compliance
Lightning Source LLC
Chambersburg PA
CBHW061515210926
51850CB00008B/1923
* 9 7 8 3 7 2 5 8 6 7 9 2 9 *